Law and Technology in a Global Digital Society

Georg Borges • Christoph Sorge
Editors

Law and Technology in a Global Digital Society

Autonomous Systems, Big Data, IT Security and Legal Tech

Springer

Editors
Georg Borges
Institute of Legal Informatics
Saarland University
Saarbrücken, Germany

Christoph Sorge
Institute of Legal Informatics
Saarland University
Saarbrücken, Germany

ISBN 978-3-030-90512-5 ISBN 978-3-030-90513-2 (eBook)
https://doi.org/10.1007/978-3-030-90513-2

This Springer imprint is published by the registered company Springer Nature Switzerland AG.
The registered company address is: Gewerbestrasse 11, 6330 Cham, Switzerland

Preface

The transformation of society into a digital society, in which essential areas of life are carried out through the exchange of data via electronic networks and machines equipped with artificial intelligence to carry out activities that were previously the preserve of humans, is taking place worldwide with great dynamics.

This transformation process involves the law in two ways: On the one hand, law has to adapt in order to provide the changed society with appropriate rules, and on the other hand, it acts as a driving factor of the transformation by setting limits to technological development while at the same time providing a framework and basis for it.

This interdependence of law and technology can only be grasped and shaped through interdisciplinary and at the same time international work. This is the approach taken by the present work, which addresses central aspects of law in this transformation process.

Data protection and IT security describe technical and legal cornerstones of the digital society. Accordingly, important aspects of data protection in the context of key technologies, such as blockchain as well as artificial intelligence (AI), are addressed on the basis of the GDPR, which is not only applicable in the European Union, but is also influencing the data protection landscape in many countries around the world. The potential of controlling technology through law will be explored using the example of IT security and exemplified by current issues.

Artificial intelligence and machines equipped with it, autonomous systems as self-controlling machines, challenge central basic assumptions by making machines actors. The question of how AI-equipped systems can be addressed by law is the subject of an overview as well as several contributions covering criminal law, liability law, and data protection law.

Legal technology, or "Legal Tech" for short, expresses the transformation of legal work through modern technologies. The potential of LegalTech is discussed from a theoretical point of view, both from a technical and from a legal perspective. Contributions on smart contracts analyze the tension between law and technology using the example of a technology with great disruptive potential for the practice of business transactions.

The concept of this work originated in the Summer School "IT Law and Legal Informatics," in which the Institute for Legal Informatics annually invites young researchers from all over the world to interdisciplinary work since 2017. The contributions have been written by lecturers and participants of the Summer School and analyze from different disciplinary and international perspectives, with a focus on European law, the challenges of law and technology in the transformation process.

Finally, on behalf of the editors, I would like to express my sincere gratitude to the staffs in the Institute of Legal Informatics, Ms. Caroline Hertwig, Mr. Max Birla, and Mr. Nikolas Hamm for their very fine work and support. Without their dedicated contribution, this book would not have been appeared.

On behalf of the editors

Saarbrücken, Germany
June 2021

Georg Borges

Contents

Part III IT Security

Part IV Legal Tech

Part I
Artificial Intelligence

A Legal Framework for Autonomous Systems

Georg Borges

Contents

Abstract Artificial intelligence and autonomous systems are a central technological basis for the global transformation of the economy and society into the so-called "digital society". This transformation process requires an adaptation of the legal framework for this technology, the scope of which goes far beyond partial amendments in matters of detail, but raises fundamental questions of law. The question of the role of machines in law is relevant in numerous areas of law and calls for new legal concepts.

The following article outlines the current legal discussion on the legal framework for autonomous systems and describes legal challenges of artificial intelligence in the field of civil and commercial law.

G. Borges (✉)
Institute of Legal Informatics, Saarland University, Saarbrücken, Germany
e-mail: georg.borges@uni-saarland.de

G. Borges, C. Sorge (eds.), *Law and Technology in a Global Digital Society*,
https://doi.org/10.1007/978-3-030-90513-2_1

1 Legal Questions Raised by Autonomous Systems

Autonomous systems, such as cars[1] or automated systems which give financial[2] or legal advice[3] are central elements of the digital revolution and will change the world as we currently know it. This will, no doubt, result in both desired consequences (such as simplification of daily life[4] and a reduction in the number of accidents[5]) as well as undesired consequences (ranging from the loss of jobs[6] to displacement in entire areas of industry,[7] loss of autonomy[8] and, of course, endangering behaviour and fatal accidents[9]). It is self-explanatory that autonomous systems will create

[1]Oppermann and Stender-Vorwachs (2020) offer an insight into numerous aspects of autonomous driving.

[2]See an overview of this topic in Phaneuf (2020), Report from 9 September 2020, available at https://www.businessinsider.com/ai-in-finance?r=DE&IR=T. Accessed 28 August 2021.

[3]A prominent example of this is the virtual intelligent agent "Watson" developed by IBM, see the IBM-Blog for a description, available at https://www.ibm.com/cloud/watson-assistant. Accessed 28 August 2021; For analysis of legal questions raised by machines ("Legal Tech") generally, see Frese (2015), pp. 2091 f.; Galetzka et al. (2021), pp. 20 ff.; Hähnchen et al. (2020), pp. 625 ff.; Kilian (2017), pp. 3048 ff.; Prior (2017), pp. 575 ff.; Römermann and Günther (2019), pp. 551 ff.; J. Wagner (2017b), pp. 898 ff.

[4]Household robots (e.g. vacuum cleaners and lawn mowers) are commercially successful with an estimated 30–40% annual increase in the number of robots sold (see the Executive Summary World Robotics 2020 Service Robots https://ifr.org/img/worldrobotics/Executive_Summary_WR_2020_Service_Robots.pdf. Accessed 28 August 2021).

[5]When the introduction of autonomous vehicles is complete, the expectation is that there will be a drastic reduction of up to 90% in the number of accidents (according to a prediction by McKinsey, summary available at https://www.mckinsey.com/industries/automotive-and-assembly/our-insights/ten-ways-autonomous-driving-could-redefine-the-automotive-world. Accessed 28 August 2021; Also, similarly for market penetration of 90% Fagnant and Kockelman (2015), pp. 167, 173). For the significance of this situation for the implementation of autonomous vehicles from an ethical and legal perspective, see the report of the German Ethics Commission on Automated and Connected driving, thesis 5 (p. 10) und thesis 18 (p. 13), available at https://www.bmvi.de/SharedDocs/DE/Publikationen/DG/bericht-der-ethik-kommission.pdf?__blob=publicationFile. Accessed 28 August 2021.

[6]The estimations differ considerably: While Frey/Osborne argue that 47% of all jobs in the USA are under threat from increasing autonomisation (Frey and Osborne (2017), pp. 254, 265), OECD states that 14% of jobs being at high risk of automation (OECD (ed.), *OECD Employment Outlook 2019: The Future of Work*, available at https://read.oecd-ilibrary.org/employment/oecd-employment-outlook-2019_9ee00155-en#page1. Accessed 28 August 2021).

[7]For an overview regarding the overall impact of digitalisation on different industries in general, see Palka (2018).

[8]Symbolic the vision that the creation exceeds the creator, cf. e.g. Kuflik (1999), p. 181 ("the creations 'outperform' their creators").

[9]See for example the report of the National Transportation Safety Board (NTSB) of 19 November 2019 on the fatal collision between a vehicle controlled by a developmental automated driving system and a pedestrian, available at, https://www.ntsb.gov/investigations/AccidentReports/Reports/HAR1903.pdf. Accessed 28 August 2021.

numerous legal questions for all areas of law, some of which being less important than others.[10]

Existing legal provisions must frequently be amended so that they apply to autonomous systems. A specific example is the amendment to the Vienna Convention on Road Traffic to include autonomous cars as road users.[11] Through its 8th[12] revision of the Road Traffic Act, the German legislature has expressly declared that the use of vehicles with high levels of automated and fully automated driving function is permissible, although this permission only applies to vehicles in which the driving function can be overtaken by a human driver at any time (compare § 1 (a) para. 2 Nr. 3 StVG).

In February 2021, the German Federal Ministry of Transport and Digital Infrastructure published the coordinated draft of a further amendment of the Road Traffic Act.[13] The draft amendment provisions go way beyond existing rules as, for the first time, the operation of fully-automated ("autonomous") vehicles (SAE level 4) would be admitted to public roads. In case of a fully automated car, the role of the driver will be replaced by a "technical supervisor" would not have to be bed that have no driver any longer, but just a Technical Supervisor with limited influence on the car's manoeuvres.

The legal questions on this subject go far beyond the conventional type of detailed regulation: in its resolution on Civil Law Rules On Robotics of 16 February 2017,[14] the European Parliament not only highlighted numerous legal questions relating to robotics but also highlighted other extremely wide-ranging questions such as the development of a new liability system, specifically, liability funds for robots—and, under the heading "ePerson", questioned whether machines should themselves

[10]For a court decision regarding the permissibility of the continuous use of a lawn mower, see AG Siegburg, NZM 2016, p. 526, **ECLI:DE:AGSU1:2015:0219.118C97.13.00**.

[11]Effective from 23 March 2016, a new paragraph 5*bis* was inserted into Art. 8 of the Convention which enables autonomous vehicles to compatible with the provisions of the convention; see Arzt and Ruth-Schumacher (2017), pp. 61 f.; von Bodungen and Hoffmann (2015), pp. 521 ff.; Jänich et al. (2015), p. 318; Lutz (2015), pp.122 ff.; different von Kaler and Wieser (2018), p. 372, who argue that a further amendment to the Convention is necessary to address Level 4 and 5 driving features.

[12]"Achtes Gesetz zur Änderung des Straßenverkehrsgesetzes" from 16 June 2017, BGBl. I 2017, pp. 1648 ff.; for the changes see for example Armbrüster (2017), pp. 83 ff.; Balke (2018), pp. 5 ff.; Berndt (2017), pp. 121 ff.; Grünvogel (2017), pp. 973 ff.; König (2017), pp. 123 ff.; Kütük-Markendorf (2017), pp. 349 ff.; Schirmer (2017), pp. 253 ff.; von Kaler and Wieser (2018), pp. 369 ff.

[13]Available at https://www.bmvi.de/SharedDocs/DE/Anlage/Gesetze/Gesetze-19/gesetz-aenderung-strassenverkehrsgesetz-pflichtversicherungsgesetz-autonomes-fahren.pdf?__blob=publicationFile. Accessed 28 August 2021.

[14]European Parliament resolution of 16 February 2017 with recommendations to the Commission on Civil Law Rules on Robotics (2015/2103(INL))—A8-0005/2017, available at http://www.europarl.europa.eu/sides/getDoc.do?pubRef=-//EP//NONSGML+TA+P8-TA-2017-0051+0+DOC+PDF+V0//DE. Accessed 28 August 2021; on the resolution of the European Parliament see also Günther (2017), p. 651; Keßler (2017), p. 589; Lohmann (2017), p. 168.

be given legal status or even legal capacity to act. This would clearly be an important step in the development of the law.

1.1 Current Discussion Regarding the Law of Automated Systems

Artificial intelligence and autonomous systems, together with other new technology trends such as "big data" or the "Internet of things", have gained the legislator's attention for a couple of years already. In several strategy papers, the European Commission highlighted the necessity for adequate liability rules in the context of new technologies—in particular with regard to the internet of things and autonomous systems. These strategy papers include the Digital Single Market Strategy for Europe[15] on using the opportunities presented by a digital single market,[16] the Strategy to build a European Data Economy,[17] as well as the Mid-Term Review on the implementation of the Digital Single Market Strategy.[18]

Remarkably, the Commission identifies as key objectives not only the strengthening of the EU's technical and industrial capacity and the diffusion of AI throughout the economy, preparing for the socio-economic changes associated with AI, but also ensuring an appropriate ethical and legal framework for AI.[19] In this context, ethical guidelines for AI should be developed, and a review of the existing regulations on the liability and security framework should be carried out.[20]

The European Commission established several expert groups who support the commission in these tasks. In the field of ethical aspects of AI, the High Level Expert Group on AI[21] developed ethical guidelines for trustworthy AI which were published in June 2018.[22] Regarding the regulation on liability and security, the Expert Group on liability and new technologies consisting of a Product Liability Directive formation and a "New Technologies" formation was established.[23] In its

[15] Communication from the Commission of 6.5.2015, COM (2015) 192 final, available at http://eur-lex.europa.eu/legal-content/EN/TXT/PDF/?uri=CELEX:52015DC0192. Accessed 28 August 2021.

[16] Digitising European Industry: Reaping the Full Benefits of a Digital single Market Communication from the Commission of 19.4.2016, COM(2016) 180 final.

[17] Communication from the Commission of 10.1.2017, COM (2017) 9 final.

[18] Communication from the Commission of 10.5.2017, COM(2017) 228 final.

[19] COM(2018) 237 final, at pp. 5 ff.

[20] *Id.* at p. 20.

[21] Detailed information on the Group available at https://ec.europa.eu/transparency/expert-groups-register/screen/expert-groups/consult?do=groupDetail.groupDetail&groupID=3591&news=1. Accessed 28 August 2021.

[22] Available at https://ec.europa.eu/newsroom/dae/document.cfm?doc_id=60419. Accessed 28 August 2021.

[23] Detailed information on the Group available at https://ec.europa.eu/transparency/expert-groups-register/screen/expert-groups/consult?do=groupDetail.groupDetail&groupID=3592. Accessed 28 August 2021.

final report, the New Technologies formation suggested amendments of the existing liability law regarding AI and autonomous systems.[24]

Based on the proposals of the expert groups, the Commission presented its "White Paper on Artificial Intelligence: A European approach to excellence and trust",[25] in February 2020 which was published together with the "Report on the safety and liability implications of Artificial Intelligence, the Internet of Things and robotics".[26]

On 20 October 2020, the European Parliament adopted three resolutions on ethics,[27] liability[28] and copyright[29] calling on the Commission to further develop the legal framework for AI. Further resolutions—in criminal matters[30] and in education, culture and the audio-visual sector[31]—are expected to be published in 2021.

The resolution on intellectual property rights for the development of AI technologies calls on the Commission to create an appropriate legal framework in the area of legal protection of products created by AI, in particular in patent and copyright law, as well as with regard to the use of data.[32] In the two other resolutions, the Parliament even presented concrete regulatory proposals. The resolution on a framework of ethical aspects of artificial intelligence, robotics and related technologies[33] contains a draft regulation on ethical principles for the development, deployment and use of artificial intelligence, robotics and related technologies. The resolution on a civil

[24] Report from the Expert Group on Liability and New Technologies—New Technologies Formation, Executive Summary, p. 3 f., available at https://www.europarl.europa.eu/meetdocs/2014_2019/plmrep/COMMITTEES/JURI/DV/2020/01-09/AI-report_EN.pdf. Accessed 28 August 2021.

[25] COM(2020) 65 final.

[26] COM (2020) 64 final.

[27] European Parliament resolution of 20 October 2020 on a framework of ethical aspects of artificial intelligence, robotics and related technologies, 2020/2012(INL), available at https://www.europarl.europa.eu/doceo/document/TA-9-2020-0275_EN.pdf. Accessed 28 August 2021.

[28] European Parliament resolution of 20 October 2020 on a civil liability regime for artificial intelligence, 2020/2014(INL), available at https://www.europarl.europa.eu/doceo/document/TA-9-2020-0276_EN.pdf. Accessed 28 August 2021.

[29] European Parliament resolution of 20 October 2020 on intellectual property rights for the development of artificial intelligence technologies, 2020/2015(INI), available at https://www.europarl.europa.eu/doceo/document/TA-9-2020-0277_EN.pdf. Accessed 28 August 2021.

[30] European Parliament Draft Report, Artificial intelligence in criminal law and its use by the police and judicial authorities in criminal matters, 2020/2016(INI), available at https://www.europarl.europa.eu/doceo/document/LIBE-PR-652625_EN.pdf. Accessed 28 August 2021.

[31] European Parliament Draft Report, Artificial intelligence in education, culture and the audiovisual sector, 2020/2017(INI), available at http://www.europarl.europa.eu/RegData/seance_pleniere/textes_adoptes/definitif/2021/05-19/0238/P9_TA(2021)0238_EN.pdf. Accessed 28 August 2021.

[32] European Parliament Draft Report, Artificial intelligence in education, culture and the audiovisual sector, 2020/2017(INI), p. 8, No. 15, available at http://www.europarl.europa.eu/RegData/seance_pleniere/textes_adoptes/definitif/2021/05-19/0238/P9_TA(2021)0238_EN.pdf. Accessed 28 August 2021.

[33] 2020/2012(INL), (*supra* note 27).

liability regime for artificial intelligence[34] contains a draft regulation on liability for the operation of artificial intelligence-systems.

In its Communication "Fostering a European approach to Artificial Intelligence" of 21 April 2021,[35] the Commission emphasised "opportunities and risks" as "the two sides of artificial intelligence".[36] The Commission identifies the existence of specific, significant risks of artificial intelligence to safety and fundamental rights, noting at the same time that these are not adequately covered by the existing legal framework.[37]

In response to this challenge,[38] the Commission presented its proposal for an "Artificial Intelligence Act"[39] on the same day. The AI Act, which comprises twelve titles with a total of 81 articles, provides for quite different measures. One element that is attracting a lot of attention is the prohibitions and restrictions on certain AI applications, such as remote biometric identification or social scoring, regulated in Title II, Art. 5. The core element of the AI Act is the regulation of so-called high-risk AI systems, to which by far the most comprehensive section, Title III with Arts. 6–51, is dedicated. Here, the duty to manage risks (Art. 9) applies with numerous individual duties (Art. 10 ff.), which are closely based on the existing product safety law, but also contain some extensions. Another new element is the transparency requirements for certain AI systems regulated in Title IV, Art. 52. The core content here is the disclosure of the fact that it is not a natural person but an AI system that is acting.

The proposed regulation, which immediately attracted worldwide attention, thus—contrary to the pretentious short title "AI Act"—by no means sets a comprehensive legal framework for artificial intelligence. Rather, the proposal relies primarily on the system of a safety check of products by the manufacturer, known from product safety law, which is, however, developed into a comprehensive risk management. Significant other areas are not covered. In particular, civil liability for AI systems, which is intended to be regulated separately, is not covered.

The importance of artificial intelligence is very much on the agenda of international organisations. For example, the United Nations Interregional Crime and Justice Research Institute (UNICRI) is overseeing the establishment of a Centre for Artificial Intelligence.[40] The report of the Development Policy and Analysis Division (DPAD) on the effects on the job market and wealth distribution of

---

[34] 2020/2014(INL), (*supra* note 28).

[35] COM(2021) 205 final.

[36] COM(2021) 205 final, p. 1.

[37] COM(2021) 205 final, p. 3 f.

[38] COM(2021) 205 final, p. 4.

[39] Proposal for a Regulation of the European Parliament and of the Council laying down harmonised rules on artificial intelligence (Artificial Intelligence Act) and amending certain Union legislative acts of 21 April 2021, COM(2021) 206 final.

[40] The Host Country Agreement with the Netherlands was concluded on 7 October 2017, see the relevant notification on the webpage of UNICRI, available at http://www.unicri.it/news/article/2017-09-07_Establishment_of_the_UNICRI. Accessed 28 August 2021.

the technology revolution brought about by AI published in July 2017 highlights the potential of the growth in productivity of technological innovation whilst, at the same time, underlines the necessity for regulation.[41]

Within the UN, several aspects of AI and related technologies have been discussed. In the 73rd session of the Third Committee of the UN General Assembly (14 November 2018) dealing with the promotion and protection of human rights, the committee expressed it's concern that technologies such as automated decision-making and machine-learning have the potential to affect human rights and recognized the need to apply international human rights law in the design, production an operation of such technologies.[42]

Among all the UN activities in the field of artificial intelligence, the work of the United Nations Economic Commission for Europe (UNECE) might have the biggest impact at present. The so-called UNECE regulations are particularly important in the context of so-called type approvals of cars which is, in many legal systems, in particular in Europe, a prerequisite for admission of vehicles to public traffic. The UNECE regulations are based on the "Agreement Concerning the Adoption of Uniform Technical Prescriptions for Wheeled Vehicles, Equipment and Parts which can be Fitted and/or be Used on Wheeled Vehicles and the Conditions for Reciprocal Recognition of Approvals Granted on the Basis of these Prescriptions", first adopted in Geneva on 20 March 1958. Revision 3 of 14 December 2016 (Rev. 03) entered into force on 13 September 2017. The Agreement aims to create uniform technical regulations for motor vehicles. The individual technical regulations are so-called "Addenda", which are adopted on the basis of the Agreement. The technical requirements are developed by the World Forum for Harmonization of Vehicle Regulations (WP. 29), formerly known as the "Working Party of experts on technical requirement of vehicles", a working group of the ECE.

At the 178th meeting in June 2019, the revised framework document on automated/autonomous vehicles was adopted,[43] which sets out central principles for highly automated driving.[44] It received great attention internationally.[45]

---

[41] The impact of the technological revolution on labour markets and income distribution, p. 3, available at https://www.un.org/development/desa/dpad/wp-content/uploads/sites/45/publication/2017_Aug_Frontier-Issues-1.pdf. Accessed 28 August 2021.

[42] Available at https://undocs.org/A/C.3/73/L.49/Rev.1. Accessed 28 August 2021.

[43] See the Report of the 178th session of the World Forum for Harmonization of Vehicle Regulations, p. 10, No. 27, available at https://unece.org/DAM/trans/doc/2019/wp29/ECE-TRANS-WP.29-1147e.pdf. Accessed 28 August 2021.

[44] The Revised Framework document on automated/autonomous vehicles, issued by the United Nations, Economic Commission for Europe (ECE), available at https://unece.org/fileadmin/DAM/trans/doc/2019/wp29/ECE-TRANS-WP29-2019-34-rev.1e.pdf. Accessed 28 August 2021.

[45] Cf. the report in Forbes, available at https://www.forbes.com/sites/lanceeliot/2019/09/06/key-insights-about-the-uns-new-framework-for-globalization-of-self-driving-cars/?sh=591a789b799b (Accessed 28 August 2021), and other journals e.g. Intelligent Transport, available at https://www.intelligenttransport.com/transport-news/87152/un-publishes-new-framework-for-autonomous-vehicles-with-safety-at-core. Accessed 28 August 2021.

Following the framework document, the individual technical rules on autonomous driving are being developed by the World Forum. For example, the Uniform provision concerning the approval of vehicles with regard to Automated Lane Keeping Systems (ALKS) came into force on 22 January 2021 as the UN Regulation No. 157.[46]

Overall, however, the United Nations do not currently play a significant role in the discussion on the legal framework for AI. In particular, global frameworks concerning human rights law chaired by the UN Human Rights Council, do not yet play a major role in AI regulation.

The Organisation for Economic Co-operation and Development (OECD) is, as part of its Going Digital Project,[47] among other things, looking at the legal questions raised by autonomous systems.[48] In May 2019, OECD member countries adopted the OECD Council Recommendation on Artificial Intelligence[49] which includes five principles for the responsible use of AI as well as recommendations for national policy and international cooperation regarding AI. The five principles for AI, are focused on growth, sustainable development and well-being, human-centred values and fairness, transparency and explainability, robustness, security and safety, as well as accountability. In June 2019, at the G20 Trade and Digital (Economy) Ministers meeting in Tsukuba City, the OECD AI Principles were adopted by the G20 resulting in the G20 AI Principles.[50]

The OECD Council Recommendation on AI, in its section on national policies, also refers to the legal framework by recommending to review and adapt the framework in order to encourage innovation and completion for trustworthy AI.[51]

National governments are taking action in the field of AI regulation or, at least, are intending to do so. In the US, a national artificial intelligence research and development strategic plan was adopted in 2016[52] and updated in 2019.[53] The

---

[46] ECE/TRANS/505/Rev.3/Add.156, available at https://unece.org/sites/default/files/2021-03/R157e.pdf. Accessed 28 August 2021.

[47] Available at https://www.oecd.org/going-digital/project/going-digital-information-note.pdf. Accessed 28 August 2021. With its project "Going Digital", OECD is striving to achieve the goal of developing a universal and comprehensive political approach in order to be able to better utilise the advantages brought about by the digital transformation.

[48] At the OECD Conference "AI: Intelligent Machines, Smart Policies", held 26 and 27 October 2017, Data Protection, IT-Security, as well as responsibility and liability for AI-systems were discussed; lectures available at http://www.oecd.org/going-digital/ai-intelligent-machines-smart-policies/conference-agenda/. Accessed 28 August 2021.

[49] Available at https://www.oecd.org/going-digital/ai/principles/. Accessed 28 August 2021.

[50] G20 Ministerial Statement on Trade and Digital Economy, available at https://www.mofa.go.jp/files/000486596.pdf. Accessed 28 August 2021.

[51] Recommendation of the Council on AI, No. 2.3. lit. b), available at https://legalinstruments.oecd.org/en/instruments/OECD-LEGAL-0449. Accessed 28 August 2021.

[52] Available at https://www.nitrd.gov/pubs/national_ai_rd_strategic_plan.pdf. Accessed 28 August 2021.

[53] Available at https://www.nitrd.gov/pubs/National-AI-RD-Strategy-2019.pdf. Accessed 28 August 2021.

2016 plan suggests seven strategies, including a strategy called "understand and address the ethical, legal and societal implications of AI". By Executive Order 13859 "Maintaining American Leadership in Artificial Intelligence" of February 2019,[54] the US Federal Government emphasised the importance of AI for economical growth and outlined it's objectives regarding AI, which include the protection of safety, security and privacy. In addition, specific legislation on AI are being made. At state level, for instance, the Artificial Intelligence Video Interview Act was passed in Illinois.[55] At federal level, a proposal for an Algorithmic Accountability Act which was firstly introduced in Congress in October 2019[56] and languished at committee level, is expected to be reintroduced in 2021.[57]

In Europe, Germany has been active in the field. In summer 2017, Heiko Maas, the then German Federal Minister of Justice and Consumer Protection, toyed with the idea of passing a law on algorithms[58] which was supposed to address automated decision making (ADM). In November 2018, the Federal Government published it's national AI strategy.[59] The Federal Government's goal is to strengthen Germany as a location for the development and use of AI, to conduct the use and development of AI responsibly and to firmly integrate AI in society.[60] As a part of the strategy, the existing legal framework should be reviewed and adapted.[61]

Several expert groups were established to investigate the influence of artificial intelligence and algorithms. In October 2019, the Data Ethics Commission (Datenethik-Kommission) presented its report along with recommendations for action for the Federal Government.[62] In October 2020, the Enquete Commission

[54] Executive Order 13859 of 11 February 2019, available at https://www.federalregister.gov/documents/2019/02/14/2019-02544/maintaining-american-leadership-in-artificial-intelligence. Accessed 28 August 2021.

[55] Available at https://www.ilga.gov/legislation/publicacts/fulltext.asp?Name=101-0260. Accessed 28 August 2021.

[56] H.R.2231—Algorithmic Accountability Act of 2019—116th Congress (2019–2020), available at https://www.congress.gov/bill/116th-congress/house-bill/2231. Accessed 28 August 2021.

[57] See the reported plan of Senator Wyden, available at https://www.meritalk.com/articles/sen-wyden-to-reintroduce-ai-bias-bill-in-coming-months. Accessed 28 August 2021.

[58] "Klarheit und Kontrolle bei Algorithmen", article from the German Federal Government of 3 July 2017, available at https://www.bundesregierung.de/Content/DE/Artikel/2017/07/2017-03-07-maas-algorithmen.html. Accessed 28 August 2021.

[59] German National AI Strategy, available at https://www.bundesregierung.de/resource/blob/975226/1550276/3f7d3c41c6e05695741273e78b8039f2/2018-11-15-ki-strategie-data.pdf?download=1. Accessed 28 August 2021.

[60] German National AI Strategy, available at https://www.bundesregierung.de/resource/blob/975226/1550276/3f7d3c41c6e05695741273e78b8039f2/2018-11-15-ki-strategie-data.pdf?download=1. Accessed 28 August 2021.

[61] German National AI Strategy, p. 38, available at https://www.bundesregierung.de/resource/blob/975226/1550276/3f7d3c41c6e05695741273e78b8039f2/2018-11-15-ki-strategie-data.pdf?download=1. Accessed 28 August 2021.

[62] Report available at https://www.bmi.bund.de/SharedDocs/downloads/DE/publikationen/themen/it-digitalpolitik/gutachten-datenethikkommission.pdf?__blob=publicationFile&v=6. Accessed 28 August 2021.

on Artificial Intelligence—Social Responsibility and Economic, Social and Ecological Potentials submitted its report to the German Parliament (Bundestag). The Federal Government published its update of the AI Strategy in December 2020.[63]

Industry is also vehemently demanding the creation of a legal framework for autonomous systems. As an example, Elon Musk's repeated appeals demanding regulation of Artificial Intelligence attracted significant attention in the worldwide media.[64]

1.2 *Artificial Intelligence and Autonomous Systems*

The subject of the current legal policy discussion largely eludes a clear-cut description.[65] In some cases, reference is made to artificial intelligence, sometimes quite specifically to machine learning or even deep learning. Other statements refer to "software agent",[66] robots[67] or cyber-physical systems.[68] On the other hand, not infrequently "new technologies" are referred to in very general terms.

The term "autonomous system" is also described in various ways.[69] The essential characteristic is the ability, brought about by software, to perceive the environment and to make and implement independent situation-based decisions.[70] Even though

---

[63] Update available at https://www.bmwi.de/Redaktion/DE/Publikationen/Technologie/strategie-kuenstliche-intelligenz-fortschreibung-2020.pdf?__blob=publicationFile&v=10. Accessed 28 August 2021.

[64] See, for example: Gibbs (2017), *Jansen*, Elon Musk warnt vor 3. Weltkrieg durch Künstliche Intelligenz, Report from 4 September 2017, available at http://www.faz.net/aktuell/wirtschaft/kuenstliche-intelligenz/elon-musk-tesla-chef-warnt-vor-kuenstlicher-intelligenz-15182958.html. Accessed 28 August 2021.

[65] See, as an example for the lack of terminological consistency, Keßler (2017), p. 589 ("Deep Learning" and "maschinelles Lernen"; "intelligente Roboter"; "künstliche Intelligenz"; "intelligente Software-Agents"); Linke (2021), p. 32 and Taeger (2016), pp. 3764 ff. ("system" and "agent" used synonymously); in respect of inventory, see Lohmann (2017), p. 169; Pieper (2016), p. 189.

[66] On the different definitions of "agents" from a technical perspective, see for example Franklin and Graesser (1996), pp. 22 ff.

[67] On the term 'robot', see Beck (2009), p. 226; on the term and classification Müller (2014), pp. 596 ff.; The used terms are slightly different between Denga (2018), pp. 69 f. ("AI robot") and Lohmann (2017), p. 169 ("autonomous robot").

[68] The term is used, for example, by Pieper (2016), p. 189; Reichwald and Pfisterer (2016), p. 208.

[69] For the term "autonomous system", see e.g. Borges (2018), p. 977; Foerster (2019), p. 422; Grützmacher and Heckmann (2019), p. 554; Kainer and Förster (2020), pp. 279 f.; Kirn and Müller-Hengstenberg (2017), *passim*; Kluge and Müller (2017), p. 24; Linke (2021), pp. 29 f.; Pieper (2016), pp. 189 f.; Reichwald and Pfisterer (2016), p. 212; Riehm (2014), pp. 113 ff.; Russell and Norvig (2010), p. 40; Schirmer (2019), pp. 716 f.; Schulz (2015), *passim*; Sosnitza (2016), p. 765; Sommer (2020), *passim*; Stiemerling (2015), p. 765; Thöne (2020), pp. 5–12; G. Wagner (2017a), pp. 707 ff.; Zech (2019), p. 199; for the term "autonomous robot", *id.*, (2016), pp. 170 f.

[70] For similar definitions, see Linke (2021), p. 32; Sommer (2020), p. 37; Sosnitza (2016), p. 765; Zech (2019), p. 198; equally for the term of "AI Systems" used by the European Commission, COM (2018) 237 final, p. 1.

software always requires a running environment and thus hardware to function, the connection of the software with a specific tangible object is not a characteristic of the autonomous system. An autonomous system could therefore also consist exclusively of software that is used in any running environment.[71]

From a legal perspective, a characteristic feature of an autonomous system is the ability to perform actions that are not controlled or predetermined from the outside, but are controlled by the system itself depending on the environment it perceives. This ultimately leads to the situation, which is relevant in numerous areas of law, that the concrete behaviour of an artefact in an specific situation cannot be attributed directly to control or programming by a human being.

2 Autonomy of Machines as a Challenge for the Law

Autonomous systems pose enormous challenges for the law because, to date, the law assumes that only humans can act autonomously and it is not designed to deal with autonomy of machines. A small, but by no means exhaustive, selection of the relevant aspects will be discussed below.

2.1 Creative Efforts of Machines

Autonomous systems create things of value in areas which were previously reserved for humans. For example, in the area of intellectual property rights law, it is necessary to discuss how the intellectual output generated by a machine should be classified.[72]

At the core of this issue is whether the output generated by the machine should be seen as an inventive step by the machine.[73] However, this approach is rejected by the majority of commentators.[74] This leads to the result that any inventive step of the machine is attributed to a person. In the case of autonomous systems, this approach is extremely questionable and, ultimately, unsuccessful.

---

[71] Thöne (2020), p. 10.

[72] See e.g. Gomille (2019), pp. 969 ff.; Lauber-Rönsberg (2019), pp. 244 ff.; Samuelson (1986), pp. 1185 ff.; Schaub (2017), pp. 346 ff.; Specht-Riemenschneider (2021), pp. 273 ff.; in the outline also Solum (1992), p. 1257 (with note 88); in the context of AI and patent law, see Nägerl et al. (2019), pp. 336 ff.

[73] Samuelson (1986), pp. 1192 ff.

[74] According to the general view held to date, only natural persons can be inventors in the sense of § 6 PatG (Kraßer and Ann (2016), § 19 at mn. 7; Melullis (2015), mn. 31; Mes (2015), mn. 10), or "Schöpfer" in the sense of § 7 UrhG (Begr. RegE zu § 7 UrhG, BT-Drucks. IV/270, p. 41; Ahlberg (2018), mn. 7; Schulze (2018), mn. 2; Thum (2019), mn. 13).

Therefore, if the protection of intellectual property rights is not to be completely disregarded,[75] it is necessary to examine the fundamental question of whether the output generated by a machine should actually be recognised as an inventive step. This would, to the extent that machines are not themselves holders of rights, be attributable to a legal person which leads to the question of which criteria the output of the machine should be measured against. Among a long list of possible recipients the following should be considered: the producer or supplier of the system, the software programmer,[76] the legal owner or the person in physical possession of the machine as well as the user or operator.[77]

This question goes well beyond Intellectual Property Rights Law because autonomous systems deliver something far more valuable than a mere "substitute" for an intellectual creation. They create experience, most noticeably through machine learning as well as knowledge, both in the sense of new insights as well as in the sense of the aggregation of that which is already known.

The legal attribution of data created by machines is being discussed under the heading "Ownership of Data" in literature[78] as well as at the level of the legislature[79] and in industry.[80] In Germany, the idea of a "property" in data has been rejected so far. The majority view expressed by the 71st "Deutsche Juristentag" held in 2016 was that the "right to one's own database" should not be protected as a so-called "absolute right", e.g. property in things, pursuant to § 823 Abs. 1 of the German Civil Code (BGB) but protection should instead be achieved though the creation of new protective norm ("Schutzgesetz") based on the model of § 303a (data manipulation) of the German Criminal Code (StGB)[81] which refers to the deletion or manipulation of data. The idea of "ownership in data" has also been rejected by the

---

[75]But see Schaub (2017), pp. 342 ff., who does not consider it necessary to develop the protection of independent autonomous systems or *de lege lata* (*ibid.*, pp. 347 ff.) nor *de lege ferenda* (*ibid* pp. 348 ff.) further.

[76]The deliberations in Schaub (2017), pp. 343 ff. are limited to the manufacturers of systems and the programmers of software.

[77]On the question of the attribution of rights to the user, see in particular Samuelson (1986), pp. 1200 ff.

[78]See e.g. Berberich (2010), *passim*; Zech (2012), *passim.*

[79]The European Parliament also posed this question in its resolution (*supra* note 14), cf. para. 18 (p. 9).

[80]For example in the area of the increasing automisation of industry (Industry 4.0), the question of the ownership of data is the focus of discussion; see e.g. IT-Sicherheit für die Industrie 4.0, Studie im Auftrag des Bundesministeriums für Wirtschaft und Energie, Final Report, pp. 103 ff., available at https://www.bmwi.de/Redaktion/DE/Publikationen/Studien/it-sicherheit-fuer-industrie-4-0.pdf?__blob=publicationFile&v=4. Accessed 28 August 2021.

[81]Resolution No. 28 of the 71st DJT (Deutscher Juristentag, Essen 2016).

Conference of the German State Ministries of Justice[82] as well as by authors who had argued for such a concept some years ago.[83]

However, the discussion on the question whether creations of machines should be protected is going on in the field of copyright and patents, raising a more general legal question: Who should these valuable goods be apportioned to?

2.2 *Declaration of Intent and Contracts*

Legal doctrines must be developed further in order to be able to classify autonomous systems appropriately. Legal doctrines, such as the declaration of intent ("Willenserklärung") which in German law as well as in many other legal traditions is considered as a declaration made by a natural person, are heavily questioned by modern technologies.[84] According to the traditional concept a declaration of intent requires, in addition to an objective element which is the perceivable declaration, there to be a subjective element, the intent, which is defined as a human intention to act.[85]

While for decades already, contracts have been made by the use of machines, many authors and court decisions continued to trace back the declaration made by a computer to human intention.[86] An illustrative example is provided by a ruling of the Frankfurt Higher Regional Court in 2003 dealing with the acceptance of an order placed in a web store that was declared automatically by the system. The court classified the automatic declaration of acceptance as a valid declaration of intent with the reasoning that the software only executes commands that are determined by human programming. Therefore, according to the court, the automatically generated declaration had its origin in a human action, which was initiated by the declarant and therefore originated from the declarant's will. With this reasoning, the court affirmed a declaration of intent by the operator of the online store.[87] Other courts followed this line of argument.[88] The court did not see or did not want to see the systematic gap in the argument that the programming was not carried out by the operator of the online store and therefore it would have to be justified how the subjective idea of the programmer should be attributed to the operator of the online store.

---

[82] Working Party "Digital Restart" of the Conference of the German State Ministers of Justice, Report from 15 May 2017, pp. 29 ff., available at https://jm.rlp.de/fileadmin/mjv/Jumiko/Fruehjahrskonferenz_neu/Bericht_der_AG_Digitaler_Neustart_vom_15._Mai_2017.pdf. Accessed 28 August 2021.

[83] See, e.g. Hoeren (2019), pp. 5 ff. revising his earlier opinion published in *id.*, (2013), pp. 486 ff.

[84] Cf. Teubner (2018), pp. 177 ff.

[85] Faust (2021), § 2 at mn. 4; Grigoleit and Herresthal (2015), mn. 6; Leipold (2019), § 10 at mn. 17.

[86] For an overview, see Borges (2008), pp. 194 f.; Borges and Sesing (2018), pp. 189 ff.

[87] OLG Frankfurt, MMR 2003, 405, 406, **ECLI:DE:OLGHE:2002:1120.9U94.02.0A.**

[88] OLG Hamm, MMR 2004, 761, 762, **ECLI:DE:OLGHAM:2004:0112.13U165.03.00**; similarly LG Köln, MMR 2003,p. 482, **ECLI:DE:LGK:2003:0416.9S289.02.00**.

Some recent monographs event represent the view that the elements of a declaration of intent can be fulfilled without the subjective element in the form of a human thought[89] which would be a remarkable step in the doctrine of a declaration of intent. On the other hand, many current publications also contain traditional explanations of the subjective elements of a declaration of intent which describe a human intention to act.[90] If this were true, machine-generated declarations could not be recognised as declarations of intent, this conclusion however is not drawn in the current literature. Instead, it is unanimously accepted that machine-generated declarations are to be recognised as declarations of intent.[91]

2.3 *Liability for Autonomous Systems*

Whether it is an autonomous car which causes an accident, a care-giving robot which injures a patient, a chatbot which makes racist comments or a system for choosing staff which discriminates on the grounds of sex, age or origin, autonomous systems will inevitably cause damage and they will breach legal provisions. There will be a multitude of cases in which the system will choose an action which, had the same action been chosen by a human, would be illegal or would even go as far as to breach the rights of third parties.

Liability for this type of damage which, from a Civil Law perspective, is the liability to pay damages, is currently at the centre of both legal discussion and legal policy considerations. The discussion is extremely controversial and the broad spectrum of suggested solutions ranges from the view that the current system is sufficient,[92] to extensive deliberations about new systems such as the possibility mentioned by the European Parliament in its 2017 resolution of the creation of liability funds.[93]

Quite often, the introduction of strict liability is suggested, as the European Parliament did in its resolution of 2017 already[94] and again with its proposal for

[89] Wiebe (2002), pp. 214 ff. (attribution of the declaration according to risk sphere); Schulz (2015), p. 104; similarly already Brehmer (1992), pp. 80 ff.

[90] Faust (2021), § 2 at mn. 4; Grigoleit and Herresthal (2015), mn. 6; Leipold (2019), § 10 at mn. 17.

[91] Arnold (2020), mn. 5a; Borges and Sesing (2018), p. 191; Cornelius (2002), pp. 355 f.; Faust (2021), § 2 at mn. 5; Kuhn (1991), pp. 70 ff.; Leipold (2019), § 10 at mn. 21; Paulus (2019), pp. 963 ff.; Paulus and Matzke (2018), pp. 440 ff.; Singer (2017), mn. 57; Spindler (2019), mn. 6; Wiebe (2002), p. 139.

[92] For example von Bodungen and Hoffmann (2016), p. 508; Grützmacher (2016), pp. 697 f.; Spindler (2015), p. 774; specifically for systems with embedded software Wagner (2019), p. 36.

[93] Resolution of the European Parliament of 16 February 2017 (*supra* note 14), para. 58, p. 15; Beck (2013), p. 256; rejected by Hanisch (2014), p. 43.

[94] European Parliament resolution of 16 February 2017, (*supra* note 14), pp. 17 f., para. 53 ff.

adjustments to the civil liability regimes for artificial intelligence.[95] In this context, the importance of insurance is highlighted by the legislator, e.g. by the European Parliament in its 2017 resolution,[96] expert groups[97] and many authors.[98] An important aspect, inter alia, here is the question whether compulsory insurance for autonomous systems should be introduced.[99]

2.4 New Legal Instruments

It is already foreseeable that the development and use of autonomous systems will lead to wide-ranging changes to the legal framework. The most striking example is the deliberation mentioned in the 2017 resolution of the European Parliament to give autonomous systems their own status with legal personality.[100] Even though this concept is, quite rightly, being met with a great deal of scepticism,[101] and gave rise to an open letter of scientists who vehemently opposed the introduction of the legal status of "an electronic person" to robots,[102] the mere existence of such deliberations highlights the need for the legal system to address autonomous systems in an entirely different way from what has been possible to date.

It will, in any event, be necessary to have legal provisions which deal with the legal aspects of the particular characteristics of decision-making by systems.

This is obvious, not only by public discussions surrounding autonomous vehicles and their choice of potential victims in the case of an unavoidable accident,[103] but

[95] European Parliament resolution of 20 October 2020 with the recommendations to the Commission on a civil liability regime for artificial intelligence (2020/2014(INL)), (*supra* note 18), p. 7, para. 14.

[96] European Parliament resolution of 16 February 2017 (*supra* note 14), p. 18, para. 57.

[97] See for example Expert Group on Liability and New Technologies—New Technologies Formation, Liability for Artificial Intelligence and other emerging digital technologies, 2019 (*supra* note 24), p. 62.

[98] Borges (2019), pp. 156 f.; cf. Hanisch (2014), p. 58; Horner and Kaulartz (2016b), p. 26; explicitly for autonomous vehicles Lutz (2015), p. 121; Spiecker gen. Döhmann (2016), p. 704; Spindler (2015), pp. 775 f.

[99] For an introduction of compulsory insurance, see e.g. Spindler (2015), p. 775; reserved support also from Horner and Kaulartz (2016a), p. 14.

[100] European Parliament resolution of 16 February 2017 (*supra* note 14), para. 59 ff., p. 15.

[101] Abbot and Sarch (2019), pp. 376 f.; Armbrüster (2017), p. 85; Denga (2018), p. 77; Bryson et al. (2017), pp. 273 ff.; cf. Kainer and Förster (2020), p. 298; Koch (2019), p. 115; Karner (2019), pp. 123 f.; Lohmann (2017), p. 171; Müller-Hengstenberg and Kirn (2014), pp. 307 ff.; Riehm (2014), Schaub (2017), p. 348; Solum (1992), pp. 1281 ff.; Spiecker gen. Döhmann (2016), p. 702; Spindler (2015), pp. 774 ff.

[102] Available at http://www.robotics-openletter.eu/. Accessed 28 August 2021.

[103] Regarding this dilemma, see Bonnefon et al. (2016), pp. 1573 ff.; On the significance of the ethical discussion for autonomous cars see for example Gerdes and Thornton (2015), pp. 87 ff.; For legal discussion on the ethical questions of autonomous driving, see for example Weber (2016),

also by the research and debate focussing on fairness, potential discrimination or other mistakes in autonomous decision making.[104]

One does not need to be a clairvoyant to be able to predict that legal development will not be determined by any large, specific change but that new legal instruments will need to be created to deal with situations on a case-by-case basis. It is highly likely that a concept which will be referred to here as "partial equality" of humans and machines whereby machines and humans will be treated as equals in certain legal provisions or sets of rules will play a central role. An example of this can be seen in the statement made by the US National Highway Traffic Safety Administration on 4 February 2016 classifying autonomous cars as "drivers" in the sense of Road Traffic legislation.[105]

In the current discussion on liability for damage caused by autonomous systems, it is sometimes proposed in German literature that the behaviour of machines should be apportioned by analogous application of § 278 BGB, which regulates the apportionment of fault by human vicarious agents.[106] If one were to accept this, one would be applying a norm that relates exclusively to human behaviour to the conduct of machines in the same way and would even do so in an aspect of behaviour (fault) that has been a traditional preserve of humans.

The analogous application of norms on vicarious liability, such as § 278 BGB, raises two different issues:[107] First, can an apportionment norm created for humans be used for the attribution of machine conduct? In principle, this question is to be answered in the affirmative. It is the essence of analogy to close a legal loophole, which is the case here, by applying a norm that was created for a different area of application.

---

pp. 249 ff. In its final report, the Ethics-Commission on Autonomous and Connected Driving set up by the German Federal Ministry for Transport and Digital Infrastructure considered that this dilemma situation could not be solved by legislation (*supra* note 5). For the details, see thesis 8 (p. 11).

[104] See, e.g. Gesellschaft für Informatik (GI), Technische und rechtliche Betrachtungen algorithmischer Entscheidungsverfahren, Studien und Gutachten im Auftrag des Sachverständigenrats für Verbraucherfragen, 2018, *passim,* available at https://www.svr-verbraucherfragen.de/wp-content/uploads/GI_Studie_Algorithmenregulierung.pdf. Accessed 28 August 2021; Hartmann (2019), pp. 421 f.; Steege (2019), pp. 715 ff.

[105] The statement was made in response to an inquiry from Google regarding the interpretation of the term "driver" or "driving seat" in the *Federal Motor Vehicle Safety Standards* (FMVSS). In its response, NHTSA assumed on the basis of numerous requirements, that the entity which has the driving control of the vehicle should be considered the driver—in case of autonomous vehicles, this would be the system itself; cf. the second paragraph in the "Discussion" section of NHTSA's letter, available at https://isearch.nhtsa.gov/files/Google%20%2D%2D%20compiled%20response%20to%2012%20Nov%20%2015%20interp%20request%20%2D%2D%204%20Feb%2016%20final.htm. Accessed 28 August 2021.

[106] Wulf and Burgenmeister (2015), p. 407; similarly Wolf (1989), pp. 901 f.; similar (depending on the degree of autonomy of the respective machine) Hacker (2018), pp. 248 ff.

[107] Hacker (2018), pp. 249 ff.

Another issue is whether fault is a concept that should be applied to the conduct of machines.[108] This question is extremely far-reaching: According to the traditional understanding of the law, the category of fault is only applicable to those humans who are capable of fault and, therefore, from whom one can expect an understanding of the required behaviour and the ability to adjust one's own behaviour to it.

Children, for example, are only expected to have this understanding to a very limited extent. Incidentally, this is probably also the reason why an analogous application of § 278 BGB or parallel norms on vicarious liability in other legal systems to animals was not considered, but rather an independent animal owner's liability was developed.

So, should machines be classified as entities able to incur fault? Apparently, this is not assumed even by advocates of analogous application of vicarious liability to machine conduct. If a "fault" of the machine were not a condition for liability of the operator, the analogous application of § 278 BGB would result in a strict liability of the operator of the autonomous system which seems to be inadmissible, even in contractual relationships.

However, the introduction of new strict liability rules is reserved for the legislator in many legal systems, such as in German law. The (analogous) application of § 278 BGB is therefore quite rightly rejected.[109] These examples show that the innocent question of applying norms created for human behaviour to machine conduct raises fundamental questions regarding the classification of autonomous systems in law.

It may be concluded that the development of an adequate legal framework for autonomous systems cannot be achieved by simply analogising existing rules designed for humans to machines.

2.5 *Law Enforcement*

Autonomous systems pose significant challenges for law enforcement. In the field of criminal law topics such as the question of whether autonomous systems can themselves be punished are being discussed.[110] The parallel question whether a human can be held liable for the actions of an autonomous system,[111] is in the focus of Civil Law discussions.

---

[108] Hacker (2018), pp. 256 ff.

[109] Freyler (2020), pp. 288 f.; Günther and Böglmüller (2017), p. 55; Horner and Kaulartz (2016a), p. 7; Müller-Hengstenberg and Kirn (2014), p. 311; in favor of a solution *de lege ferenda* Klingbeil (2019), pp. 723 ff.

[110] Gless et al. (2016), p. 416; Hilgendorf (2012), pp. 128 ff.; Joerden (2013), pp. 203 ff.; Ziemann (2013), pp. 183 ff.

[111] Beck (2009), pp. 227 ff.; Grützmacher (2016), p. 696; Hilgendorf (2013), pp. 11, 16 ff.; Lohsse et al. (2019), p. 20; Müller (2014), p. 605; from a philosophical perspective Matthias (2010), pp. 72 ff.; Spiecker gen. Döhmann (2016), p. 701; Spindler (2015), pp. 766 f.

If civil law liability is used as an instrument to steer behaviour, it is immediately clear that the incentives created by the thread of liability is only as effective as the possibility of the victim to succeed with damages claims in a civil court proceeding.[112]

A great deal of further research is required on this issue. A central aspect in this respect is the attribution of the burden of proof. The person injured by an autonomous system will often have no knowledge of the necessary facts to support a claim for damages, especially in cases where the claim for damages presupposes fault on the part of the manufacturer or operator of the system or a defect in the system. Where fact-finding instruments such as the pre-trial discovery of US law are not available, as is the case in continental European jurisdictions, the injured party runs the risk of not being able to enforce a claim for damages due to lack of evidence.

This problem was highlighted, for example, by the New Technologies Formation of the EU Commission's Expert Group on Liability and New Technologies in its final report which called for the reversal of the burden of proof.[113] This demand was taken up by the European Parliament in its resolution of 20 October 2020.[114] Its proposal for a regulation on the liability of operators of AI systems provides for a reversal of the burden of proof at the expense of the operators.[115] Even though the Parliament's regulatory proposal is still rather immature in this respect, it rightly points to the need to regulate the question of the burden of proof as a central aspect of the enforceability of claims for damages in the case of damage caused by autonomous systems.

3 Conclusion

Autonomous systems are key elements of the current formation of a digital society. The current legal framework cannot deal adequately with the autonomy of machines and must therefore be developed further. The development cannot be achieved by simple analogy to norms which were designed for human behaviour.

The further development of a legal framework for autonomous systems must take place quickly as the development and implementation of the systems are forging ahead at great speed. The effort must be international, preferably with global coverage, as the development, production and implementation of the systems crosses national borders.

[112] Cf. Förster (2021), mn. 9; Kötz and Wagner (2021), chap. 4, II at mn. 32; Wagner (2020), mn. 46.

[113] Expert Group on Liability and New Technologies—New Technologies Formation, Liability for Artificial Intelligence and other emerging digital technologies, 2019 (*supra* note 24), p. 52.

[114] European Parliament resolution of 20 October 2020 with recommendations to the Commission on a civil liability regime for artificial intelligence (2020/2014(INL)) (*supra* note 28), p. 8, para. 6.

[115] European Parliament resolution of 20 October 2020 with recommendations to the Commission on a civil liability regime for artificial intelligence (2020/2014(INL)) (*supra* note 28), p. 5, para. 5.

Moreover, the further development of the legal framework requires cross-disciplinary research. Legal regulation of the complex technologies and their economic and social embedding can only be successful if the technology and its global social and economic importance is sufficiently taken into account. To support and guide the development of a legal framework for autonomous systems is a task assigned to legislature. The necessary cross-disciplinary research will not take place without sufficient support and, without goal-oriented guidance, key aspects for legislature will not be achievable within a short period of time. The challenge to create an appropriate legal framework for autonomous systems is one which must be overcome by politics and academia working together.

References

Abbot, R., & Sarch, A. (2019). Punishing artificial intelligence: Legal fiction or science fiction. *UC Davis Law Review, 53*(1), 323–384.

Ahlberg, H. (2018). § 7. In H. Ahlberg & H.-P. Götting (Eds.), *Beck online-Kommentar-Urheberrecht* (25th ed.). C.H. Beck.

Armbrüster, C. (2017). Automatisiertes Fahren – Paradigmenwechsel im Straßenverkehrsrecht? *Zeitschrift für Rechtspolitik (ZRP), 50*(3), 83–86.

Arnold, A. (2020). Vorb. § 116 BGB. In H. P. Westermann, B. Grunewald, & G. Maier-Reimer (Eds.), *Erman BGB Kommentar* (16th ed.). Otto-Schmidt-Verlag.

Arzt, C., & Ruth-Schumacher, S. (2017). Zulassungsrechtliche Rahmenbedingungen der Fahrzeugautomatisierung. *Neue Zeitschrift für Verkehrsrecht (NZV), 30*(2), 57–62.

Balke, R. (2018). Automatisiertes Fahren – Begriffsbestimmungen und haftungsrechtliche Fragestellungen im Zusammenhang mit dem automatisierten Fahren. *Straßenverkehrsrecht (SVR), 18*(1), 5–8.

Beck, S. (2009). Grundlegende Fragen zum rechtlichen Umgang mit der Robotik. *Juristische Rundschau (JR)*, 225–230.

Beck, S. (2013). Über Sinn und Unsinn von Statusfragen – zu Vor- und Nachteilen der Einführung einer elektronischen Person. In E. Hilgendorf & J.-P. Günther (Eds.), *Robotik und Gesetzgebung* (pp. 239–262). Nomos.

Berberich, M. (2010). *Virtuelles Eigentum*. Mohr Siebeck.

Berndt, S. (2017). Der Gesetzentwurf zur Änderung des Straßenverkehrsgesetzes – ein Überblick – Das automatisierte Fahren auf dem Vormarsch – oder doch nur ein Schnellschuss? *Straßenverkehrsrecht (SVR), 17*(4), 121–127.

Bonnefon, J., Shariff, A., & Rahwan, I. (2016). The social dilemma of autonomous vehicles. *Science, 352*(6293), 1573–1576.

Borges, G. (2008). *Verträge im elektronischen Geschäftsverkehr*. Nomos.

Borges, G. (2018). Rechtliche Rahmenbedingungen für autonome Systeme. *Neue Juristische Wochenschrift (NJW), 71*(14), 977–982.

Borges, G. (2019). New liability concepts: The potential of insurance and compensation funds. In S. Lohsse, R. Schulze, & D. Staudenmayer (Eds.), *Liability for artificial intelligence and the internet of things* (pp. 145–164). Nomos.

Borges, G., & Sesing, A. (2018). Der Irrtum des Kühlschranks und andere Fragen zu Willenserklärungen von Maschinen: Eine Bestandsaufnahme. In A. Matusche-Beckmann & T. Sato (Eds.), *Saar-Tage 2016: Rechtsprobleme der Informationsgesellschaft* (pp. 179–201). Verlag Alma Mater.

Brehmer, N. (1992). *Wille und Erklärung: Zu Geltungsgrund, Tatbestand und Zurechnung der Willenserklärung*. Nomos.

Bryson, J. J., Diamantis, M. E., & Grant, T. D. (2017). Of, for, and by the people: The legal lacuna of synthetic persons. *Artificial Intelligence and Law, 25*(3), 273–291.

Cornelius, K. (2002). Vertragsabschluss durch autonome elektronische Agenten. *Multimedia und Recht (MMR), 5*(6), 353–358.

Denga, M. (2018). Deliktische Haftung für künstliche Intelligenz. *Computer und Recht (CR)*, 69–78.

Fagnant, D. J., & Kockelman, K. (2015). Preparing a nation for autonomous vehicles: Opportunities, barriers and policy recommendations. *Transportation Research Part A: Policy and Practice, 77*(C), 167–181.

Faust, F. (2021). *Bürgerliches Gesetzbuch Allgemeiner Teil* (7th ed.). Nomos.

Foerster, M. (2019). Automatisierung und Verantwortung im Zivilrecht. *Zeitschrift für die gesamte Privatrechtswissenschaft (ZfPW), 5*(4), 418–435.

Förster, C. (2021). § 823. In R. Poseck & W. Hau (Eds.), *Beck-Online Kommentar BGB* (57th ed.). C.H. Beck.

Franklin, S., & Graesser, A. (1996). In J. Müller, M. J. Wooldridge, & N. R. Jennings (Eds.), *Intelligent agents III. Agent theories, architectures, and languages – ECAI' 96 workshop (ATAL), Budapest, Hungary, 12–13 August 1996. Proceedings* (pp. 21–35). New York: Springer.

Frese, Y. (2015). Recht im zweiten Maschinenzeitalter. *Neue Juritische Wochenschrift (NJW), 68*(29), 2090–2092.

Frey, C. B., & Osborne, M. A. (2017). The future of employment: How susceptible are jobs to computerisation? *Technological Forecasting and Social Change, 114*(C), 254–280.

Freyler, C. (2020). Robot-Recruiting, Künstliche Intelligenz und das Antidiskriminierungsrecht. *Neue Zeitschrift für Arbeitsrecht (NZA), 37*(5), 284–290.

Galetzka, C., Garling, S., & Partheymüller, J. (2021). Legal tech – "Smart law" oder Teufelszeug? – Übersicht der aktuellen Rechtsprechung zu legal tech zwischen Rechtspolitik und technischer Entwicklung. *Multimedia und Recht (MMR), 24*(1), 20–26.

Gerdes, J. C., & Thornton, S. M. (2015). Implementable ethics for autonomous vehicles. In M. Maurer, J. C. Gerdes, B. Lenz, & H. Winner (Eds.), *Autonomes Fahren – Technische, rechtliche und gesellschaftliche Aspekte* (pp. 87–102). Springer Vieweg.

Gibbs, S. (2017). *Elon Musk: Regulate AI to combat 'existential threat' before it's too late*. Retrieved August 28, 2021, from https://www.theguardian.com/technology/2017/jul/17/elon-musk-regulation-ai-combat-existential-threat-tesla-spacex-ceo

Gless, S., Silverman, E., & Weigend, T. (2016). If robots cause harm, who is to blame? Self-driving cars and criminal liability. *New Criminal Law Review, 19*(3), 412–436.

Gomille, C. (2019). Kreative künstliche Intelligenz und das Urheberrecht. *Juristen Zeitung (JZ)*, 969–975.

Grigoleit, H. C., & Herresthal, C. (2015). *BGB Allgemeiner Teil* (3rd ed.). C.H. Beck.

Grünvogel, T. (2017). Das Fahren von Autos mit automatisierten Funktionen. *Monatsschrift für deutsches Recht (MDR)*, 973–975.

Grützmacher, M. (2016). Die deliktische Haftung für autonome Systeme – Industrie 4.0 als Herausforderung für das bestehende Recht? *Computer und Recht (CR)*, 695–698.

Grützmacher, M., & Heckmann, J. (2019). Autonome Systeme und KI – vom vollautomatisierten zum autonomen Vertragsschluss? – Die Grenzen der Willenserklärung. *Computer und Recht (CR)*, 553–561.

Günther, J., & Böglmüller, M. (2017). Künstliche Intelligenz und Roboter in der Arbeitswelt. *Betriebs-Berater (BB)*, 53–58.

Günther, J.-P. (2017). Europäische Regelungen im Bereich Robotik – alles noch Science Fiction? *Der Betrieb (DB)*, 651–653.

Hacker, P. (2018). Verhaltens- und Wissenszurechnung beim Einsatz von Künstlicher Intelligenz. *Rechtswissenschaft (RW)*, 243–288.

Hähnchen, S., Schrader, P., Weiler, F., & Wischmeyer, T. (2020). Legal Tech – Rechtsanwendung durch Menschen als Auslaufmodell? *Juristische Schulung (JuS), 60*(7), 625–635.

Hanisch, J. (2014). Zivilrechtliche Haftungskonzepte für Robotik. In E. Hilgendorf (Ed.), *Robotik im Kontext von Recht und Moral* (pp. 27–63). Nomos.

Hartmann, F. (2019). Diskriminierung aus der Black Box – Neue Herausforderungen durch KI-gestützte Personalentscheidungen. *Europäische Zeitschrift für Arbeitsrecht (EuZA), 12*(4), 421–422.

Hilgendorf, E. (2012). Können Roboter schuldhaft handeln? Zur Übertragbarkeit unseres normativen Grundvokabulars auf Maschinen. In S. Beck (Ed.), *Jenseits von Mensch und Maschine* (pp. 119–132). Nomos.

Hilgendorf, E. (2013). Das Recht vor den Herausforderungen der modernen Technik. In E. Hilgendorf & S. Hötitzsch (Eds.), *Das Recht vor den Herausforderungen der modernen Technik* (pp. 11–40). Nomos.

Hoeren, T. (2013). Dateneigentum – Versuch einer Anwendung von § 303a StGB im Zivilrecht. *Multimedia und Recht (MMR), 16*(8), 486–491.

Hoeren, T. (2019). Datenbesitz statt Dateneigentum – Erste Ansätze zur Neuausrichtung der Diskussion um die Zuordnung von Daten. *Multimedia und Recht (MMR), 22*(1), 5–8.

Horner, S., & Kaulartz, M. (2016a). Haftung 4.0 Verschiebung des Sorgfaltsmaßstabs bei Herstellung und Nutzung autonomer Systeme. *Computer und Recht (CR)*, 7–19.

Horner, S., & Kaulartz, M. (2016b). Haftung 4.0. *Zeitschrift für Innovation und Technikrecht (InTeR)*, 22–27.

Jänich, V. M., Schrader, P. T., & Reck, V. (2015). Rechtsprobleme des autonomen Fahrens. *Neue Zeitschrift für Verkehrsrecht (NZV), 28*(7), 313–318.

Joerden, J. C. (2013). Strafrechtliche Perspektiven der Robotik. In E. Hilgendorf & J.-P. Günther (Eds.), *Robotik und Gesetzgebung* (pp. 195–211). Nomos.

Kainer, F., & Förster, L. (2020). Autonome Systeme im Kontext des Vertragsrechts. *Zeitschrift für die gesamte Privatrechtswissenschaft (ZfPW), 6*(3), 275–303.

Karner, E. (2019). Liability for robotics: Current rules, challenges, and the need for innovative concepts. In S. Lohsse, R. Schulze, & D. Staudenmayer (Eds.), *Liability for artificial intelligence and the internet of things* (pp. 117–124). Nomos.

Keßler, O. (2017). Intelligente Roboter – neue Technologien im Einsatz – Voraussetzungen und Rechtsfolgen des Handelns informationstechnischer Systeme. *Multimedia und Recht (MMR), 20*(9), 589–594.

Kilian, M. (2017). Die Zukunft der Juristen – Weniger, anders, weiblicher, spezialisierter, alternativer – und entbehrlicher? *Neue Juristische Wochenschrift (NJW), 70*(42), 3043–3050.

Kirn, S., & Müller-Hengstenberg, C. D. (2017). *Rechtliche Risiken autonomer und vernetzter Systeme – Eine Herausforderung*. DeGruyter Oldenbourg.

Klingbeil, S. (2019). Schuldnerhaftung für Roboterversagen: Zum Problem der Substitution von Erfüllungsgehilfen durch Maschinen. *Juristen Zeitung (JZ)*, 718–725.

Kluge, V., & Müller, A.-K. (2017). Autonome Systeme. *Zeitschrift zum Innovations- und Technikrecht (InTeR)*, 24–31.

Koch, B. (2019). Product liability 2.0 – Mere update or new version? In S. Lohsse, R. Schulze, & D. Staudenmayer (Eds.), *Liability for artificial intelligence and the internet of things* (pp. 97–116). Nomos.

König, C. (2017). Die gesetzlichen Neuregelungen zum automatisierten Fahren. *Neue Zeitschrift für Verkehrsrecht (NZV), 30*(3), 123–128.

Kötz, H., & Wagner, G. (2021). *Deliktsrecht* (14th ed.). Verlag Franz Vahlen.

Kraßer, R., & Ann, C. (2016). *Patentrecht – Lehrbuch zum Deutschen und Europäischen Patentrecht und Gebrauchsmusterrecht* (7th ed.). C.H. Beck.

Kuflik, A. (1999). Computers in control: Rational transfer of authority or irresponsible abdication of autonomy? *Ethics and Information Technology, 1*, 173–184. Retrieved August 28, 2021, from https://doi.org/10.1023/A:1010087500508

Kuhn, M. (1991). *Rechtshandlungen mittels EDV und Telekommunikation*. C.H. Beck.

Kütük-Markendorf, M. (2017). IT und Software. Die hoch- oder vollautomatisierte Fahrfunktion als Vorstufe zum autonomen Fahren. *Compuer und Recht (CR)*, 349–355.

Lauber-Rönsberg, A. (2019). Autonome "Schöpfung" – Urheberschaft und Schutzfähigkeit. *Gewerblicher Rechtschutz und Urheberrecht (GRUR), 121*(3), 244–253.
Leipold, D. (2019). *BGB I, Einführung und Allgemeiner Teil* (10th ed.). Mohr Siebeck.
Linke, C. (2021). *Digitale Wissensorganisation – Wissenszurechnung beim Einsatz autonomer Systeme*. Nomos.
Lohmann, M. F. (2017). Ein europäisches Roboterrecht – überfällig oder überflüssig? *Zeitschrift für Rechtspolitik (ZRP), 50*(6), 168–171.
Lohsse, S., Schulze, R., & Staudenmayer, D. (2019). Liability for artificial intelligence. In S. Lohsse, R. Schulze, & D. Staudenmayer (Eds.), *Liability for artificial intelligence and the internet of things* (pp. 11–23). Nomos.
Lutz, L. S. (2015). Autonome Fahrzeuge als rechtliche Herausforderung. *Neue Juristische Wochenschrift (NJW), 68*(3), 119–124.
Matthias, A. (2010). *Automaten als Träger von Rechten. Plädoyer für eine Gesetzesänderung* (2nd ed.). Logos Verlag.
Melullis, K.-J. (2015). § 6. In G. Benkhard, (founder) *Benkard Patentgesetz* (11th ed.). C.H. Beck.
Mes, P. (2015). § 6. In P. Mes (Ed.), *Patentgesetz, Gebrauchsmustergesetz, Kommentar* (4th ed.). C.H. Beck.
Müller, M. F. (2014). Roboter und Recht. *Aktuelle juristische Praxis – Pratique juridique Actuelle (AJP)*, 595–608.
Müller-Hengstenberg, C. D., & Kirn, S. (2014). Intelligente (Software-)Agenten: Eine neue Herausforderung unseres Rechtssystems – Rechtliche Konsequenzen der "Verselbstständigung" technischer Systeme. *Multimedia und Recht (MMR), 17*(5), 307–313.
Nägerl, J., Neuburger, B., & Steinbach, F. (2019). Künstliche Intelligenz: Paradigmenwechsel im Patentsystem. *Gewerblicher Rechtsschutz und Urheberrecht (GRUR), 121*(4), 336–341.
Oppermann, B., & Stender-Vorwachs, J. (2020). *Autonomes Fahren, Rechtsfolgen, Rechtsprobleme, technische Grundlagen* (2nd ed.). C.H. Beck.
Palka, A. (2018). *Digitalisierung gefährdet Millionen von Jobs – welche besonders betroffen sind.* Retrieved August 28, 2021, from https://www.handelsblatt.com/unternehmen/management/digitaletransformation/oecd-studie-zur-zukunft-des-arbeitsmarktes-digitalisierunggefaehrdet-millionen-von-jobs-welche-besonders-betroffen-sind/21217278.html
Paulus, D. (2019). Die automatisierte Willenserklärung. *Juristische Schulung (JuS)*, 960–965.
Paulus, D., & Matzke, R. (2018). Smart Contracts und das BGB – Viel Lärm um nichts? *Zeitschrift für die gesamte Privatrechtswissenschaft (ZfPW), 4*(4), 431–465.
Phaneuf, A. (2020). *Artificial intelligence in financial services: Applications and benefits of AI in finance*. Retrieved August 28, 2021, from https://www.businessinsider.com/ai-in-finance
Pieper, F.-U. (2016). Die Vernetzung autonomer Systeme im Kontext von Vertrag und Haftung. *Zeitschrift zum Innovations- und Technikrecht (InTeR)*, 188–194.
Prior, P. (2017). Legal Tech – Digitalisierung der Rechtsberatung. *Zeitschrift für Anwaltspraxis (ZAP)*, 575–581.
Reichwald, J., & Pfisterer, D. (2016). Autonomie und Intelligenz im Internet der Dinge. *Computer und Recht (CR)*, 208–212.
Riehm, T. (2014). Von Drohnen, Google-Cars und Software-Agenten – Rechtliche Herausforderungen autonomer Systeme. *Der IT-Rechtsberater (ITRB)*, 113–115.
Römermann, V., & Günther, T. (2019). Legal tech als berufsrechtliche Herausforderung – Zulässige Rechtsdurchsetzung mit Prozessfinanzierung und Erfolgshonorar. *Neue Juristsiche Wochenschrift (NJW), 72*(9), 551–555.
Russell, S., & Norvig, P. (2010). *Artificial intelligence: A modern approach* (3rd ed.). Prentice Hall.
Samuelson, P. (1986). Allocating ownership rights in computer generated works. *University of Pittsburgh Law Review, 47*, 1185–1228.
Schaub, R. (2017). Interaktion von Mensch und Maschine – Haftungs- und immaterialgüterrechtliche Fragen bei eigenständigen Weiterentwicklungen autonomer Systeme. *Juristen Zeitung (JZ)*, 342–349.

Schirmer, J.-E. (2017). Augen auf beim automatisierten Fahren! Die StVG-Novelle ist ein Montagsstück. *Neue Zeitschrift für Verkehrsrecht (NZV), 30*(6), 253–257.

Schirmer, J.-E. (2019). Von Mäusen, Menschen und Maschinen – Autonome Systeme in der Architektur der Rechtsfähigkeit. *Juristen Zeitung (JZ)*, 711–718.

Schulz, T. (2015). *Verantwortlichkeit bei autonom agierenden Systemen – Fortentwicklung des Rechts und Gestaltung der Technik*. Nomos.

Schulze, G. (2018). § 7. In T. Dreier & G. Schulze (Eds.), *Urheberrechtsgesetz, Verwertungsgesellschaftengesetz, Kunsturhebergesetz Kommentar* (6th ed.) C.H. Beck.

Singer, R. (2017). Vorb. §§ 116–144. In N. Habermann (Ed.), *Staudinger Kommentar zum BGB* (Vol. 1). DeGruyter.

Solum, L. B. (1992). Legal personhood for artificial intelligences. *North Carolina Law Review, 70*, 1231–1287.

Sommer, M. (2020). *Haftung für autonome Systeme – Verteilung der Risiken selbstlernender und vernetzter Algorithmen im Vertrags- und Deliktsrecht*. Nomos.

Sosnitza, O. (2016). Das Internet der Dinge – Herausforderung oder gewohntes Terrain für das Zivilrecht? *Computer und Recht (CR)*, 764–772.

Specht-Riemenschneider, L. (2021). Urheberrechtlicher Schutz für Algorithmenerzeugnisse? – Phasenmodell de lege lata, Investitionsschutz de lege ferenda? *Wettbewerb in Recht und Praxis (WRP)*, 273–278.

Spiecker gen. Döhmann, I. (2016). Zur Zukunft systemischer Digitalisierung – Erste Gedanken zur Haftungs- und Verantwortungszuschreibung bei informationstechnischen Systemen. *Computer und Recht (CR)*, 698–704.

Spindler, G. (2015). Roboter, Automation, künstliche Intelligenz, selbst-steuernde Kfz – Braucht das Recht neue Haftungskategorien? – Eine kritische Analyse möglicher Haftungsgrundlagen für autonome Steuerungen. *Computer und Recht (CR)*, 766–776.

Spindler, G. (2019). Vorb. §§ 116 ff. BGB. In G. Spindler & F. Schuster (Eds.), *Recht der elektronischen Medien – Kommentar* (4th ed.). C.H. Beck.

Steege, H. (2019). Algorithmenbasierte Diskriminierung durch Einsatz von Künstlicher Intelligenz – Rechtsvergleichende Überlegungen und relevante Einsatzgebiete. *Multimedia und Recht (MMR), 22*(11), 715–721.

Stiemerling, O. (2015). "Künstliche Intelligenz" – Automatisierung geistiger Arbeit, Big Data und das Internet der Dinge – Eine technische Perspektive. *Computer und Recht (CR)*, 762–765.

Taeger, J. (2016). Die Entwicklung des IT-Rechts im Jahr 2016. *Neue Juristische Wochenschrift (NJW), 69*(52), 3764–3770.

Teubner, G. (2018). Digitale Rechtssubjekte? Zum privatrechtlichen Status autonomer Softwareagenten. *Archiv für die civilistische Praxis (AcP), 218*(2-4), 155–205.

Thöne, M. (2020). *Autonome Systeme und deliktische Haftung. Verschulden als Instrument adäquater Haftungsallokation?* Mohr Siebeck.

Thum, D. (2019). § 7. In A.-A. Wandtke & W. Bullinger (Eds.), *Praxiskommentar zum Urheberrecht* (5th ed.). C.H. Beck.

von Bodungen, B., & Hoffmann, M. (2015). Belgien und Schweden schlagen vor: Das Fahrsystem soll Fahrer werden! *Neue Zeitschrift für Verkehrsrecht (NZV), 28*(11), 521–526.

von Bodungen, B., & Hoffmann, M. (2016). Autonomes Fahren – Haftungsverschiebung entlang der Supply Chain? (2. Teil). *Neue Zeitschrift für Verkehrsrecht (NZV), 29*(11), 503–509.

von Kaler, M., & Wieser, S. (2018). Weiterer Rechtsetzungsbedarf beim automatisierten Fahren. *Neue Zeitschrift für Verwaltungsrecht (NVwZ), 37*(6), 369–373.

Wagner, G. (2017a). Produkthaftung für autonome Systeme. *Archiv für die civilistische Praxis (AcP), 217*(6), 707–765.

Wagner, J. (2017b). Legal Tech und Legal Robots in Unternehmen und den diese beratenden Kanzleien. *Betriebs-Berater (BB)*, 898–905.

Wagner, G. (2019). Roboter als Haftungssubjekte? Konturen eines Haftungsrechts für autonome Systeme. In F. Faust & H.-B. Schäfer (Eds.), *Zivilrechtliche und rechtsökonomische Probleme des Internet und der künstlichen Intelligenz* (pp. 1–39). Mohr Siebeck.

Wagner, G. (2020). Vorb. § 823. In F. J. Säcker, R. Rixecker, H. Oetker, & B. Limperg (Eds.), *Münchner Kommentar zum BGB* (Vol. 8, 8th ed.). C.H. Beck.
Weber, P. (2016). Dilemmasituationen beim autonomen Fahren. *Neue Zeitschrift für Verkehrsrecht (NZV), 29*(6), 249–254.
Wiebe, A. (2002). *Die elektronische Willenserklärung*. Mohr Siebeck.
Wolf, M. (1989). Schuldnerhaftung bei Automatenversagen. *Juristische Schulung (JuS)*, 899–902.
Wulf, H. M., & Burgenmeister, C. (2015). Industrie 4.0 in der Logistik – Rechtliche Hürden beim Einsatz neuer Vernetzungs-Technologien. *Computer und Recht (CR)*, 404–412.
Zech, H. (2012). *Information als Schutzgegenstand*. Mohr Siebeck.
Zech, H. (2016). Zivilrechtliche Haftung für den Einsatz von Robotern – Zuweisung von Automatisierungs- und Autonomierisiken, Intelligente Agenten und das Recht. In S. Gless & K. Seelmann (Eds.), *Intelligente Agenten und das Recht* (pp. 165–204). Nomos.
Zech, H. (2019). Künstliche Intelligenz und Haftungsfragen. *Zeitschrift für die gesamte Privatrechtswissenschaft (ZfPW), 5*(2), 198–218.
Ziemann, S. (2013). Wesen, Wesen, seid's gewesen? Zur Diskussion über ein Strafrecht für Maschinen. In E. Hilgendorf & J.-P. Günther (Eds.), *Robotik und Gesetzgebung* (pp. 183–195). Nomos.

“Moralizing Technology” and Criminal Law Theory

Tatsuhiko Inatani

Contents

Abstract The conventional criminal law theory usually relies on the idea of humanism, where human agent, who is independent of his/her external circumstances, can unilaterally control non-human agent through his/her free will. However, AIs and emerging technologies gradually reveal that we will not be able to sustain this assumption, due to the “autonomous” nature of these artifacts. It becomes more and more obvious that these artificial objects influence our mode of existence in various ways. Therefore, it is important to focus on and to analyze interpenetrative relationship between human subject and artificial object, in order to reconstruct the criminal law which can accompany with the recent technological development. The future criminal law should not exclusively target the will of human agent, rather it should consider the nature of hybrid of subject/object and

T. Inatani (✉)
Kyoto University, Kyoto, Japan
e-mail: inatani.tatsuhiko.3w@kyoto-u.ac.jp

G. Borges, C. Sorge (eds.), *Law and Technology in a Global Digital Society*,
https://doi.org/10.1007/978-3-030-90513-2_2

distribute the criminal responsibility to human and non-human thorough structured democratic process for elaborating our "good life."

1 Introduction

"Guns kill people" is a slogan of those who try to control the unrestricted sale of guns. To which the National Rifle Association replies with another slogan, "Guns don't kill people; *people* kill people[1]". According to Bruno Latour, pro-gun controls are "materialists" in terms that they claim that on account of gun the law-abiding citizen becomes dangerous; while the NRA is a "moralist" as they say the gun is a tool, a medium, a neutral carrier of human will. However both camps are wrong because the relationship between a man and a gun does not have the essential nature of object and subject, but is rather an intertwining relation[2] in which "You are different with a gun in your hand; the gun is different with you holding it.[3]" What we ought to be looking at is how the mode of this hybrid of human and non-human should be and where it takes us.[4] To focus on the human-technology hybrid and its consequence has become more and more important today as technological development—signified by artificial intelligence—has shaken the foundations of our ways of thinking and the state of society.

According to Peter=Paul Verbeek, "Contrary to what many people intuitively think, these technologies are not simply neutral instruments that facilitates our existence, ...they give shapes to what we do and how we experience the world. And doing so they contribute actively the ways we live our lives... Cars, for instance, do not only take us from A to B. They also lengthen the radius enclosing our most frequent social contacts. They help to determine how far we live from where we work. And they organize how we design cities and neighborhoods.[5]"

This consequence of technology may be subject to moral evaluation. "(S)ome low-hanging overpasses on parkways in Long Island (New York) giving access to the beach is a good example. Architect Robert Moses deliberately built these overpasses so low that buses cannot use the parkways, implicitly limiting access to the beach for African Americans who could not afford cars of their own.[6]" Nevertheless, it is difficult to assume a complete control of the results that technology provokes. As phones and typewriters originally invented for the deaf and blind unexpectedly gave a massive impact to society, technology may, in the context of usage, be given a role

---

[1] Latour (1999), p. 176.

[2] Verbeek (2011), p. 4.

[3] Latour (1999), p. 179.

[4] See Latour (1999), pp. 179 f.

[5] Verbeek (2011), p. 1.

[6] Verbeek (2011), p. 5.

that the developer could not foresee. If technology is beyond our control, but actively affecting our morality and making us what we are, in order to understand "the ethical dimensions and relevance of technology," we have to cross the divide between "human and nonhuman.[7]"

However, according to Verbeek, these claims are a major challenge to the conventional ethics. "Material objects do not have minds or consciousness, they lack free will and intentionality and cannot be held responsible for their actions; therefore they cannot be fully fledged parts of the moral community, the argument goes[8]".

Given this kind of "humanism", embodied by "modern" thinkers who argued essential division between subject and object based on the dualism of mind and body, it can be said that modern criminal law theory, founded on the premise of human "autonomy" derived from the concept of "free will", is also based on "humanism." But to reduce the gun violence, it is not necessary to use criminal law based on this kind of humanism.

For example, as suggested by Latour, let's take a look at the human-nonhuman hybrids in the context of gun murder. The problem here is the undesirable state of human-nonhuman hybrids in which a minor dispute could immediately turn into critical gun violence. Then it might be apparent that we have other options rather than punishment of the perpetrator, like redesigning the gun or to changing the relationship between human and nonhuman. Of course, this possibility does not necessarily mean ethically/legally censuring a gun murderer is wrong. To perceive these censures as essentially wrong would be the same mistakes committed by "materialists" whom Latour criticizes. Instead, what I do want to clarify is that to pay close attention to the human-nonhuman hybrids allows us, criminal law scholars, to depart from the perpetuate tendency to think that appealing to the gun user's autonomy, thorough punishment, is the almost only desirable way to solve or deter the crime, constructively, fundamentally and ethically. Criminal law scholars who are often humanists, usually concentrate to condemn the human perpetrator and overlook to analyze the mode of existence of the nonhuman that affected the consequence of hybrids, otherwise we might find equally or even more promising ways to fix the problem.

Today, as the interpenetrative relation between human and nonhuman has become more and more intricate and emerging "autonomous" technology such as artificial intelligence has trembled the fundamental division between human and nonhuman more and more, an approach, focusing on the human-nonhuman hybrids, which Latour and Verbeek suggest—so called the post-phenomenology by Verbeek—becomes more and more important for legal theory to *accompany*[9] technology. Under this approach, not only human but also nonhuman has morality; we should face the reality that nonhuman—technology and object—is not neutral

[7] See Verbeek (2011), p. 6.

[8] Verbeek (2011), p. 2.

[9] See Verbeek (2011), p. 95.

but actively intervenes in the formation of human experiences and practices, hence it affects our moral judgements, or comes to make moral judgements in the place of humans. As a result, deontology and value theory based on essentialized human nature is relativized, and a better mode of existence of society or human is pursued by the strategic combination of actions towards both human and nonhuman. As will be mentioned later, adopting this approach has an important meaning in studying why and how we should allow the presence of nonhuman that could cause, in a sense, uncontrollable danger to the society such as autonomous cars (hereinafter, AC).

This paper examines the conventional criminal law theory from a viewpoint of post-phenomenology and sketch the alternative way to *accompany* technology. This examination will be done in the context of artificial intelligence, particularly of AC, because "uncontrollable" nonhuman may demonstrate the limitation of conventional criminal law theory.

The first section will start with an overview of the issues of humanism and will show how conventional criminal law theory inherit these problems as they are based on humanism. I will demonstrate the limitation of conventional theory by application it to AC settings. The second section will investigate the post-phenomenological approach and demonstrate how this approach overcomes the problems of conventional criminal law theory. Then, the third section will essay a new regime of criminal law based on the post-phenomenology, especially in the context of AC. Finally, the closing remark will raise some remaining issues and conclude the paper.

2 Humanism and Criminal Law Theory

2.1 The Humanism and Its Problems

According to Verbeek, the most important problem of humanism, with its limitation of moral agency to human, is that it makes difficult to conceptualize the actual moral role of nonhuman in the world.[10]

As Latour points out, our world is full of artifacts with a moral role. This is to say, seatbelts that make unpleasant sounds when the vehicle starts to move without its passenger wearing it and speed-bumps that force vehicles to slow down also influence the mode of human existence.[11]

The example of obstetrical ultrasound, which Verbeek repeatedly mentioned in his book, is even more striking, because obviously "this technology transcends the mere functionality of making visible an unborn child in the womb.[12]" By making it

[10] See Verbeek (2011), p. 22.
[11] See Latour (1992), pp. 163–166.
[12] See Verbeek (2011), p. 23.

seem as if the baby is floating freely, independent from the mother's body, it shows the baby as an independent living organism and contributes to its personification—especially for the father's perception. In addition, the technology shows the baby from the perspective of various medical parameters, especially that of genetic diseases' risks, therefore converting pregnancy to a medical state which needs monitoring and professional supervision. It also converts congenital diseases to an avoidable difficulty, hence irreversibly changing pregnancy from an encounter to unpredictable risks to a process of ethical choice. It makes the parents to become a decision maker of the unborn child's life. In this way, "(u)ltrasound inevitably and radically changes the experience of being pregnant and the interpretations of unborn life.[13]" In addition to this striking situation, we must also note that this technology was, of course, not developed to raise such ethical questions. This example demonstrates the fact that technology forms an important object of our moral evaluation in an uncontrollable manner and we are obliged to make them.

However, it suddenly becomes difficult to evaluate these moral roles of obstetrical ultrasound if we persist in the conventional humanist ethics, that premise the free will and autonomy as essential character of human in order to evaluate human behavior ethically and exclude nonhuman's influence from ethical evaluation. Because under this approach, if technology shapes human decisions, our decisions are no longer free or autonomous. Namely, it is intrinsically impossible to include these decisions as an object of ethical evaluation from the viewpoint of conventional ethics, that only targets free and autonomous decisions for normative evaluation.

Of course, there may be arguments that it is possible to make normative evaluation as long as there is at least a certain level of free will. However, in that case, the required level of free will has to be made clear and justified; and if the ground for normative evaluation is derived from some kinds of essentialized view of human, that is precisely humanism itself, provoking the vexing problem here.

In addition, if we choose some strands of humanism, super plasticity of our body senses formed through interactions with external circumstances and of our experience of existence based on those senses could also be a grave problem. The absence of essential character of human as subject and nonhuman as object, although humanist ethics heavily rely on this division, rather clearly shows that humanism cannot serve as a foundation of our normative theory. According to Andy Clark, "(f)or what is special about human brains, and what best explains the distinctive features of human intelligence, is precisely their ability to enter into deep and complex relationship with nonbiological constructs, props, and aids.[14]" Therefore a lot of our tools have become as "proper parts of the computational apparatus that constitutes our mind.[15]" Ever since Descartes' advocacy of mind-body dualism, our mind has been considered as the core of modern subject and our body as the core of

[13] Verbeek (2011), p. 27.

[14] Clark (2003), p. 5.

[15] Clark (2003), p. 6.

modern object, but these two are not separated or independent in the first place. Issues so called "expanded mind" through "embodiment" makes this point clear.

For example, adults don't need to give much attention to the weight of the pen, the tension of the paper and the shape of symbols when writing symbols with a pen and paper. However, children who try to write symbols for the first time have to think about the force they put on the pen, the tension of the paper, the shape of the symbol etc. in order to carry out this task. Here, we make paper, pen and symbols embodied. The process in which we acquire paper, pen and symbols and come to be able to use them unconsciously is fairly similar to the process of learning body control such as walking. We become able to leave the exercise to our legs when walking without consciously controlling them. In the same way, we become able to leave the complicated information processing to the paper, pen and symbols by their embodiment. This entrustment is critically important as a lot of our information processing is done unconsciously.

The revealed vague boundaries of mind/body/artifacts mean that modern dualism, which divides subject/object, cannot be essential, and that we rather exist as hybrids (although it is debatable whether we should call this status a cyborg as Clark does) constantly interpenetrating subjects and objects, and continuously divide and form subjects and objects depending on the situation and the goal of each moment. Thus we cannot properly handle the morality of nonhuman unless we find an ethical theory that avoids the essentialization of human existence and is not based on humanism. As Latour observed, we are unable to appropriately handle the world fulfilled with human-nonhuman hybrids because of the modern way of thinking that strictly divides subjects and objects.[16]

2.2 Humanism and Criminal Law Theory

The field of criminal law is full of humanism. Modern criminal law theories are affected by deontological ethics and/or consequentialism ethics;[17] twins from the strict modern division between subject and object. On the one hand deontological ethics exclusively focused on the behavior of human as a subject with free will, and exclude the behavior affected by objects from the object of ethical evaluation. On the other hand, consequentialism ethics exclusively focus on the object affected by the subject with free will and exclude the human behavior itself from the object of ethical evaluation. These are twins because they both essentialize the way of human existence based on the modern humanism and cannot leave a room for evaluating any interpenetrating relationship between subjects and objects.[18]

[16]See Latour (1999), pp. 193–215.

[17]In Japanese criminal law theory of wrongfulness (Rechtswidrigkeit in German) is defined in the manner of deontology and/or consequentialism. See Yamaguchi (2016), pp. 104 ff.

[18]See Verbeek (2011), pp. 30 ff.

These features of humanism might face serious problems in the field of criminal negligence. This is because criminal negligence is, by its definition, not caused by a definite intention, and the influences from nonhuman tend to be salient.

For example, in a car accident, the failure to control the car (which is obviously nonhuman) could be understood as the failure to embody the car, hence we may pursue the criminal liability of the driver. A sort of condemnation like that the driver should have practiced more to become able to control the car better may be a blunt expression of her criminal liability. But at the same time, as it is obvious from the fact that some artifacts are easier to be embodied than others (for example chopsticks made for children's practice), it may also be understood as a failure of the design of the artifact that affects to humans. As we have seen with the gun control and gun crimes, these two ways of pursuing liability may have equal value for our society.

However, it should be impossible to normatively evaluate behavior affected by artifacts when we stand on humanism. Therefore, pursuit of legal liability in this context should at least be on an assumption that the danger could have been controlled without being affected by the technological artifact. Then the way we can choose is either to charge against the driver who failed to control, or to charge against the designer/manufacturer of the artifact who should have only made controllable artifacts. Either ways, charges are determined in relation to the ideal human behavior. Then, most importantly, to pursue the responsibility of the artifact itself, which may be done through improving or modifying its mode of existence and its influence on human, is left outside the scope of interest of criminal law scholars, at least directly.

The fact that leading criminal law theories about criminal negligence are grounded on the perpetrator's foreseeability of the consequence precisely shows this. According to these theories, those who should have foreseen the dangerous consequence deserve to be criminally liable for letting the consequence occur.[19] Here, we can clearly see a humanist approach in that the subject who failed to control the danger-generating object should be criminally punished.

This trend is even more clear in recent theories which focus on the danger rather than the consequences.[20] This approach perceives protection of legal goods through criminal sanction as the role of criminal law and stands on the idea that we should have a strict enough regulation for acts that generate danger. It even brings forth the idea of due care violations for all of those who are involved in generating danger. From this way of thinking, more emphasis is put on the subject's control on the object compared to conventional approaches that require foreseeability of a specified outcome.

[19] See Yamaguchi (2016), pp. 241–256.

[20] See Higuchi (2014), pp. 228 f. But I must mention that, as the author suggested, recent Japanese supreme court cases may indicate the quasi-danger-based approach in the context of criminal negligence, although the court superficially repeated the due care approach.

Furthermore, the tendency of toughening the law for criminal negligence in the course of driving is another sign that present criminal law think the human (subject) control over artifacts (object) is important.[21]

2.3 Problems Caused by Humanism in Criminal Law Context

The same problems as those of philosophy and ethics based on humanism face emerge in the context of criminal law based on humanism. By essentializing the mode of human existence, the interpenetration of humans and nonhuman are overseen. That results in the inability to handle the actual moral role that artifacts play in the way of forming human existence and her normative evaluation.

However, the fact that criminal law is related to the exercise of punishment—the most severe form of state power—may aggravate the matter. Firstly, there might be excessive normalization of humans by forcing an essentialized figure of humans ultimately through severe punishment. This movement should conflict with the core principle of Japanese constitutional law, the individual autonomy[22] and ultimately constrains human creativity because it always trembles the previous order of society and the fixed figure of humans. Of course, we need an elaborated concept of autonomy in this context, because the concept in conventional ethics, that is perfectly independent of external circumstances with free will, itself is problematic here. I will return this point soon. Secondly, once the criminal law is executed despite the first problem, it lacks the idea of pursuing the responsibility of artifacts itself, so it does not have the means to improve the mode of existence of artifacts. As long as enforcement of criminal law based on humanism is continued, it is difficult to create a society where humans and nonhumans can coexist better, despite the sacrifice of creativity.

In order to tackle the first problem, we must start by recognizing that neither the *raison d'être* of present criminal law theory—retribution, protection of legal goods and danger management—nor the types of punishment—fines, imprisonment and education—have a substantial essence. As Michel Foucault observed, the triangle of Knowledge = Power relationship = Mode of constitution of subject (hereinafter K=P=S), where each distinctive element constitutes each other without absorbing each other, continuously forms ourselves and our reality in multiple ways and levels.[23] The reason, why we tend to perceive the rightness of our criminal theories from the age of enlightenment and our practices of the punishment such as fines and imprisonments to be essential, is because the triangle of K=P=S, which emerged in the age of enlightenment, is being reproduced through the institution and technology

---

[21] In Japan legislator pressed by popular sentiment repeatedly has enacted the tougher law on criminal negligence of driving car. See Kawamoto (2015).

[22] See Art. 13 of Japanese constitutional law.

[23] About the triangle, see Foucault (2009), p. 10.

for discipline, such as military, school and prison, unceasingly forming ourselves and (at least some of) our reality.[24]

Then if we essentialize the mode of human existence, establish the legal goods to be protected and the danger to be controlled derived from the essentialization, and set the *raison d'être* of criminal law as the protection and the control itself, the fantasy of "transparent communication[25]" may turn into the actual nightmare. Because human creativity, that allows us to imagine the other world than it is, is the core of autonomy and this very human ability inevitably trembles the status quo of human and social existence, perfectly ordered world without friction in the ideal means completely oppressed world without autonomy in the real. As I mentioned above, the enforcement of criminal law in such a manner will violently conflict with the Japanese constitution that situate the individual autonomy in the core of its value. If we want to make our criminal justice system truly compatible with the constitution, there needs to be some space to be left for the autonomous individuals to continuously challenge and redefine the current mode of human existence and social order for the purpose of coexistence with others. But at the same time, we must tackle this problem without appealing for the concept of the conventional concept of autonomy, that is the mode of human existence completely independent of the external circumstances with her free will, because this concept its self inevitably causes the problem of essentialism and the fantasy of "transparent communication."

Here, Foucalt's interpretation of Kant's "What is enlightenment" gives us a suggestive implication in looking at the relationship between the autonomy of individuals and the achievement of a better society that allows coexistence with others avoiding the essentialization of mode of human existence.[26] It hints, with following lectures of *parrêsia*, the possibility of understanding Kant's "enlightenment" as a practice that improves our reality more desirable through the constitution of subject by following the truth she speaks by herself frankly (*parrêsia*)—an autonomous subject by "self-normalization"—and jumping into the power relationship that forms her and others at the very moment; the incident that inevitably changes the power relationship—better or worse—between her and the others at her risk.[27] The autonomy of individuals and legal norms for coexistence can only be harmonized once each human is permanently allowed to engage in the practice of *parrêsia*, that is to speak her own truth frankly at her own risk and to lead herself and others through the constitution of her own subject by following her own truth, that is based on the strategy and technique of governmentality aimed at destructing the

---

[24] See generally Foucault (1975). Of course I don't have any intention to say that our criminal law theory exclusively relies on such kind of K=P=S. As represented by Law and Economics school, initially led by Gary Becker's canonical work, so called Biopolitique and its K=P=S permeates widely in the discourse of criminal law. See Becker (1968); Foucault (2004), pp. 245–294 (Leçon du 21 et 28 Mars 1979).

[25] See Foucault (2001), pp. 1545 f.

[26] See Foucault (2008), pp. 3–39 (Leçon du 5 Janvier 1983).

[27] See Hirose (2011), pp. 266–297.

dominance of current truth that hinders the coexistence among humans.[28] In short re-reading the concept of autonomy of the Art. 13 of Japanese constitution as described above, we can find the way to get rid of the essentialization of the mode of human existence in the context of criminal law.

In addition, we should not forget that this kind of leading can be derived from not only humans but also nonhuman, as seen in the case of vehicles, phones and typewriters. For example, if the mode of existence of a certain nonhuman is determined with the sole goal of diminishing "danger" in a certain value system, there is a risk that the potential possibility of the nonhuman contributing to realize the coexistence with others in our society.

Therefore, humanism-based criminal law should face double challenges in the area of criminal negligence.

However, this does not mean that the normalization of humans and nonhumans by criminal justice system itself is undesirable. For the socially vulnerable who cannot protect themselves on their own to live an autonomous life, it is critical that the rule of law restricts those who tend to dominate from being oppressive. It is also obvious that difficult-to-handle and dangerous artifacts need to be improved and altered to a better mode of existence.

The real problem is the risk of being forcibly dominated, without even noticing, by the existing social order as if it is the ideal order which allows autonomy and coexistence to be accomplished, as the current mode of human existence is essentialized, ideas such as legal goods or danger is defined by the essentialization and the protection of the legal goods or the control of danger is naturalized as the *raison d'être* of criminal intervention.

Thus, when enforcing the criminal law, it is necessary to have an opportunity to reflect whether the object and methods of criminal law are sincerely aimed at the achievement of a society where autonomous individuals co-exist as the constitution expects. In order to do this, we must construct a legal theory which allows both human and nonhuman to keep their creativity without essentializing the mode of human existence.

By organizing the first problem like this, the second problem also becomes clear. That is, from the perspective of achieving a society where autonomous individuals can co-exist, charging the user or developer, compared to directly improving the mode of nonhuman existence itself, may have a high cost of side effect and low benefit. This is to say, the strategy of criminal law based on humanism is to maintain

[28] See Foucault (2001), p. 1547; Of course I don't have any intention to argue that Foucault admitted any substance of legal norms. Conversely it is supposed he doesn't admit any substance of legal norms, accordingly we should start our thinking from our liberty not our legal status in the current political institution in order to shake the current power relationship, because to find the space in the context of current law, that is inevitably formed through some arts of governmentality, means just to obey the current power relationship in logic. See Foucault (2001), p. 1548. But if we understand the character of Foucault's law—"responsiveness and necessary vacuity"—properly, law may become the strong and necessary tool to realize the world of coexistence among humans. See generally Golder and Fitzpatrick (2009).

legal goods and social order by sanctioning humans so as to give them an incentive to control their artifacts so that they are controlled well enough by humans, but this strategy does not ensure the improving the mode of existence of artifacts at all. Upon a humanist-based criminal law, if a certain artifact is evaluated as being controllable under current social value and the danger derived from it is thought to be caused by the user, even if there is a chance that its mode of existence can be improved—for example by making it easier to be embodied—it can be left as it is. This is especially the case when there is an uncertainty, or danger, in relation to the improvement and that risk to be punished may fall onto the developer.

Illustrating the criminal law theory of AC may clarify the limits of humanism-based criminal law on this point. As it is well known, to develop AC, a public road experiment that involves certain danger is necessary for the artificial intelligence's learning because its program, so called deep learning, needs massive data to improve its function up to the practical level. Even after the AC is put into practice, that artificial intelligence is, in a sense, always halfway in its learning/development, and it is impossible to completely predict the operation of these types of artificial intelligence,[29] so even if it is expected to be of trifling chance, fatal accidents caused by AC cannot be avoided.

In this case, according to the dominant interpretation of criminal law, the AC developer will fulfill the "Tatbestand[30]" of professional negligence resulting in death. This is to say, the result of death from the public road experiments or actual operation is obviously foreseeable, because of the inherent nature of the artificial intelligence, and if they hadn't done these things, the result could have been unmistakably avoided.

Of course, this does not immediately mean the AC developer is criminally liable. For example, there are judicial doctrines such as that of permissible danger, which rejects the wrongfulness or the fulfillment of "Tatbestand" due to its social benefit even if the action is dangerous, and "Seito-gyomu-koi 正当業務行為", which justifies the undesirable result from the professional work due to its social benefit. It may also be the case that the government gives a special approval not to punish the AC developer. These judicial doctrines and political considerations may help avoid the benefits that AC brings to society as a whole—namely sustaining life in an ageing village due to better transportation, ensuring movement of the severely disabled, reducing the number of car accidents, increasing employment due to the creation of a new industry etc.—from being restrained by punishments.

[29] Of course this kind of unpredictability has already occurred in the case of logic programming, if the branches of the logic is multiplied and its function is based on certain probability. But, the inherent character of unpredictability of deep learning should be demonstrated to perceive the problem of humanist criminal law theory in this context.

[30] I do know that this technical term, signifying the elements of behavior the law describes about particular criminal behavior, is unique in German criminal law theory (and also in Japanese one affected heavily by German theory), but I cannot avoid to use this concept because of the convenience of description.

However, it is difficult to justify these judicial doctrines and political considerations from the conventional criminal law theories based on humanism. According to humanism-based criminal law, it should in itself be banned to create an uncontrollable technological artifact in a society in logic—especially if it generates grave danger like injuries and deaths. For example, leading criminal law scholars poses questions about the possibility to justify the critical result, such as death, from manufacturing the automobile based on the calculation of social its benefit, because, for them, the danger of automobile is controlled by drivers if they properly use it. It means that manufacturing the automobile generates "no" danger, even if it is well known that some driver cannot control them in reality, because the danger "should" be confined by the driver with free will, normatively. This argument is completely understandable from the point of view of humanism-based criminal law as they claim all artifacts must be controlled by humans with their free will. However, adopting the same argument in the context of AC would inevitably actualize the negative aspect of humanism-based criminal law. The creativity of the developer, which holds a potential to a realize the better society of coexistence, must be sacrificed, and the automobiles, which causes more injuries and deaths than AC, is basically left as it is.

However, easily taking the conclusion of not condemning the AC developer due to the social benefit of AC without a sufficient normative theory to support why and how an uncontrollable and dangerous technological artifact can exist in society, while leaving humanism as a base of its theory, will inevitably cause other dangers such as the enforcement of punishments on the grounds of rough utilitarianism or unfounded political decisions.

In the end, to answer the question when/how/why technological development such as AC is acceptable, it is necessary to fundamentally switch from conventional humanist ways of thinking, and to restructure criminal law theories based on a normative theory that can appropriately capture the mutual penetration of human and nonhuman.

3 The Post-Phenomenology and Its Potential

3.1 Ethical Theory Based on the Analytical Framework of Composite Intentionality

Facing the reality that there are influences from the side of artifacts as mentioned earlier in Sect. 2.1—they have a great impact on our mode of existence and ethical evaluation or make moral judgement on our behalf—Verbeek put forward an analytical framework employing the concept of "composite intentionality".

As it is well known, the concept of intentionality in terms of phenomenology was developed by Edmund Husserl and it basically means that human consciousness is always aimed at a target. From this point of view, it is derived that there is no consciousness unrelated to the world or a world unrelated to consciousness. By focusing on the idea of intentionality, it becomes distinct that human experience is given by relationships with the world and "(r)ather than separating the humans and world, the concept makes visible the inextricable connection between them.[31]"

Verbeek first observes the many way how artifacts are involved in our intentionality—glasses, thermometers, automobiles, telephones, air conditioners, MRI scanners. Then, he states that human experiences, mediated by these technologies and artifacts, are given by the function of intentionality in which human and artifacts collaborate with each other.

In addition, Verbeek points out that these functions of technology and artifacts have an "active" and "autonomous" nature which cannot always be explained by the designer or user's intention. For example, as mentioned before, ultrasonic devices have actively changed the birth of children with congenital disease from an issue of coincidence to an issue of ethical choice, unrelated to the designer or user's intention.

Considering this situation, Verbeek introduces the idea of "composite intentionality[32]". By using the term "composite" it is made clear how artifacts have an impact on the formation of human experience that cannot be explained only by human intentionality. On the other hand, by leaving the term "intentionality" it is also clarified that these artifacts' influence basically appears through human as an agent. Due to this concept, what has been recognized as an issue of human "free will" can now be analyzed from the perspective that allows us to divide intentionality of human themselves and influence by artifacts. The analytical framework of "composite intentionality" has opened the way to ethically analyze the interpenetrative state of human and nonhuman by restructuring the basic concept of humanism ethics.

3.2 *"Composite Intentionality" and the Later Foucault's Ethics*

After preparing a theoretical framework to appropriately analyze matters that humanism failed to target, Verbeek moves on to the next step—the introduction of the core concept for his ethics.

Here, Verbeek looks at the reevaluation of virtue ethics from the later Foucault's work. As is well known, the later Foucault tried to advocate an ethical theory different from simple deontology or consequentialism theory. Paying attention to the concept of "*the good life*" or "*self-practice*" to make one's life as beautiful artwork, Foucault investigated the ethics as "*self-practice*" directed to "*the good life*"

[31] See Verbeek (2011), p. 55.

[32] Verbeek (2011), p. 58.

and the constitution of subject as a trial based on the aesthetic concerns about herself in the ancient Greek.[33]

Taking this into account, Verbeek observed "(h)umans are not only the objects of power here but also subjects that create their own existence against the background of and in confrontation with these powers.[34]" That is, although "(b)y mediating our actions and perceptions, technologies form a structure of power, disciplining, organizing and normalizing the subject,[35]" but considering the fact that Foucault demonstrated the importance of freedom to constitute a better subject through power rather than freedom from power, he argues that "(r)ather than separating the human domain from the domain of technology, we need to ask ourselves in what ways we want both domains to interface. Rather than trying to save humanity from technology, the aim of our moral self-constitution should be to let humanity and technology blend in desirable ways[36]"

However, Verbeek's argument does not completely negate the value of conventional ethical theory.[37] Even on the premise of interpenetrative relationship between human and artifacts, under circumstances such as where the influence of artifacts is relatively weak and humans' intentionality seems strongly, it cannot be denied that it may be better to approach the mode of human existence by ethical theories based on humanism's strong point—"free will". What is important is to recognize that in order to improve the mode of human existence, it should not be essentialized, and both modes of existence of human and nonhuman, as well as their relationship, should be continuously refined to reach the desirable world where the hybrids, as many as possible, can coexist with each other.

3.3 *Morality of Designing*

If there is a certain morality of nonhuman itself, and a desirable mode of human existence should be structured through the mode of nonhuman existence, the act of creating artifacts—design—also inevitably involves with the dimension of ethical practice. Designing should be regarded as "materializing morality.[38]" "All technology-in-design will eventually mediate human actions and experience, thus helping to form our moral decisions and the quality of our lives.[39]" Therefore Verbeek remarks that "the inevitability of technological mediation urges ethics to deal with these mediations in a responsible way, informing design practices and

[33] See generally, Foucault (1984a, b).

[34] Verbeek (2011), p. 67.

[35] Verbeek (2011), p. 70.

[36] Verbeek (2011), p. 84.

[37] Verbeek (2011), p. 84.

[38] Verbeek (2011), p. 90.

[39] Verbeek (2011), p. 90.

contributing to the development of technologies with morally justifiable mediating capacities.[40]" The ethics of technology should aim to *accompany* technological development.

Verbeek then asserts that it is necessary to anticipate the result of technological mediation in the practice of design, and presents three ways to do this, taking into consideration how technology mediates ethical practice in different ways under different contexts of usage but cannot be completely predicted due to the autonomy of artifacts/technology itself. The first is the designer's moral imagination, the second is, so called, Constructive Technology Assessment (hereinafter CTA) and the third is scenario-based product design.[41] Some explanations for the second and the third may be required. "CTA is based on an evolutionary view of technology development." That consists of systematical "feedings back assessments from all relevant actors – users, lobbies, designers, companies, and like – into the design process. It does so by organizing meetings of these actors in which the aim is to reach consensus about design of the technology that is 'constructively assessed.' This. . .is called 'constructive' because it does not assess technologies after they have been developed but during their development, so that the assessments can be used to modify the original design." However, CTA tends to focus too much on human actors and not on nonhuman entity, that is the center of technology-in-design, Verbeek suggests some ameliorations. He argues that the participants in CTA are required to anticipate possible mediating roles of technology in design.[42] The method of scenario-based product design aims to design products from the point of view of the way they are used, therefore it forces designers to anticipate the use context of the products they are designing. Then faced with the difficulties to develop meaningful scenarios in this methodology, Verbeek implies the application of CTA or usage of virtual reality for potential users.[43]

According to Verbeek, after the step of anticipating mediations, designers will also have to assess the quality of the anticipated mediations toward responsible technology design. Here he introduces the methodology of stakeholder analysis, in this method "all relevant moral arguments regarding the technology-in-design are gathered and balanced, taking the perspective of all stakeholders involved. . .The aim of stakeholder analysis is to lay bare all moral arguments that are relevant to a given ethical problem by making an inventory of all stakeholders involved all arguments that are relevant from their point of view. . .By weighing all these arguments against each other, we can reach an informed conclusion about the moral quality of a decision.[44]"

However, to expand its scope to technology-in-design itself, Verbeek modifies and expands the methodology that can reflect the elements as follows. "First, there is

---

[40] Verbeek (2011), p. 95.

[41] See Verbeek (2011), pp. 97–105.

[42] Verbeek (2011), pp. 102 f.

[43] See Verbeek (2011), pp. 104 f.

[44] Verbeek (2011), pp. 105 f.

the moral quality of the *intended mediations* that are deliberately inscribed in the technology. Second, there are the implicit mediations, insofar as they can be anticipated by means of moral imagination, constructive technology assessment, and scenarios and simulations. Third, the *forms of mediations* are relevant. . ." there are the forms of force, persuade and seduce. "Fourth the eventual outcome of the technological mediations. . .need to be morally assessed.[45]"

After introducing the assessment method, Verbeek makes comments on three distinguished important points: responsibility, freedom and democracy. Firstly, in the field of responsibility, Verbeek, while expanding the methodology of ethical evaluation to nonhuman entities, cautions that "it does not make sense to consider technologies full-fledged moral agents in the way human beings are moral agents. . .Rather the moral questions themselves and the various options available for answering them are coshaped by these technologies; it is an amalgam of human and technologies that acts morally and bears moral responsibility. . .This blurring of boundaries between humans and technologies does not make human being less responsible; rather it opens up a new realm of responsibility" for users and designers. Therefore, he mentions that "including technology in the realm of responsibility makes visible how both users and designers can have moral responsibility for technologically mediated actions.[46]"

Secondly, in the field of freedom, Verbeek, influenced by Foucauldian approach to freedom, claims that as all human actions are mediated by artifacts and technology, we should not use the concept of absolute freedom and complete autonomy, but rather understand the concept of freedom as a "relational freedom. . .(that) comes about in the relations people have with what helps to shape their subjectivity.[47]" He then makes caveats concerned with freedom that "(w)hen technological mediations do not leave any room for 'relational freedom' for human beings to constitute their (moral) subjectivity, they oppress and limit human subjects and must not be permitted to function as a basis for generating forms of subjectivity.[48]"

Lastly, regarding the field of democracy, given the fact that some "technologies influence human behavior and install visions of the *good life* without users' being explicitly aware of this or even being able to opt out. . ." Verbeek concludes, partly relying on Latourian concept of "public things,[49]" that moralization of technology does need a more democratic structure like CTA.[50]

[45] Verbeek (2011), pp. 106 f.

[46] Verbeek (2011), pp. 108 f.

[47] Verbeek (2011), p. 111.

[48] Verbeek (2011), p. 111.

[49] See Latour (2005).

[50] See Verbeek (2011), pp. 112 f.

3.4 Possibility of Application to Criminal Law Theory

The baseline of Verbeek's argument may be summarized as follows. First, human intentionality, always mediated by nonhuman, is composite intentionality. Second, therefore the relationship between humans and nonhuman must be continuously refined towards a desirable mode of human existence, through the constitution of subjectivity based on relational freedom. Third, in the course of design, it is necessary to look at the morality of users and designers and to undertake the democratic procedure properly designed to anticipate and assess the mediations of technology-in-design, especially from the viewpoint of responsibility, freedom and democracy.

I believe that the application of the argument to the problems we face in the context of criminal law is very fruitful. Because, as we have seen in Sect. 2.3, the problems of current criminal law, demonstrated in the context of AC, are caused by humanism that cannot capture nor handle the interpenetrative relationship between human and nonhuman. A restructuring the criminal law theory towards the desirable world where human and others can coexist, can be accomplished through making it based on the argument proposed by Verbeek.

Of course, to do so, it is inevitable to make some modifications to his theory. And in the course of this process, the concept of autonomy and freedom occupy the critical status, because as seen above, Verbeek proposes Foucauldian "relational freedom" to investigate the alternative ethics, instead of the concept of "autonomy" that is often problematic in this context.

First of all, I do clarify that I share the same concern that the concept of autonomy has tendency to essentialize the mode of human existence as a human figure emerged in philosophy of enlightenment and hide the importance of mediations of nonhuman that in fact forms our mode of existence in many ways, as I mentioned in Sect. 2.3. But to employ and secure the "relational freedom," we must constitute our subjectivity based on our own truth, derived from the practice of aesthetics of life, at our own risk. In other words, to divide human and nonhuman from the hybrids, we have to figure out the particular type or mode of human and nonhuman discretionally, even though they are different from the conventional ones. And, better or worse, this practice of constitution of subjectivity inevitably affects the power relationship, that changes the world and in turn starts to have a risk to dominate others, at the very moment of its emergence. Therefore, the usage of "relational freedom" is inherently connected to self-normalization based on frankly truth telling (*parrêsia*), a kind of practice of autonomy, that inevitably affects the others and must provoke another *parrêsia* consecutively, due to its inherent tendency to dominate. Then I do want to remain a nuance of arbitrary self-normalization involved with autonomy and of its risk to leave the room for the permanent game of power relationship in the context of technology, because as Verbeek properly demonstrates the critical capacity for technology-in-design is imagination of other mode of human existence and the world. But to imagine the other reality, we must obey the truth of our own at first, the other truth of the reality than it is. In short, I do continue to use the concept of autonomy, but of course elaborated and not complete one, inspired by Foucault's Kant reading in his lecture, mentioned in Sect. 2.3, to employ the "relational freedom" in a manner that Verbeek clarifies. With this little modification to his argument, I will explore the alternative criminal law theory inspired by him, in the next section.

4 "The Moralizing Technology" and Criminal Law

4.1 Purpose and Strategy

For the ethics of Verbeek inspired by Foucault, it is critical to continuously challenge the reality through constitution of subjectivity in order to enable a more desirable mode of human existence.

However, since law itself is technology, it is necessary to reflect its mode of existence in the first place, to consider whether the law genuinely contributes to the *good life* of people, when analyzing the law. If we recognize the actual practice of law, that itself combines the multiple sorts of science and technology through the mobilization of limited resource, we should avoid to essentialize the law, because once we find the essential substance of law, we will trigger the problems represented by "transparent communication" as mentioned in Sect. 2.3. Rather we have to keep its character of "responsiveness and necessary vacuity[51]", to combines any type of truth to our reality and make our world more desirable for coexistence with others.[52]

Therefore, to achieve the goal of more desirable mode of human existence, we should rather strategically combine various knowledge and maintain or build the networks that allows to connect human with nonhuman, and continuously reflect and challenge the current world fulfilled with hybrids in a democratic manner, as Latour suggested.[53] In short, we need to explore not only dogmas but also knowledges that inevitably influence the legal dogmas, and in order to achieve more desirable mode of human existence, we introduce multiple knowledge, sometimes combined, that is strategically useful to achieve the goal—like economics, sociology, and cognitive science etc. Of course we should also be cautious about the dominance that may be caused whenever the constitution of subjectivity based on other truth emerges, as I mentioned in Sect. 3.4.

---

[51] See Golder and Fitzpatrick (2009).

[52] We can find this sort of function in Latour's exploration of law. See Latour (2009).

[53] See Latour (2004).

4.2 Risks Should Be Allowed

We will start the argument with the concept of the risk concerned with criminal negligence. Under humanism based criminal law, we should focus solely on human consciousness, to confine the risk of her activity through criminal punishment. But once we rebuild our theory on the concept of composite intentionality, this is meaningless to target human exclusively, because his intentionality is always mediated by nonhuman, for example automobile. And if we consult the useful knowledges, like economics or cognitive science, they may suggest that the cost of punishment in particular case is higher than redesign or some design of automobile makes it difficult for human to embody it. In these cases punishment may not be carried out to achieve our *good life*, because redesigning nonhuman will bring us more fruitful result and constitutes our subjectivity more desirable. Of course, to redesign nonhuman, we should take course of democratic process of anticipation and assessment of its influence—like CTA or stakeholder analysis—but what is clear is that there is a risk that should not be led to punishment to achieve more desirable mode of our existence.

This argument is especially important in the context of AC, because this kind of technology has huge potential to ameliorate our quality of life, but inevitably concerned with risk to be suppressed in the existing value. Or in more legal term, the constitution which permits us to engage in a certain kind of practice of autonomy, prohibits the punishment on the creative activity which may lead us to the other world where more hybrids will coexist each other than now. I denote this "risk" as "risk should be allowed" because this risk is not only able to be allowed for the social benefit of the action that generates it but should be allowed for the inherent creativity of the action that generates it.

4.3 The Alternative Regime of Criminal Law

Then what kind of criminal justice system we should envisage to deal with the "risk should be allowed" or the redesign of nonhuman matters. I will sketch the alternative system as below.

According to Verbeek, it does not make sense to treat nonhuman as full-fledged moral agent as human. To handle nonhuman as exactly same as human is certainly a kind of "fetish" as Latour criticized,[54] and ultimately should be trapped by the risk of essentialization.

Therefore, it cannot be denied that designers and developers bear a certain ethical responsibility for the artifacts' impact and way of existence. However, to immediately associate this ethical responsibility to penal sanction, as we have repeatedly seen, could eventually lead to a world of "transparent communication" without

[54] See Latour (1999), ch. 9.

creativity—although it cannot be denied that there are cases where it is desirable to immediately impose penal sanction such as when they intentionally left an obviously dangerous artifact as it is.

But to suspend the responsibility of human and to make them to engage in redesigning nonhuman, seemingly radical idea, in fact not new for our criminal justice system. So called DPA (Deferred Prosecution Agreements) employed by U.S. Department of Justice in the context of corporate crime is good model for envisaging the alternative system, because by this procedure, prosecutor tries to change the culture or structure of the corporation through agreements based on negotiation—mode of nonhuman existence—backed up by the severe punishment.[55] In this sense, this procedure not only targets human, but also nonhuman, reflecting the reality that nonhuman inevitably influences human mode of existence in a manner that the designer cannot anticipate. Of course there is acute criticism to this procedure due to its opaque process and lack of feedback system. But we do know how to complement this procedure in technological setting, thanks to Verbeek. I mean applying the CTA and stakeholder analysis will improve the DPA based procedure aimed at nonhuman.

I will sketch this alternative system, taken AC cases as example as follows. First, when AC makes accident, we should determine whether we punish the manufacture or redesign the AC itself backed up by potential criminal sanction. But to do so, we need enough information about artificial intelligence mounted on the AC, for example. Then we need a strong leverage to extract the information from manufacture, as well known in the context of DPA. I propose the substantive criminal law that penalizes any activity concerned with the accident of AC, and once manufacture or its employee is prosecuted, we allow the administrator to exclude the products from the market. This leverage makes manufacture to collaborate with authority and to give necessary information to them because there is strong incentive to avoid the prosecution. Logically, as long as manufacture collaborate, prosecutor should not prosecute the manufacture and its employee on the ground of the accident caused by AC.

Second, after the acquisition of enough information, prosecutor and stakeholder, including representative of the corporation, concerned administrator, independent engineer, independent scientist, representative of consumer and victim etc., should anticipate and assess the mode of existence of AC, in the light of desirable mode of human existence. In the course of assessment, they should, if possible, figure out the statistically undesirable programming or data processing used for building AC. After the collection of undesirable behavior, legislator or delegated administrator penalize them, and make it another leverage for manufacture. I mean if employee is prosecuted on the ground of this violation, administrator should exclude the manufacture from the market. This kind of leverage with normal type of DPA will lead

[55] See e.g. Garrett (2007). I should notice that this kind of procedure now spread to United Kingdom or France.

manufacture to build proper corporate governance in the context of developing AC or artificial intelligence in general.

Third, prosecutor and the other stakeholders should decide whether to redesign AC. It should be noticed that in order to decide this, they should pay enough attention to the character of artificial intelligence. That is, they exclusively focus on the statistical danger caused by the AC, because we can only control artificial intelligence in a statistical manner in logic. Otherwise, for example, to try to fix the AC on the ground that the accident seems idiot for human may aggravate the situation, because the quality of artificial intelligence based on their "otherness." They may surpass our capacity because they process the information differently from us. Therefore making them understandable may worsen their quality and ultimately be trapped by "fetishism." In short, prosecutors and stakeholders should decide to redesign the AC, only if its accident rate exceeds the certain level that is not compatible with the world to be realized for their "*good life*." It is desirable to make this level public not to deter manufacture too much.

Finally, if they decide to redesign the AC, the process and the method for monitoring it should be articulated and agreed, through democratic process, like CTA or stakeholder analysis. Once these agreements are made, the prosecution is deferred.

The criminal justice system as I sketched above, will distribute the criminal responsibility to human and nonhuman based on the structured democratic process for elaborating our mode of existence. This kind of system is necessary, because the interpenetrative relationship between humans and nonhuman is increasingly intensified and "autonomous" artificial intelligence will influence our society more. As Verbeek observed,[56] because it becomes clearer and clearer that we form our own existence and experiences under the influence of technology, it is time to reflect the issues that were addressed in various contexts under the name of autonomy, freedom and liability—and to take a new step on our criminal justice system.

At least, the reason why it is acceptable to let an uncontrollable and dangerous artifact such as AC to exist should not be found in the discourse that simply it is beneficial to society, but rather in the belief, constructed through the democratic process, that there is a institutionally assured possibility for it to contribute to our "*good life*."

There may remain criticisms against my approach urging that it is a way of sacrificing an indispensable individual life for the social benefit. However, as we have seen so far, Verbeek and Latour—and Foucault who prepared the foundation of their theory to some extent—have made clear that no human exists without any influence of others, including nonhuman. In other words, to examine humans independently of others' influence under the name of complete individual autonomy means no other than unintentional essentialization of the mode of human existence, formed through existing value judgment. If so, to ignore or deny the possibility that technological artifacts hold, on the grounds of an essentialized human existence and

[56] See Verbeek (2011), p. 138.

the concept of danger derived from this, cannot mean more than making the maintenance of the existing order its own goal in the end. For example, regarding ACs when simply evaluating the danger to the existing life rather than the opened possibility to future human existence, does advocate unintentionally essentialize his value judgement[57]? In our society, we ultimately cannot avoid situations where humans conflict each other. However that is the very reason why we should continue to seek a solution that allows us to respect human autonomy in a sense that trembles the existing social and legal order.

5 Conclusion

This paper is an essay on various issues, aiming to build an alternative criminal law theory that is able to accompany technology, like artificial intelligence, that often fundamentally changes our mode of existence in an unpredictable manner. I have to admit that there remain many points to be improved. Especially, at least in Japanese context, I have to figure out concrete transitional procedure form the current regime to the alternative regime, because we even don't have an effective enough system designed to treat the corporate crime, like DPA, I mentioned above. But I think it is important to demonstrate the interpenetrative relationship between human and nonhuman, and to examine our justice system in the light of the fact to avoid too much normalization. I hope this short essay may contribute to future discussion about criminal law in the context of autonomous nonhuman like artificial intelligence.

References

Becker, G. S. (1968). Crime and punishment: An economical approach. *Journal of Political Economy, 76*(2), 169–217.
Clark, A. (2003). *Natural-Born Cyborgs: Minds, technologies, and the future of human intelligence*. Oxford University Press.
Foucault, M. (1975). *Surveiller et punir: Naissance de la prison*. Gallimard.
Foucault, M. (1984a). *Histoire de la sexualité II: L'usage des plaisirs*. Gallimard.
Foucault, M. (1984b). *Histoire de la sexualité III: Le souci de soi*. Gallimard.
Foucault, M. (2001). L'éthique de souci de soi comme pratique de la liberté. In M. Foucault (Ed.), *Dits et écrits II. 1976-1988* (pp. 1527–1548). Gallimard.
Foucault, M. (2004). *Naissance de la biopolitique: Cours au collège de France. 1978-1979*. Seuil.
Foucault, M. (2008). *Le Gouvernement de soi et des autres. Cours au Collège de France. 1982-1983*. Seuil.

[57] Because it is also arbitrary judgement that normal automobile is not dangerous. In other words, we don't have any essential meaning about the concept of danger as well as human.

Foucault, M. (2009). *Le Courage de la vérité: Le gouvernement de soi et des autres II. Cours au Collège de France. 1984.* Gallimard/Seuil.

Garrett, B. J. (2007). Structural reform prosecution. *Virginia Law Rev, 93*(4), 853–957.

Golder, B., & Fitzpatrick, P. (2009). *Foucault's law*. Taylor & Francis.

Higuchi, R. (2014). Chuui-gimu-no-naiyo-kakutei-kijyun. In K. Takayama & S. Shimada (Eds.), *Yamaguchi Atsushi-sensei-kentei-ronbunshu* (pp. 197–261). Seibundo. (in Japanese only).

Hirose, K. (2011). *Kouki Foucault*. Seidosha. (in Japanese only).

Kawamoto, T. (2015). *Koutsu-hanzai-taisaku-no-kenkyu (in Japanese only)*. Seibundo.

Latour, B. (1992). Where are the missing Masses? The sociology of a few Mundane Artifacts. In W. E. Bijker & J. Law (Eds.), *Shaping Technology/Building Society* (pp. 225–259). Republished in D.G. Johnson, J.M. Wetmore (Eds.), *Technology and Society: Building our Sociotechnical Future* (2008, pp. 151–180). Cambridge, MA: MIT Press.

Latour, B. (1999). *Pandora's hope: Essays on the reality of science studies*. Harvard University Press.

Latour, B. (2004). *Politics of nature*. Harvard University Press.

Latour, B. (2005). From Realpolitik to Dingpolitik, or how to make things public. In B. Latour & P. Weibel (Eds.), *Making things public: Atmospheres of democracy* (pp. 14–41). MIT Press.

Latour, B. (2009). *The making of law: An ethnography of the Conseil d'Etat*. Polity.

Verbeek, P.-P. (2011). *Moralizing technology: Understanding and designing the morality of things*. University of Chicago Press.

Yamaguchi, A. (2016). *Criminal law* (3rd ed., (in Japanese only)). Yuhikaku.

Liability for Autonomous Systems

Georg Borges

Contents

Abstract In the legal discussion on autonomous systems, the question of responsibility for damage caused by these systems is a major issue. This article analyzes the current law and the ongoing discussion on liability for damages caused by autonomous systems in European and German law with the aim of providing impulses for the development of an adequate legal framework for liability. It focuses on the question of the appropriate concept and addressees of liability.

The article comes to the conclusion that the traditional fault-based liability de lege ferenda has to be supplemented by a system of strict liability and in this respect advocates a strict liability in the sense of a causal liability. The addressee of strict liability should not only be the operator of the system, but also its producer. With regard to causal liability of the producer, however, a differentiation is required: The producer should only be the addressee of causal liability if, instead of the operator of the system, he plays the central role in controlling the risks emanating from the autonomous system.

G. Borges (✉)
Institute of Legal Informatics, Saarland University, Saarbrücken, Germany
e-mail: georg.borges@uni-saarland.de

G. Borges, C. Sorge (eds.), *Law and Technology in a Global Digital Society*,
https://doi.org/10.1007/978-3-030-90513-2_3

1 Introduction

Autonomous systems, such as self-driving cars, are not only a fascinating technical development; they also raise numerous legal questions. Complex legal questions result not least in relation to liability for damage caused by such systems. Again, autonomous vehicles attract most public attention. In February 2018, an Uber car which was driving autonomously caused the death of a pedestrian.[1] The news of this event was greeted with shock and intensified the debate about the safety of autonomous cars.[2] It is obvious, that not only self-driving cars, but any kind of autonomous system may cause damage: A care robot may injure a patient, an industrial robot as well as a household robot may hit bystanders. Even pure software such as a chatbot, may infringe personality rights or other protected goods.

The question of who is liable for damage which is caused by autonomous systems is, due to its immense practical relevance, justifiably at the centre of current legal discussion. Different positions are taken both in the questions of whether the current liability system needs modification in the light of autonomous systems and in which direction the liability regime for autonomous systems should be developed. These questions are discussed in this article with a focus on European and German law.

2 Damage Caused by Autonomous Systems as a Challenge for Tort Law

2.1 Gaps in the Current Liability System

In the current legal policy debate, the question arises as to which goals the legal framework for autonomous systems should achieve in this respect. The rules on civil liability traditionally aim to compensate the victim for damage suffered to protected legal assets.[3] At the same time, the threat of liability also serves to steer behavior by providing an incentive to avoid damage.[4] However, the risk of liability can also trigger undesirable strategies of liability avoidance. In the current debate, for example, a possible chilling effect of liability risks to the development and market launch of products equipped with artificial intelligence is often mentioned.

With regard to the objectives of the liability regime for artificial intelligence, there is a broad consensus that compensation for damage caused by autonomous systems

[1] Wakabayashi (2018).

[2] Several articles dealing with the safety have been published since then, see for example Kerr (2018), Financial Times (2018), Naughton (2018), and O'Donnell (2018).

[3] Schwarze (2019), mn. A2; Hager (2017), mn. 9; Shavell (2004), pp. 259 f.; Wagner (2020b), mn. 43; Schäfer and Ott (2012), pp. 166, 565.

[4] Hager (2017), mn. 10; Schäfer and Ott (2012), p. 169; Shavell (2004), p. 187; Wagner (2020b), mn. 45–47.

to protected legal interests should be ensured as far as possible.[5] It is also widely stated that the current legal system has gaps in this respect.[6]

The EU commissions Expert Group on Liability and New Technologies—New Technology Formation, for example, concludes in its final report that the current liability regime in Europe does not guarantee compensation for damage caused by new technologies in all cases where this would be desirable, and infers from this that there is a need for further development of liability law.[7]

The same position is taken by the European Parliament in its resolution of October 2020, which calls for the introduction of new liability rules for autonomous systems.[8]

The reasons for such liability gaps are manifold. From a systematic perspective, it is important that the traditional liability concept of tort law which focuses on fault as a basis for liability reaches its limits in the case of autonomous systems: To the extent that a system is permitted to act autonomously, and in fact does so, there is no basis for fault on the part of a natural person regarding damage caused by the system.[9]

Even if an interim result of the discussion can be drawn to the effect that a further development of the liability law with regard to autonomous systems is necessary, it is still very unclear and highly controversial which steps should be taken in this respect. At the forefront of the debate is the question of whether new legislation on strict liability should be introduced. This is clearly illustrated by the resolution of the European Parliament of October 2020, which proposes a statutory strict liability for operators of "High-risk" AI systems.[10]

2.2 *Roles and Participants*

Any consideration of liability for autonomous systems must distinguish between the various roles and participants that are part of the liability system. In contract liability, the roles of seller and buyer of the system are paramount. On the other hand, tort

---

[5] Bräutigam and Klindt (2015), p. 1139; explicitly just for tort law Schaub (2017), p. 348; cf. Wagner (2019), p. 30.

[6] Borges (2018), p. 982; Hanisch (2013), p. 120; Spiecker gen. Döhmann, I. (2016), p. 704; Wagner (2020a), p. 734.

[7] Report from the Expert Group on Liability and New Technologies—New Technologies Formation, pp. 3 f., 35 f., available at https://www.europarl.europa.eu/meetdocs/2014_2019/plmrep/COMMITTEES/JURI/DV/2020/01-09/AI-report_EN.pdf. Accessed 28 August 2021.

[8] European Parliament, resolution with recommendations to the Commission on a civil liability regime for artificial intelligence, Brussels 2020-10-20, 2020/2014(INL), available at https://www.europarl.europa.eu/doceo/document/TA-9-2020-0276_EN.html. Accessed 28 August 2021.

[9] Explicitly for autonomous driving Borges (2016b), p. 273; Riehm (2014), p. 114; cf. Spindler (2015), p. 768.

[10] See Article 14 of the proposed resolution *supra* note 8.

liability for damage caused by artifacts distinguishes between two other spheres of potentially responsible persons: the manufacturer (and supplier) of the artifact and the user (or legal owner, operator, or provider).[11] However, within the current liability system, it seems necessary to differentiate between three roles instead of two when considering autonomous systems: firstly, the manufacturer of the product, then the operator of the product who controls its use generally and finally, on the side of the operator, the actual person who controls the specific use and who will be referred to as 'user'. This three-way split becomes obvious in the case of vehicles where, typically, distinctions are drawn between the manufacturer, registered keeper and driver of the vehicle with different liability systems being applicable to each of them.

Very recently, an interesting distinction regarding the role of the operator has been suggested. The EU Commission's Expert Group on Liability and New Technologies—New Technology Formation distinguishes, in its final report, between frontend operator and backend operator.

2.3 Fault-Based Liability, Strict Liability and Objective Liability

Civil law recognises several different liability models with differing characteristics[12] and a deciding factor in the discussion regarding the liability for autonomous systems is which of these liability models should be adopted.[13]

The traditional liability model in Civil Law is fault-based liability[14] pursuant to which liability is only established when it can be proven that an offender breached a specific duty of care[15] and that this breach was the actual cause of the damage.[16] An example for this liability concept is § 823 para. 1 German Civil Code (BGB) which reads: "A person who, intentionally or negligently, unlawfully injures the life, body, health, freedom, property or another right of another person is liable to make compensation to the other party for the damage arising from this."[17] Together with

[11] In addition, Zech (2016), pp. 177 ff., differentiating between the manufacturer on the one hand and the user on the other hand.

[12] See also hereto in overview Kötz and Wagner (2016), pp. 2 f. *et passim*; Jansen (2003), pp. 42 ff.; Koziol (2019), p. 377; Owen (2010), p. 571; Zech (2016), pp. 180 ff.

[13] Borges (2018), p. 979; *id.* (2019), p. 149; Sommer (2020), pp. 447 ff.; Spindler (2019), pp. 134 ff.; Thöne (2020), pp. 187 ff.

[14] Deutsch (1996), mn. 5; Owen (2010), p. 580; Lahe (2005), p. 61; Gray (1999), p. 293; Karner (2019), p. 117; for French tort law cf. Brüggemeier (2020), pp. 334 ff.

[15] So for example Kötz and Wagner (2016), mn. 110, 113; particularly on § 823 Sec. 1 BGB see also Brüggemeier (1986), mn. 178; Watson (2013), p. 3; Goldberg and Zipursky (2001), p. 658.

[16] Wright (1985), p. 1737; Gray (1999), p. 94.

[17] Translation published by the German Federal Ministry of Justice and Consumer Protection available at, http://www.gesetze-im-internet.de/englisch_bgb/englisch_bgb.html#p3489. Accessed 28 August 2021.

§ 276 para. 2 BGB which defines negligence as the breach of a duty of care, this provision clearly reflects the concept of fault-based liability.

This liability concept which can be found in most if not all jurisdictions is contrasted with so-called "strict liability". However, it is often difficult to establish which liability model is being referred to by the term "strict liability".[18] To simplify matters, two liability models will be distinguished from one another: one is objective liability[19] which apportions responsibility as a result of the fact that damage has been caused; it requires to be a responsibility for a specific sphere of risk (for example, the operation of a vehicle) but neither requires the breach of a duty to act nor additional criteria such as the existence of a defective product. Objective liability mirrors the legal concept of strict liability[20] in the form as put forward in economic discussion[21] and often referred to 'strict liability in the narrow sense'.[22] As an example, the liability of the registered keeper of a vehicle pursuant to § 7 para. 1 of the German Road Traffic Act (Straßenverkehrsgesetz, StVG)[23] may be mentioned. This provision reads as follows: "If, during the operation of a motor vehicle or a trailer intended for use with a motor vehicle, a person suffers death, the body or health of a person is injured or an item of property is damaged, the vehicle holder has an obligation to pay compensation to the injured person for the resulting damage."[24]

The term 'strict liability' is also used to refer to many other liability models,[25] all of which have additional requirements which must be fulfilled such as the breach of a duty of care in the form of a breach of a duty to act or the defect of a product. The key difference from traditional fault-based liability lies in the different apportionment of the evidentiary burden.[26]

An example of this is the liability pursuant to European Product Liability Law. The European Product Liability Directive of 1985[27] (PLD) holds the producer liable

---

[18] In part, numerous distinctions are drawn between different types of strict liability, a summary can be found in Deutsch (1992), pp. 73 ff.

[19] This term is used, for example, in the German Federal Court's (Bundesgerichtshof, BGH) decision of 20.10.1961, Az. VI ZR 41/61, VRS Bd. 22, p. 10 ff.; OLG Düsseldorf, NJW-RR 1999, p. 1623; Borges (2016b), p. 274; *id.* (2018), p. 980; Deutsch (1996), p. 54 ("Kausalhaftung als enge Gefährdungshaftung").

[20] See Jorgensen (1970), p. 46; Ramon (1982), pp. 1692 ff.

[21] Schäfer and Ott (2012), p. 229.

[22] Deutsch (1996), mn. 661; Deutsch (1992), p. 75.

[23] Explicitly OLG Düsseldorf, NJW-RR 1999, p. 1623; Borges (2016b), p. 274; Böhmer (1970), p. 1725 ("Verursachungshaftung").

[24] Translation published by the German Federal Ministry of Justice and Consumer Protection, available at https://www.gesetze-im-internet.de/englisch_stvg/englisch_stvg.html#p0014. Accessed 28 August 2021.

[25] Deutsch (1992), pp. 75 ff.

[26] Borges (2016b), p. 278; *id.* (2018), p. 981.

[27] Council Directive of 25 July 1985 on the approximation of the laws, regulations and administrative provision of the Member States concerning liability for defective Goods (85/374/EEC), O.J. No. L 210/29.

for damages caused by a defect of his product (compare Art. 1 PLD). The liability according to Art. 1 PLD is generally referred to as "strict liability".[28] However, as the liability according to Art. 1 PLD requires the product to have had an identifiable defect at the time the product was put into circulation (compare Art. 1 PLD), which is to be proven by the plaintiff (Art. 4 PLD), the "strict liability" according to the European Product Liability Directive results, at least in some jurisdictions, in producing a similar outcome to fault-based liability.[29]

The boundaries between the different liability models remain unclear. In particular, evidentiary rules can lead to there being, in effect, little difference between fault-liability and strict liability.[30] A further method which is keenly used by jurisprudence is the formulation of wide-ranging duties of care, which, in practice, has the same effect as strict liability due to the fault-based liability. However, it should be borne in mind that fault-based liability and strict liability are, in essence, very different liability concepts and so the question of whether fault-based liability or strict liability should apply to autonomous systems remains fundamentally important.

3 Liability for Autonomous Cars

The liability for autonomous systems has been particularly focusing on the example of autonomous cars and will be briefly discussed below at the example of German law.

As it has already been mentioned, § 7 StVG holds the registered keeper of a vehicle, with the exception of joyriding and cases of Acts of God, comprehensively responsible for damage caused by the use of such vehicle. The liability of the driver is entirely secondary. The result does not come from the provisions of Tort Law which do, of course, apply but from the special liability regulations of the Road Traffic Act and the German Compulsory Insurance Act (Gesetz über die Pflichtversicherung für Kraftfahrzeughalter—Pflichtversicherungsgesetz, PflVG). In its § 18, the Road Traffic Act also provides for liability of the driver. However, while both the keeper and the driver of the vehicle are liable as regards to the external relationship with the victim, in practice, insurance will cover the damage.

According to § 1 of the German Compulsory Insurance Act, the registered keeper of the car is obliged to take out and maintain insurance cover for any personal injury,

---

[28]Taschner (1986), p. 612; Dielmann (1986), p. 1392; European Commission (2018), p.7.

[29]Wagner (2019), pp. 35 f.; cf. for the German law Oechsler (2018b), mn. 42.; cf. also for the ProdHaftG Seibl (2021), mn. 24.

[30]Borges (2018), p. 981; Spickhoff (2016), p. 866; Thöne (2020), pp. 137 f.; to the extent that a relaxation of evidentiary burden to a partial reversal of the burden of proof is accepted in the context of § 823 para. 1 BGB, liability pursuant to § 823 para. 1 BGB could certainly be described as strict liability, compare Borges (2016b), p. 278; see also Deutsch (1992), p. 77; Grundmann (2016), mn. 44 (for presumed fault); see also von Bar (1980), pp. 273 ff.

property damage and other financial losses caused by the use of the vehicle. § 115 of the German Insurance Contract Act (Gesetz über den Versicherungsvertrag—Versicherungsvertragsgesetz, VVG) grants a claim against the insurer to the victim. For this reason, in practice, the compensation for damage arising out of car accidents is managed by insurance companies. Being even more important for the apportionment of liability, § 116 para. 1 sentence 1 VVG reads: "As regards the relationship between the joint and several debtors under § 115 (1), fourth sentence, the insurer shall be solely liable insofar as he is obligated to indemnify the policyholder based on the insurance agreement."[31] The effect of this rules of the Insurance Contracts Act, together with the Compulsory insurance Act is, that usually only the compulsory insurance taken out by the registered keeper will be responsible for paying for damages arising out of car accidents. The existing liability of the driver is crowded out,[32] the liability of the car's producer, e.g. under product liability law, is usually not considered in practice.

§ 7 StVG is also applicable to autonomous cars so that, pursuant to the legal system currently in force, the registered keeper of an autonomous car is comprehensively liable for all damage caused by such car.[33] In particular, liability is not excluded pursuant to § 7 para. 2 StVG because the failure of technical equipment would not be considered to be an act of God.[34]

However, the application of these provisions to autonomous cars is not convincing. The comprehensive liability of the vehicle's registered keeper is based not least on the idea that the registered keeper is in the position of being able to comprehensively control the vehicle.[35] He can decide who may drive the car and if he does not drive the car himself, he can control the degree of risk to a great extent through his choice and guidance of the driver. In addition, evidentiary deliberations point towards holding the registered keeper rather than the driver of the vehicle liable as the identity of the registered keeper at the time of the accident is much easier to establish than the one of the driver.[36]

In the case of autonomous vehicles, the registered keeper does not in fact exercise control because he has no influence on the specific control of the vehicle. This role is

[31] Translation published by the German Federal Ministry of Justice and Consumer Protection, available at https://www.gesetze-im-internet.de/englisch_vvg/englisch_vvg.html#p0482. Accessed 28 August 2021.

[32] Recourse against the registered keeper and the driver is only possible under exceptional circumstances; see Burmann and Hess (2019), mn. 1–13.

[33] Armbrüster (2017), p. 84; von Bodungen and Hoffmann (2016a), pp. 450 f.; Borges (2016b), p. 274; Fleck and Thomas (2015), p. 1394; Gless and Janal (2016), p. 571; Jänich et al. (2015), p. 315; Lutz et al. (2013), p. 61; Schrader (2015), pp. 3537 ff.; Sosnitza (2016), p. 768; Wagner (2017), p. 758.

[34] See Jänich et al. (2015), p. 315; Schrader (2015), p. 3538; Sosnitza (2016), p. 768; Wagner (2017), p. 758.

[35] Borges (2016b), p. 279; Greger (2014), mn. 23.

[36] Borges (2016b), p. 279.

taken by the vehicle manufacturer.[37] Therefore, as has been justified in a different context,[38] objective liability for the manufacturer of autonomous vehicles should be introduced to incentivise such manufacturers to produce autonomous cars which are as safe as possible.[39]

Literature, in the main part, rejects such a development of the legal framework.[40] In practice, there is a great deal of support for liability of the manufacturer of autonomous systems. In 2017. Bitkom, Germany's digital industry association, published a survey on how liability of damage caused by autonomous vehicles should be apportioned. Interestingly, 73% of 1.006 respondents aged 14 and older believed that vehicle manufacturers (35%) or software providers (38%) should be liable, whereas 0% thought that the registered keeper should be liable. Out of 177 companies in the automotive industry, only 12% would hold the registered keeper of the vehicle liable compared to 60% (Software-Providers (41%), vehicle manufacturers (19%)) for the liability of the manufactures side.[41]

As a conclusion it can be said that, according to the survey, a broad majority both of ordinary people and of companies in the automotive industry, considers that liability for damage caused by autonomous cars should be attributed to the car producers rather than to the registered keeper. This view can also be found in the public discussion in Germany,[42] while other voices consider the existing liability scheme to be sufficient.[43]

4 Liability for Autonomous Systems in General

The discussion on liability for autonomous vehicles highlights the central questions that arise in general with regard to liability for autonomous systems. In particular, there is the question whether objective liability for autonomous systems should be introduced and who should be the addressee of such liability.

---

[37] Borges (2018), p. 981.

[38] Borges (2016b), pp. 277 ff.; *id.* (2016a), pp. 617 ff.

[39] In this direction also Kütük-Markendorf (2017), p. 354; more reservedly by Sosnitza (2016), p. 772.

[40] Implied in Schaub (2017), p. 348; in favour of comprehensive strict liability of an undefined "operator" Gless and Janal (2016), p. 574; in favour of the sufficiency of the existing liability concepts von Bodungen and Hoffmann (2016b), p. 508; the same result also in Sosnitza (2016), p. 772 (changes to the existing liability regime are not necessarily required) Wagner and Goeble (2017), p. 266.

[41] Cf. the press release of Bitkom e.V., available at https://www.bitkom.org/Presse/Presseinformation/Wer-haftet-fuer-mein-selbstfahrendes-Auto.html. Accessed 28 August 2021.

[42] E.g. Prantl (2018).

[43] See e.g. Leibfritz (2019).

4.1 *Liability of the Operator of Autonomous Systems*

As part of any consideration of general liability for autonomous systems, it is necessary to consider the liability of the operator. In contrast to the regulations for vehicles, most jurisdictions do not provide any specific liability rule applicable to operators of autonomous systems in general so far. In German law, e.g., liability is solely governed by the general statutory regulation of Tort Law set out in § 823 para. 1 BGB.[44]

4.1.1 Liability for Animals as a Traditional Form of "Operator Liability for Autonomous Systems"

However, objective liability of the operator is not entirely foreign to Civil Law. On the contrary, one can even find objective liability of the operator for what can be called an "autonomous system". Following Roman legal tradition, several national laws contain provisions establishing objective liability of the keeper of an animal.[45] As an example, § 833 sentence 1 BGB holds the "keeper" of an animal liable for damage caused by such animal. This provision reads: "If a human being is killed or if the body or health of a human being is injured or a physical object is damaged by an animal, the person who keeps the animal is obliged to compensate the injured person for the damage arising from this."

§ 833 sentence 1 BGB only applies to so-called "purely domestic animals" (luxury animals). In the case of animals which serve an occupational or economic purpose for the keeper, a different liability regime according to § 833 sentence 2 BGB comes into play.[46]

Specific rules on the liability of the keeper of an animal can be found in many jurisdictions.[47] According to Art. 1243 of the French Civil Code (Code Civil), the owner of an animal as well as the person who uses the animal are liable regardless of fault.[48] This liability occurs regardless of the type of animal.[49]

[44] Borges (2018), p. 981.

[45] von Bar (1996), mn. 206 f., mn. 210 f., mn. 354; Wagner (2020c), mn. 4.

[46] Eberl-Borges (2018), mn. 6.

[47] See, for an overview, von Bar (1996), mn. 206 ff.; Spickhoff (2020), mn. 18–29.

[48] RGZ 14, 316 (317 ff.); von Bar (1996), mn. 208, 210, 213; Spickhoff (2020), mn. 6; Cour de cassation, Chambre civile 2, 24.05.1991, n° 90-12912; Cour de cassation, Chambre civile 2, 18.10.1995, n° 93-17.277.

[49] Spickhoff (2020), mn. 6; Cour de cassation, Chambre civile 2, 06.05.1970, n° 69-11.121; Cour de cassation, Chambre civile 2, 08.11.1984, n° 83-12.643; Cour de cassation, Chambre civile 2, 09.01.1991, n° 89-15.489.

English law contains a detailed regulation on liability for damage caused by animals in the Animals Act of 1971,[50] which differentiates between dangerous and non-dangerous animals. In the case of dangerous animals, § 2 para. 1 provides for strict liability of the animal keeper.[51] In U.S. law, the animal owner liability distinguishes according to the wildness of the animal, providing strict liability of the animal owner for wild animals.[52]

This example shows that the legislature has, for many years, used objective liability to apportion liability to those who operate dangerous objects in order to protect third parties from the risk associated with such objects.

4.1.2 Liability for Animals as a Model for Liability for Autonomous Systems?

The possibility of using the liability for the keeping of animals as a model for the liability for autonomous systems has not yet been discussed in any great depth. Whilst there are, in part, parallels being drawn,[53] admittedly without explanation of their significance, a transfer of the differentiation set out in § 833 BGB to the creation of rules for autonomous systems is also taken into consideration.[54] Others completely reject any such transfer,[55] including similarities to other regulations, such as liability for buildings pursuant to § 836 BGB for example.[56]

It can come as no surprise that a simple application of § 833 BGB or a parallel regulation on the liability for animals to autonomous systems is not suggested. As it already has been mentioned, in the case of § 833 BGB, the law makes a vital distinction between so-called purely domestic animals and animals which serve an occupational or economic purpose for the keeper.[57] The liability for keepers of

[50] Animals Act 1971, available at https://www.legislation.gov.uk/ukpga/1971/22. Accessed 28 August 2021. See also *Lord Denning M. R.* in Cummings v. Granger [1977] Q. B. 397, 404 described § 2 (2) as "very cumbrously worded" and giving rise to "several difficulties"; *Slade L. J.* in Curtis v. Betts [1990] 1 W. L. R. 459, 462 agreeing.

[51] von Bar (1996), mn. 357; Spickhoff (2020), mn. 25; without mentioning the respective rule, Goodhart (1941), pp. 269 f.

[52] Goodhart (1941), pp. 269 f.; Hay (2020), pp. 159 f. at mn. 392.

[53] Brunotte (2017), pp. 585 ff.; Sosnitza (2016), p. 772; contradicting for example Grützmacher (2016), p. 698; reservedly Horner and Kaulartz (2016a), p. 14; *the same* but more in favour of *de lege ferenda* (2016b), pp. 24 ff.

[54] It is explained in such a way in Schaub (2017), p. 348, who distinguishes between private and commercial operators.

[55] Börding et al. (2017), p. 140; Grützmacher (2016), p. 698; Zech (2016), p. 196.

[56] See Grützmacher (2016), p. 698 by reference to a similar problem regarding the proof of the cause of the damage; on the general purpose of § 836 BGB in solving evidential problems, see Bernau (2018), mn. 3; Wagner (2017), mn. 2.

[57] For the reasons behind the creation of § 833 BGB in its current form, see Eberl-Borges (2018), mn. 1 ff.

so-called purely domestic animals is objective liability.[58] In the case of animals which serve an occupational or economic purpose for the keeper, however, liability is excluded if the keeper can prove that he duly supervised the animal. As a result, the liability for animals used in a business set forth in § 833 sentence 2 BGB can be characterised as fault based liability combined with a reversal of the burden of proof regarding fault.[59]

As for the application of § 833 BGB by analogy to autonomous systems, it is not even clear what part of § 833 BGB should be applied. From a legal perspective, it would be possible in theory to apply the differentiation of purpose set out in § 833 BGB in order to distinguish between private and commercial use.[60]

But with a commercial purpose typically being used a justification for more stringent liability, there is doubt on whether to apply to "useful robots", i.e. robots of a specific (commercial) purpose, a less stringent liability than to "luxury robots" however defined.

Moreover an even more difficult question arises when it comes to the field of application of such liability in general. It will be extremely difficult to define the application area of such a strict liability system for "robots" or "autonomous systems". It will not be possible to find highly selective and, at the same time, general criteria for the distinction between "normal" and "autonomous" machines.[61]

These deliberations show that the development of a specific liability regime for "robots" or "autonomous systems" or for "software" cannot be achieved through a transfer of the regime governing the liability for the keepers of animals.[62]

4.1.3 What Can Be Learned from Traditional Rules on Operators' Liability?

However, traditional rules of tort law provide valuable reference points of the further development of a liability regime for autonomous systems. It is important to take into account not only the liability for keepers of animals but also other situations giving rise to liability set out in §§ 832 et seq. BGB. Of particular interest here is the liability of parents for damage caused by their children pursuant to § 832 BGB. It is noteworthy that for parents the liability concept is the same as for the keepers of animals used for occupational or economic purposes according to § 833 sentence 2 BGB.

It then becomes clear that, in the case of liability for beings acting independently (§§ 832, 833 BGB), the legislature is traditionally more likely to rely on fault-based

[58] Deutsch (1992), p. 75 ("enge Gefährdungshaftung"); similarly also BGH, NJW 1976, p. 2130 as well as Kruse (2012), p. 1362.

[59] Borges (2018), p. 981.

[60] In this direction see for example, Schaub (2017), p. 348.

[61] Borges (2018), p. 981.

[62] Borges (2018), p. 981.

liability with a reversal of a burden of proof, as well as other ways of reducing the evidentiary burden, and adopts objective liability only in limited exceptional circumstances.

The liability for the keepers of animals points towards two essential criteria for a differentiated liability regulation. The first point of differentiation is the degree of risk which is posed by the object in question. The less stringent liability pursuant to § 833 sentence 2 BGB, for instance, only applies to animals held domestically as well as to tame animals or animals which are typically tame,[63] and which, in contrast to wild animals (compare § 960 BGB), pose less danger to a third party.[64]

Secondly, the public interest in the implementation of a dangerous system is to be considered.[65] The German law may serve as an example again: The opportunity for exculpation in relation to non-domestic animals was introduced in 1908 explicitly to avoid an excessive liability risk for farmers and so to protect the efficiency of the agricultural sector.[66]

In particular, the principle behind the rules on strict liability such as the §§ 832 to 834 BGB or parallel norm in other legal systems should lead the way by establishing a specific liability regime for those actions of independently acting beings which are neither entirely predictable nor controllable actions.[67] Unforeseen or uncontrollable events can often not be avoided, at least not with reasonable effort. So, it would not be possible to formulate a fault-based liability regime based on duties to act (duties to ensure the safety of the public) without significant gaps.

This gap in liability can be partially closed through a reversal of the burden of proof particularly as, in practice, the reversal of the burden of proof frequently acts as de facto objective liability in cases where it is not possible to establish the cause of the damage. However, in the key area of uncontrollable actions, gaps in liability remain and the risk falls upon the damaged party. According to the key principle of § 833 BGB, this allocation of risk is only acceptable in cases where the risky activity is of absolute necessity to society.

4.1.4 New Liability Rules for Operators of High-Risk AI Systems?

The legal policy discussion in the European Union points in the direction of a strict liability of operators of autonomous systems with a particular risk. In its final report, the Expert Group recommends the introduction of strict liability for the operator of an AI system.[68]

---

[63] Spindler (2017), mn. 26; Wagner (2017), mn. 46.

[64] Borges (2018), pp. 981 ff.

[65] Borges (2018), p. 982.

[66] See detailed commentary in Engelmann (1929), Anm. 2; on the ratio of § 833 para.-2 BGB see also Eberl-Borges (2018), mn. 8 ff.; on the constitutionality of the differentiation see BGH, NJW 2009, p. 3234.

[67] Borges (2018), p. 982.

[68] *Supra* note 7, p. 6.

Picking up this ball, the European Parliament proposes in its resolution of October 2020 a regulation on liability for the operation of systems equipped with artificial intelligence.[69] The core of the proposed regulation is the introduction of specific liability for the operators of so-called AI systems. In Art. 3 lit. a), the proposed regulation defines the characteristic feature of the AI system as that it "takes action with a certain degree of autonomy in order to achieve certain goals". Thus, the proposed regulation is about systems that are to be regarded as autonomous systems in the sense of the terminology used here.

The proposed regulation contains an interesting distinction between "high risk" AI systems and other AI systems. According to Art. 4 para 1 of the proposed regulation, the operators of high-risk AI systems are to be subject to strict liability. The proposed article reads: "The operator of a high-risk AI-system shall be strictly liable for any harm or damage that was caused by any physical or virtual activity, device or process driven by that AI-system." Causal liability in the sense of the terminology used here is thus proposed. In contrast, the draft regulation wants traditional fault-based liability to be applied to other AI systems, as Art. 8 para. 1 of the proposed regulation clarifies, but with a significant special feature: According to Art. 8 para. 2, a reversal of the burden of proof with regard to fault is to apply.

According to Art. 4 para 2, strict liability of operators of high-risk AI systems should apply exclusively to the systems listed in an annex. The EU Commission should be given the power to include certain AI systems in the annex or to delete them. This approach of the EU Parliament is convincing in principle: The planned introduction of a legal causal liability of the operators of autonomous systems with special risk would be a suitable step to ensure compensation for damages resulting from the operation of AI systems. The limitation of liability to a positive list is necessary to ensure legal clarity. The relevance of such a rule in practice is currently not foreseeable as it depends solely on how the EU Commission makes use of the power to subject AI systems to strict liability through delegated acts.

Even more interesting is Art. 8 of the proposed regulation, which would apply to all AI systems that are not classified as high-risk AI systems. This rule, which would have a very wide scope of application due to the broad definition of AI system in Art. 3 of the same regulation, subjects the operator of such an AI system to "fault-based" liability in its paragraph 1. At first glance, this norm seems superfluous, since operators of AI systems are subject to general tort liability anyway. The meaning of the provision can only be understood in the context of paragraph 2, sentence 1, which reads as follows:

> 2. The operator shall not be liable if he or she can prove that the harm or damage was caused without his or her fault, relying on either of the following grounds':

[69] Proposal for a Regulation of the European Parliament and of the Council on liability for the operation of Artificial Intelligence-systems, (2020/2014(INL)), available at https://www.europarl.europa.eu/doceo/document/TA-9-2020-0276_EN.pdf. Accessed 28 August 2021.

(a)
the AI-system was activated without his or her knowledge while all reasonable and necessary measures to avoid such activation outside of the operator's control were taken, or

(b)
due diligence was observed by performing all the following actions: selecting a suitable AI-system for the right task and skills, putting the AI-system duly into operation, monitoring the activities and maintaining the operational reliability by regularly installing all available updates.

[...]

In its first sentence, para. 2, the draft regulates a reversal of the burden of proof at the expense of the operator of the AI system, which justifies an aggravation of liability. At the same time, however, the norm aims at limiting liability considerably by enumerating the relevant obligations of the operator. It may be assumed that this list, which covers the typical duties of the user of a machine, is intended to be exhaustive.

Liability under Art. 8 para. 2 has a very similar structure to liability for human agents, which is regulated in a very similar way in German law, for example: According to § 831 BGB, the principal is liable for damage caused by a human assistant if he cannot prove that he exercised the necessary care in selecting and supervising the agents.[70] Liability under § 831 BGB is not particularly strict in practice because case law does not impose strict requirements on proof of careful selection and supervision. Whether this is true in the same way for the use of devices would only become apparent in judicial practice. At his moment, it can therefore not be assumed that Art. 8 of the proposed regulation would actually result in a significant expansion of the liability of operators of AI systems.

A significant relief of liability in favour of the operator would result if the first sentence of paragraph 2 was also to exclude liability under general tort law of the Member States. Even if the draft regulation is silent on this, such a primacy might result from the regulation's general primacy as lex specialis. Liability for human agents however, for example according to § 831 BGB, always applies alongside liability according to general rules. This should also apply to Art. 8 of the proposed regulation, i.e. it should not displace general tort law.

As a result, with regard to the liability concept of the proposed regulation, it thus becomes apparent that the relevance of the regulation crucially depends on which systems are classified as high-risk AI systems by the EU Commission. If the annex was to remain as empty as it is in the draft, the effect of the regulation would be

[70] § 831 para. 1 BGB reads: "A person who uses another person to perform a task is liable to make compensation for the damage that the other unlawfully inflicts on a third party when carrying out the task. Liability in damages does not apply if the principal exercises reasonable care when selecting the person deployed and, to the extent that he is to procure devices or equipment or to manage the business activity, in the procurement or management, or if the damage would have occurred even if this care had been exercised.", available at https://www.gesetze-im-internet.de/englisch_bgb/englisch_bgb.html#p3510. Accessed 28 August 2021.

limited to an appeal to the users of AI systems to observe the basic obligations in dealing with machines.

It is not foreseeable at this moment whether the EU Parliament's proposal for a regulation has a chance of becoming law. The EU Commission has announced a comprehensive legislative proposal for AI for 2021, which will probably determine the further legislative process at EU level.

4.2 Liability of the Manufacturer of Autonomous Systems

An essential aspect of the liability regime for autonomous systems concerns the question of which liability regime applies to their manufacturers in this respect.

4.2.1 General Tort Law

General tort law with its traditional fault-based liability is applicable. In some jurisdictions, case law and doctrine have derived specific rules for producer liability from this. In German law, for example, the so-called producer liability, which is based on the general tort clause of § 823 para. 1 BGB, is of great practical importance. In some areas, this producer liability according to general tort law even goes beyond product liability under the European Product Liability Directive, in that it contains obligations regarding the monitoring of products and obligations to recall products.[71] This can even apply in cases where the product is defect-free on its own, but can create a hazard in combination with other products.[72]

4.2.2 Product Liability

The focus of interest regarding the responsibility of the producer is the question of the extent to which so-called product liability, which is based on special legislation in many states worldwide and often imposes strict liability on the producer, can be applied to autonomous systems and whether this leads to appropriate results.

Product liability was introduced in many countries in the second half of the twentieth century, starting in the U.S. which is considered the "birthplace" of product liability.[73] The transition from fault based liability to strict liability of the manufacturer for defective products was heavily influenced by the 1963 landmark decision of the California Supreme Court in *Greenman v Yuba Power Products.*[74]

---

[71] Borges (2020b), mn. 83; Wagner (2020d), mn. 62; von Westphalen (2020), p. 250.

[72] BGHZ 99, pp. 172 ff.

[73] Howells (1993), p. 201.

[74] *Greenman v. Yuba Power Products, Inc.* (1963), 59 Cal. 2d 57.

This approach found its way into § 402A of the Restatement Second of Torts from 1966 and was later accepted in most states of the US.[75]

In Europe, the Product Liability Directive was enacted in 1985 in the countries of the then European Economic Community. Article 1 of the Directive provides for strict liability of the manufacturer of a product for damage caused by a defect in the product. The central prerequisite for liability is the existence of a defect in the product,[76] which, according to Art. 6 PLD, exists if the product does not offer the safety that could reasonably be expected taking into account all circumstances at the time it was placed on the market. The burden of proving the existence of the defect and its causality for the damage is on the injured party according to Art. 4 PLD.

Liability under the Product Liability Directive applies to products as defined in Art. 2 PLD. Accordingly, products are all "movables", even if they form part of another movable or immovable object. Insofar as autonomous systems consist of a combination of software and hardware, as in most cases, they are undoubtedly subject to product liability.[77] The only question is whether such systems, which can be operated purely software-based in any runtime environment, fall under the concept of a product.

Even before the Product Liability Directive came into force, it was discussed whether software falls under the concept of a product and thus is subject to product liability.[78] This controversial question has still not been conclusively answered. The European Court of Justice has not yet ruled on this question.

However, in some jurisdictions, such as in German law, there is agreement in the current literature that software is subject to product liability comprehensively, i.e. even if it is not part of a movable object.[79] However, there is a lack of case law on this aspect, so that there is legal uncertainty in this respect.

As it had already been mentioned, the EU Commission had set up an "Expert Group on liability and new technologies" with two formations in 2018. The "Product Liability Directive formation" had the task of advising on the revision of the Product Liability Directive.[80] This revision could, not least, legally regulate product liability

[75] Cf. Fairgrieve et al. (2016), Teil 1, p.19 at mn. 1; Reimann (2015), p. 251.

[76] Fairgrieve et al. (2016), Teil 1, p. 50 at mn. 77.

[77] Schulz (2015), p. 163; Sommer (2020), p. 220; for autonomous vehicles cf. Berz et al. (2000), p. 549; Borges (2016b), p. 275; Horner and Kaulartz (2016a), pp. 10 f.; Jänich et al. (2015), p. 316; Lutz et al. (2013), p. 61; Schrader (2015), p. 3538.

[78] See, as an excerpt of this discussion, the written question (No. 706/88) of *Gijs de Vries* to the Commission of the European Communities and the related answer given by *Lord Cockfield* on behalf of the Commission, OJEC No C 114, p. 42 as of 8.5.1989.

[79] Borges (2020a), mn. 43; Cahn (1996), p. 2904; Fairgrieve and Rajneri (2019), p. 27; Rebin (2021), mn. 56; Redeker (2020), mn. 878; Reese (1994), p. 1124; Reusch (2020), Ch. 4.1 at mn. 173 ff.; Taeger (1996), p. 261; Wagner (2017), pp. 716 ff.; a.A. Oechsler (2018a), mn. 65; Frietsch (1990), mn. 22.

[80] See the description of tasks of the Product Liability Directive formation within the Call for Applications, p. 4, available at https://ec.europa.eu/transparency/expert-groups-register/core/api/front/calls-application/12065/download. Accessed 28 August 2021.

with regard to software. However, the formation has not yet submitted a final report and has not commented on this aspect.

As an interim result, it can be stated that autonomous systems are generally subject to liability under the Product Liability Directive.

The main problems in relation to autonomous systems are the requirement of a product defect and the burden of proof of the injured party for such defect Determining a is product defect often difficult in the case of new technologies, as the question of what level of safety "can be expected" from a novel product is often not clarified. Moreover there is often a lack of information about the technical characteristics of the system and about the technical processes that led to the damage. In this respect, it is of decisive importance that the instrument of pre-trial discovery of US law is not available in European legal systems. German procedural law, for example, does not grant the injured party the right to obtain information or evidence from the producer about the damage or the product. Insofar as a defect in the product is not apparent from the external circumstances, for example in the case of a spontaneously explosive glass bottle, the injured party is often in lack of evidence.

This aspect has a particular impact in the case of autonomous systems. The problem already arises as to whether the behaviour of an autonomous system in a certain situation can be used to infer the existence of a defect of the system. This must be rejected in principle. The behaviour of a system in a particular situation as such is not a property of the system.[81] However, the ability of a system to behave in a certain way in a certain situation or not to show a certain behaviour can be regarded as a property.[82] By way of example, a desired property of a highly automated vehicle is the ability to drive "correctly" in a certain situation.[83]

It is important to note that product liability law does not provide for liability for conduct, not even for the property of behaving correctly in future unforeseeable situations.[84] Therefore, only the ability to behave correctly in a spectrum of foreseeable situations can be described as a property of an AI system in the sense of product liability, and only to the extent that the system is not equipped with such ability there is a defect which triggers liability.

In this context the author of this article has recently uttered a presumption of the existence of a defect in the autonomous system if the system has behaved "incorrectly". It is then incumbent for the manufacturer to prove that the fault could not

[81] Borges (2021), p. 35, sub 4.1; different by von Westphalen (2018), p. 194 (misbehaviour of a system as a defect of the product).

[82] Borges (2021), p. 35, sub 4.1.

[83] That means, at least, that an automated vehicle must be able to obey traffic rules; this requirement is, for example, set forth in a current draft for an amendment to the German Road Traffic Act (Straßenverkehrsgesetz; StVG), compare § 1e para. 2 No. 3 StVG-E, available at https://www.bmvi.de/SharedDocs/DE/Anlage/Gesetze/Gesetze-19/gesetz-aenderung-strassenverkehrsgesetz-pflichtversicherungsgesetz-autonomes-fahren.pdf?__blob=publicationFile. Accessed 28 August 2021.

[84] Borges (2021), p. 35, sub 4.1; von Westphalen (2020), p. 251; Zech (2019), p. 205.

have been detected with proper design and testing.[85] However, it remains to be seen whether the courts will follow this approach.

4.2.3 Strict Liability of Manufacturers of Autonomous Systems?

With regard to the limitations of product liability law in the case of autonomous systems, the question of whether objective liability for the manufacturers of autonomous systems should be introduced must be posed.

It is clear that the introduction of a general objective liability applicable to any kind of autonomous systems would not make much sense. The introduction of objective liability would only be appropriate if the risk is controllable by the person who is being held liable.[86] In the case of autonomous cars, the manufacturer fulfils this requirement. However, in many other situations, this would not be the case. This becomes immediately obvious if one thinks about objective liability for software. The risk to the rights of third parties associated with the use of the software cannot be foreseen by the manufacturer and certainly cannot be limited by him if he does not control the implementation of the software.

Therefore, one can assume that objective liability for the manufacturer of autonomous systems can only be considered in cases where there are limited possibilities of implementation with clearly foreseeable risks.[87] Candidates for such systems might, besides vehicles, be found amongst products which are used in public spaces which cannot be controlled by an operator, e.g. drones.

It is obvious that the introduction of any kind of objective liability requires a careful analysis of the question which party is best in place of controlling the risks associated with the system. In the case of household robots, for instance, the environment of the robot's activities and the associated risks, are outside the control of the robots manufacturer. Here, one might assume that the operator be the risks. However, in many cases where a robot is used in a private household to support its owner, the operator himself is the party who should be protected as he is the one who is facing the risks. To impose strict liability solely on him might be highly counterproductive. This situation might occur for a wide range of autonomous systems designated to serve their owner such as care robots, household robots or health apps et cetera. Wherever malfunction of such a system constitutes a significant hazard for the user, it seems that objective liability addressing the producer of such system would provide both an adequate protection for the potential victim and an incentive to develop safe products.

---

[85] Borges (2021), p. 37, sub 4.4.

[86] The controllability of the risk is a leading idea for the introduction of objective liability, see Borges (2016b), p. 278; also for an explanation of the term strict liability in the sense of the liability model described as objective liability Brüggemeier (1986), mn. 30; Hehl (1999), p. 90; Larenz (1965), p. 374; Müller-Erzbach (1910), p. 413; Rümelin (1910), p. 46; contradictory opinions expressed by Blaschczok (1993), p. 66; Köndgen (1976), p. 32.

[87] Borges (2018), p. 981.

5 Conclusion: The Necessity for a Differentiated Liability Regime

It is possible to draw the conclusion that a specific liability regime is necessary to deal with the liability for autonomous systems, inasmuch as such systems pose a risk of damage through unforeseeable actions.

It will not be possible to establish a single liability regime which can be uniformly applied to all kinds of "autonomous systems". Instead a regime of differentiated regulation is required which will probably need to be combined with a reversal of the evidential burden, with a wide scope of application and with objective liability (including compulsory insurance and limited liability etc.) for specific areas (such as autonomous vehicles).[88]

Therefore, a great deal of more research into liability gaps and appropriate liability regulation is required before the specific form of a liability regime for autonomous systems can be established.

References

Armbrüster, C. (2017). Automatisiertes Fahren – Paradigmenwechsel im Sraßenverkehrsrecht? *Zeitschrift für Rechtspolitik (ZRP), 50*(3), 83–86.

Bernau, F. (2018). § 836 BGB. In J. Staudinger (Founder), *Staudingers Kommentar zum BGB* (latest version 1 Nov 2020). Sellier/de Gruyter.

Berz, U., Dedy, E., & Granich, C. (2000). Haftungsfragen bei dem Einsatz von Telematik-Systemen im Straßenverkehr. *Deutsches Autorecht (DAR)*, 545–554.

Blaschczok, A. (1993). *Gefährdungshaftung und Risikozuweisung*. Heymanns.

Böhmer, E. (1970). Keine Gefährdungshaftung des Kraftfahrzeugfahrers. *Neue Juristische Wochenschrift (NJW), 23*(39), 1724–1725.

Börding, A., Jülicher, T., Röttgen, C., & v. Schönfeld, M. (2017). Neue Herausforderungen der Digitalisierung für das deutsche Zivilrecht. *Computer und Recht (CR), 33*(2), 134–140.

Borges, G. (2016a). Haftung für selbstfahrende Autos. In E. Schweighofer, F. Kummer, W. Hötzendorfer, & G. Borges (Eds.), *Netzwerke, Tagungsband des 19. Internationalen Rechtsinformatik Symposions (IRIS)* (pp. 611–619). OCG.

Borges, G. (2016b). Haftung für selbstfahrende Autos. *Computer und Recht (CR), 4*, 272–280.

Borges, G. (2018). Rechtliche Rahmenbedingungen für autonome Systeme. *Neue Juristische Wochenschrift (NJW), 71*(14), 977–982.

Borges, G. (2019). New liability concepts: The potential of insurance and compensation funds. In S. Lohsse, R. Schulze, & D. Staudenmayer (Eds.), *Liability for artificial intelligence and the Internet of Things, Münster Colloquia on EU law and the digital economy IV* (pp. 145–165). Nomos.

Borges, G. (2020a). § 2 ProdHaftG. In G. Borges & M. Hilber (Eds.), *Beck'scher Online-Kommentar IT-Recht* (1st ed.). C.H.Beck.

Borges, G. (2020b). § 3 ProdHaftG. In G. Borges & M. Hilber (Eds.), *Beck'scher Online-Kommentar IT-Recht* (1st ed.). C.H.Beck.

[88] Borges (2018), p. 982.

Borges, G. (2021, June). AI systems and product liability. In *ICAIL '21: Proceedings of the eighteenth international conference on artificial intelligence and law* (pp. 32–39). Retrieved August 28, 2021, from https://doi.org/10.1145/3462757.3466099

Bräutigam, P., & Klindt, T. (2015). Industrie 4.0, das Internet der Dinge und das Recht. *Neue Juristische Wochenschrift (NJW), 68*(16), 1137–1142.

Brüggemeier, G. (1986). *Deliktsrecht*. Nomos.

Brüggemeier, G. (2020). The civilian law of delict: A comparative and historical analysis. *European Journal of Comparative Law and Governance, 7*, 339–383.

Brunotte, N. (2017). Virtuelle Assistenten – Digitale Helfer in der Kundenkommunikation. *Computer und Recht (CR), 33*(9), 583–589.

Burmann, M., & Hess, R. (2019). Chapter 7 Sektion K. In U. Berz & M. Burmann (Eds.), *Handbuch des Straßenverkehrsrechts* (43rd ed.). C.H.Beck.

Cahn, A. (1996). Produkthaftung für verkörperte geistige Leistungen. *Neue Juristische Wochenschrift (NJW), 49*(44), 2899–2905.

Deutsch, E. (1992). Das neue System der Gefährdungshaftungen: Gefährdungshaftung, erweiterte Gefährdungshaftung und Kausal-Vermutungshaftung. *Neue Juristische Wochenschrift (NJW), 45*(2), 73–77.

Deutsch, E. (1996). *Allgemeines Haftungsrecht* (2nd ed.). Heymanns.

Dielmann, H. (1986). The European Economic Community's Council Directive on product liability. *International Lawyer, 20*(4), 1391–1400.

Eberl-Borges, C. (2018). § 833 BGB. In J. Staudinger (Founder), *Staudingers Kommentar zum BGB* (latest version 6 Aug 2018). Sellier/de Gruyter.

Engelmann, T. (1929). § 833 BGB. In J. Staudinger (Founder), in Gemeinschaft mit T. Loewenfeld, E. Riezler, P. Mayring, K. Kober, T. Engelmann, F. Herzfelder und J. Wagner (Eds.), *Kommentar zum Bürgerlichen Gesetzbuche für das Deutsche Reich nebst Einführungsgesetz* (8th ed. 1929). Schweitzer.

European Commission. (2018). Commission Staff Working Document, SWD(2018) 157 final. Resource Document. Retrieved August 28, 2021, from https://ec.europa.eu/transparency/regdoc/rep/10102/2018/EN/SWD-2018-157-F1-EN-MAIN-PART-1.PDF

Fairgrieve, D., Howells, G., Møgelvang-Hansen, P., Straetmans, G., Verhoevens, D., Machnikowski, P., Janssen, A., & Schulze, R. (2016). Product liability directive. In P. Machnikowski (Ed.), *European product liability, an analysis of the state of the art in the era of new technologies* (pp. 17–111). Intersentia.

Fairgrieve, D., & Rajneri, E. (2019). Is software a product under the product liability directive? *Zeitschrift für Internationales Wirtschaftsrecht (IWRZ), 4*(1), 24–27.

Fleck, J., & Thomas, A. (2015). Automatisierung im Straßenverkehr. *Neue Juristische Online-Zeitschrift (NJOZ), 37*, 1393–1397.

Frietsch, E. (1990). § 2 ProdHaftG. In H. Taschner & E. Frietsch (Eds.), *Produkthaftungsgesetz und EGProdukthaftungsrichtlinie* (2nd ed.). C.H.Beck.

Gless, S., & Janal, R. (2016). Hochautomatisiertes und autonomes Autofahren – Risiko und rechtliche Verantwortung. *Juristische Rundschau (JR), 10*, 561–575.

Goldberg, J., & Zipursky, B. (2001). The restatement (third) and the place of duty in negligence law. *Vanderbilt Law Review, 54*, 657–750.

Goodhart, A. (1941). Restatement of the Law of Torts, Volume III: A comparison between American and English Law. *University of Pennsylvania Law Review and American Law Register, 89*(3), 265–305.

Gray, C. (1999). *The philosophy of law: An encyclopedia*. Garland Publishing.

Greger, R. (2014). § 1. In R. Greger & M. Zwickel (Eds.), *Haftungsrecht des Straßenverkehrs* (5th ed.). DeGruyter.

Grundmann, S. (2016). § 276. In F. J. Säcker, R. Rixecker, H. Oetker, & B. Limperg (Eds.), *Münchener Kommentar zum Bürgerlichen Gesetzbuch* (Vol. 2). C.H.Beck.

Grützmacher, M. (2016). Die deliktische Haftung für autonome Systeme – Industrie 4.0 als Herausforderung für das bestehende Recht? *Computer und Recht (CR), 32*(10), 695–698.

Hager, J. (2017). Vorbemerkungen zu §§ 823. In J. Staudinger (Founder), *Staudingers Kommentar zum BGB*. Sellier/de Gruyter.
Hanisch, J. (2013). Zivilrechtliche Haftungskonzepte für Roboter. In E. Hilgendorf & J. Günther (Eds.), *Robotik und Gesetzgebung* (pp. 109–123). Nomos.
Hay, P. (2020). *US-Amerikanisches Recht* (7th ed.). C.H.Beck.
Hehl, S. (1999). *Das Verhältnis von Verschuldens- und Gefährdungshaftung*. Roderer.
Horner, S., & Kaulartz, M. (2016a). Haftung 4.0, Verschiebung des Sorgfaltsmaßstabs bei Herstellung und Nutzung autonomer Systeme. *Computer und Recht (CR), 32*(1), 7–14.
Horner, S., & Kaulartz, M. (2016b). Haftung 4.0. *Zeitschrift zum Innovations- und Technikrecht (InTeR), 1*, 22–27.
Howells, G. (1993). *Comparative product liability*. Dartmouth.
Jänich, V., Schrader, P., & Reck, V. (2015). Rechtsprobleme des autonomen Fahrens. *Neue Zeitschrift für Verkehrsrecht (NZV), 28*(7), 313–318.
Jansen, N. (2003). *Die Struktur des Haftungsrechts*. Mohr Siebeck.
Jorgensen, S. (1970). The decline and fall of the law of torts. *The American Journal of Comparative Law, 18*(1), 39–53.
Karner, E. (2019). Liability for robotics: Current rules, challenges, and the need for innovative concepts. In S. Lohsse, R. Schulze, & D. Staudenmayer (Eds.), *Liability for artificial intelligence and the Internet of Things, Münster Colloquia on EU law and the digital economy IV* (pp. 117–125). Nomos.
Kerr, D. (2018, March 21). Are driverless cars safe? Uber fatality raises questions. *CNET*. Retrieved August 28, 2021, from www.cnet.com/news/uber-self-driving-car-fatality-in-arizona-has-people-asking-how-safe-are-driverless-cars/
Köndgen, J. (1976). *Haftpflichtfunktionen und Immaterialschaden am Beispiel von Schmerzensgeld bei Gefährdungshaftung*. Duncker & Humblot.
Kötz, H., & Wagner, G. (2016). *Deliktsrecht* (3rd ed.). Vahlen.
Koziol, H. (2019). Die Sicherstellungshaftung – eine weitere Spur im Haftungsrecht? *Archiv für die civilistische Praxis (AcP), 219*(3-4), 376–419.
Kruse, C. (2012). Zu den Voraussetzungen der Tierhalterhaftung bei ungeklärter Verursachung. *Versicherungsrecht (VersR), 63*(31), 1360–1364.
Kütük-Markendorf, M. (2017). Die hoch- oder vollautomatisierte Fahrfunktion als Vorstufe zum autonomen Fahren. *Computer und Recht (CR), 33*(6), 349–355.
Lahe, J. (2005). Forms of liability in the law of delict: Fault-based liability and liability without fault. *Juridica international, 10*, 60–70.
Larenz, K. (1965). Die Prinzipien der Schadenszurechnung. Ihr Zusammenspiel im modernen Schuldrecht. *Juristische Schulung (JuS), 5*, 373–379.
Leibfritz, C. (2019, July 24). Autonomes Fahren kommt. Wer haftet aber im Schadensfall? *EFahrer*. Retrieved August 28, 2021, from https://efahrer.chip.de/news/autonomes-fahren-kommt-wer-haftet-aber-im-schadensfall_101019
Lutz, L., Tang, T., & Lienkamp, M. (2013). Die rechtliche Situation von teleoperierten und autonomen Fahrzeugen. *Neue Zeitschrift für Verkehrsrecht (NZV), 26*(2), 57–63.
Müller-Erzbach, R. (1910). Gefährdungshaftung und Gefahrtragung. *Archiv für die civilistische Praxis, 106*(3), 309–476.
Naughton, K. (2018, March 27). Just how safe is driverless car technology, really? *Bloomberg*. Retrieved August 28, 2021, from www.bloomberg.com/news/articles/2018-03-27/just-how-safe-is-driverless-car-technology-really-quicktake
O'Donnell, B. (2018, April 8). How safe should we expect self-driving cars to be? *USA Today*. Retrieved August 28, 2021, from https://eu.usatoday.com/story/tech/columnist/2018/04/08/how-safe-should-we-expect-autonomous-cars/451494002/
Oechsler, J. (2018a). § 2 ProdHaftG. In J. Staudinger (Founder), *Staudingers Kommentar zum BGB* (latest version 28 Feb 2020). Sellier/de Gruyter.
Oechsler, J. (2018b). Einl. ProdHaftG. In J. Staudinger (Founder), *Staudingers Kommentar zum BGB*. Sellier/de Gruyter.

Owen, D. (2010). Bending nature, bending law. *Florida Law Review, 62*, 569–617.

Prantl, H. (2018, March 20). Haftung für den Roboter. *Sueddeutsche Zeitung (SZ)*. Retrieved August 28, 2021, from https://www.sueddeutsche.de/politik/autonome-autos-haftung-fuer-den-roboter-1.3913628

Ramon, F. (1982). Strict liability. *Louisiana Law Review, 42*(5), 1679–1707.

Rebin, I. (2021). § 2 ProdHaftG. In B. Gsell, W. Krüger, S. Lorenz, & C. Reymann (Eds.), *Beck'scher Online-Großkommentar zum Zivilrecht*. C.H.Beck.

Redeker, H. (2020). *IT-Recht* (7th ed.). C.H.Beck.

Reese, J. (1994). Produkthaftung und Produzentenhaftung für Hard– und Software. *Deutsches Steuerrecht (DStR), 32*(31), 1121–1127.

Reimann, M. (2015). Product liability. In M. Bussani & A. Sebok (Eds.), *Comparative Tort Law* (pp. 250–279). Edward Elgar.

Reusch, P. (2020). Chapter 4.1. In M. Kaulartz & T. Braegelmann (Eds.), *Rechtshandbuch Artificial Intelligence und Machine Learning*. C.H. Beck, Vahlen.

Riehm, T. (2014). Von Drohnen, Google-Cars und Software-Agenten. *IT-Rechtsberater (ITRB), 5*, 113–115.

Rümelin, M. (1910). *Schadensersatz ohne Verschulden: Rede gehalten bei der akademischen Preisverteilung am 6. November 1910*. J.C.B. Mohr.

Schäfer, H., & Ott, C. (2012). *Lehrbuch der ökonomischen Analyse des Zivilrechts* (5th ed.). Springer Gabler.

Schaub, R. (2017). Interaktion von Mensch und Maschine. *JuristenZeitung (JZ), 72*(7), 342–349.

Schrader, P. (2015). Haftungsrechtlicher Begriff des Fahrzeugführers bei zunehmender Automatisierung von Kraftfahrzeugen. *Neue Juristische Wochenschrift (NJW), 68*(49), 3537–3542.

Schulz, T. (2015). *Verantwortlichkeit bei autonom agierenden Systemen*. Nomos.

Schwarze, R. (2019). Vorbemerkungen zu §§ 280–285. In J. Staudinger (Founder), *Staudingers Kommentar zum BGB*. Sellier/de Gruyter.

Seibl, M. (2021). § 1 ProdHaftG. In B. Gsell, W. Krüger, S. Lorenz, & C. Reymann (Eds.), *Beck'scher Online-Großkommentar zum Zivilrecht*. C.H.Beck.

Shavell, S. (2004). *Foundations of economic analysis of law*. Belknap Press.

Sommer, M. (2020). *Haftung für autonome Systeme, Verteilung der Risiken selbstlernender und vernetzter Algorithmen im Vertrags- und Deliktsrecht*. Nomos.

Sosnitza, O. (2016). Das Internet der Dinge – Herausforderung oder gewohntes Terrain für das Zivilrecht? *Computer und Recht (CR), 11*, 764–772.

Spickhoff, A. (2016). Die Grundstruktur der deliktischen Verschuldenshaftung. *Juristische Schulung (JuS), 56*(10), 865–871.

Spickhoff, A. (2020). § 833 BGB. In W. Krüger, B. Gsell, S. Lorenz, & C. Reymann (Eds.), *Beck'scher Online-Großkommentar zum Zivilrecht*. C.H.Beck.

Spiecker gen. Döhmann, I. (2016). Zur Zukunft systemischer Digitalisierung - Erste Gedanken zur Haftungs- und Verantwortungszuschreibung bei informationstechnischen Systemen. *Computer und Recht (CR), 10*, 698–704.

Spindler, G. (2015). Roboter, Automation, künstliche Intelligenz, selbst-steuernde Kfz – Braucht das Recht neue Haftungskategorien? *Computer und Recht (CR), 12*, 766–776.

Spindler, G. (2017). § 833 BGB. In W. Hau & R. Poseck (Eds.), *Beck'scher Online-Kommentar BGB* (57th ed.). C.H.Beck.

Spindler, G. (2019). User liability and strict liability in the Internet of Things and for robots. In S. Lohsse, R. Schulze, & D. Staudenmayer (Eds.), *Liability for artificial intelligence and the Internet of Things, Münster Colloquia on EU law and the digital economy IV* (pp. 125–145). Nomos.

Taeger, J. (1996). Produkt- und Produzentenhaftung bei Schäden durch fehlerhafte Computerprogramme. *Computer und Recht (CR)*, 257–271.

Taschner, H. (1986). Die künftige Produzentenhaftung in Deutschland. *Neue Juristische Wochenschrift (NJW), 39*(10), 611–616.

The Financial Times View. (2018, March 20). Driverless cars and the imperative of safety. *Financial Times*. Retrieved August 28, 2021, from www.ft.com/content/df4b266e-2c2f-11e8-9b4b-bc4b9f08f381

Thöne, M. (2020). *Autonome Systeme und deliktische Haftung*. Mohr Siebeck.

von Bar, C. (1980). *Verkehrspflichten*. Heymanns.

von Bar, C. (1996). *Gemeineuropäisches Deliktsrecht Band 1: Die Kernbereiche des Deliktsrechts*. C.H.Beck.

von Bodungen, B., & Hoffmann, M. (2016a). Autonomes Fahren – Haftungsverschiebung entlang der Supply Chain? (1. Teil). *Neue Zeitschrift für Verkehrsrecht (NZV), 29*(10), 449–454.

von Bodungen, B., & Hoffmann, M. (2016b). Autonomes Fahren – Haftungsverschiebung entlang der Supply Chain? (2. Teil). *Neue Zeitschrift für Verkehrsrecht (NZV), 29*(11), 503–509.

von Westphalen, F. (2018). Tastende Versuche, Rechtsrisiken künstlicher Intelligenz einzuhegen. *Zeitschrift für Internationales Wirtschaftsrecht (IWRZ), 3*(5), 193–194.

von Westphalen, F. (2020). Produkthaftungsrechtliche Erwägungen beim Versagen Künstlicher Intelligenz (KI) unter Beachtung der Mitteilung der Kommission COM(2020) 64 final. *Verbraucher und Recht (VuR), 35*(7), 248–254.

Wagner, G. (2017). Produkthaftung für autonome Systeme. *Archiv für die civilistische Praxis (AcP), 6*, 707–764.

Wagner, G. (2019). Robot liability. In S. Lohsse, R. Schulze, & D. Staudenmayer (Eds.), *Liability for artificial intelligence and the Internet of Things, Münster Colloquia on EU Law and the Digital Economy IV* (pp. 27–63). Nomos.

Wagner, G. (2020a). Verantwortlichkeit im Zeichen digitaler Techniken. *Versicherungsrecht (VersR), 12*, 717–741.

Wagner, G. (2020b). Vor § 823 BGB. In F. J. Säcker, R. Rixecker, H. Oetker, & B. Limperg (Eds.), *Münchener Kommentar zum Bürgerlichen Gesetzbuch* (Vol. 7). C.H.Beck.

Wagner, G. (2020c). § 833 BGB. In F. J. Säcker, R. Rixecker, H. Oetker, & B. Limperg (Eds.), *Münchener Kommentar zum Bürgerlichen Gesetzbuch* (Vol. 7). C.H.Beck.

Wagner, G. (2020d). § 1 ProdHaftG. In F. J. Säcker, R. Rixecker, H. Oetker, & B. Limperg (Eds.), *Münchener Kommentar zum Bürgerlichen Gesetzbuch* (Vol. 7). C.H.Beck.

Wagner, B., & Goeble, T. (2017). Freie Fahrt für das Auto der Zukunft? *Zeitschrift für Datenschutz (ZD), 7*(6), 263–269.

Wakabayashi, D. (2018, March 19). Self-driving Uber car kills pedestrian in Arizona, where robots roam. *New York Times*. Retrieved July 20, 2020, from www.nytimes.com/2018/03/19/technology/uber-driverless-fatality.html

Watson, D. (2013). Style over substance: A comparative analysis of the English and French approaches to fault in establishing tortious liability. *University of Manchester - Student Law Review, 2*(1), 1–14.

Wright, R. (1985). Causation in Tort Law. *California Law Review, 73*, 1735–1828.

Zech, H. (2016). Zivilrechtliche Haftung für den Einsatz von Robotern – Zuweisung von Automatisierungs- und Autonomierisiken. In S. Gless & K. Seelmann (Eds.), *Intelligente Agenten und das Recht* (pp. 163–205). Nomos.

Zech, H. (2019). Künstliche Intelligenz und Haftungsfragen. *Zeitschrift für die gesamte Privatwissenschaft (ZfPW), 5*(2), 198–219.

Machine Learning Decision-Making: When Algorithms Can Make Decisions According to the GDPR

Giulia Del Gamba

Contents

Abstract Not all legal bases can be useful for automated decision-making, eventually not even the ones provided by Article 22 GDPR. The method used to make automated decisions and to profile data subjects using Machine Learning technologies is still somehow obscure to the data protection community—and beyond. Strengths and weaknesses of automated decisions should dispel some myths and encourage a better understanding of this complex phenomena and its functioning.

G. Del Gamba (✉)
University of Turin, Turin, Italy

G. Borges, C. Sorge (eds.), *Law and Technology in a Global Digital Society*,
https://doi.org/10.1007/978-3-030-90513-2_4

1 Introduction

This paper aims at explaining the circumstances in which decisions based on automated processing are legitimate according to the EU General Data Protection Regulation (GDPR)[1] and to analyse the more feasible (or infeasible) legal bases for Automated Decision-Making (ADM).

The first part of the paper intends to clarify the method used for automated decisions and to explain how the profiling activity can improve the decision process. The analysis will be developed with a focus on the use of Machine Learning (ML) technologies to carry out ADM. The intention is to point out benefits and challenges of the use of this technology and understand whether—and where—there is the need to further investigate the legal regime of automated decisions.

In the second part, the legal bases that allow ADM will be presented. Through this comprehensive analysis, the intention is to highlight strengths (such as accuracy) and weaknesses (such as opacity) of automated decisions. Without deepening the issue already, this paper supports some authors' observations concerning the possibility of automated decisions to be equally—or more—accurate than decisions made by humans (HDM).

Considering that ML automated decisions are bringing efficiencies to a wide range of areas (such as medicine, cyber security, finance, etc.) there is a need to deeply understand the phenomena and its functioning. The author does not assume that all innovations are good, however intends to stress that this complex revolution needs to be properly addressed and regulated. Instead of banning the scientific research and development, the present analysis proposes to go beyond what is merely considered a technological threat and evaluate whether, despite the complexity, it is possible to minimise risks and ensure an adequate protection of individuals' rights. The outcome of this paper is to encourage an open discussion on whether it is acceptable that data protection provisions—like explainability—would impact model assessment, precluding the use of some approaches—even if more accurate.

2 ADM in General

ADM is an extremely broad concept and does not include just one type of decisions. Examples can range from issuing a line of credit by banks, to administrative or judicial decisions. More complex decisions are taken with the support of algorithms[2]

---

[1] Regulation (EU) 2016/679 of the European Parliament and of the Council of 27 April 2016, on the protection of natural persons with regard to the processing of personal data and on the free movement of such data and repealing Directive 95/46/EC (General Data Protection Regulation), OJ 2016 L 119/1.

[2] "*Algorithm: a process or set of rules to be followed in calculations or other problem-solving operations, especially by a computer.*" Future of Privacy Forum (2018).

but, as will be explained later on, algorithmic intervention is not always indispensable. In theory, during the ADM process, machines take decisions without involving human participation.

There are two main ways to carry out fully automated decisions. The first one does not imply high "intelligence", however, in the second scenario decisions are taken after a preliminary data processing analysis, called profiling. For convenience, it will be referred to the two categories as "*ADM with or without profiling*".

2.1 ADM Without Profiling

The typical example of this type of decision process concerns the automated imposition of speeding fines achieved only by evidence from speed cameras.[3] Since, for instance, the driving habits of the individual are not monitored and not taken into account to assess the amount of the fine, this decision is taken without profiling—and without even processing personal data.

In general, it is believed that when ADM occurs, and involves personal data—unlike the example, profiles will be created from such a process. For this reason, it has been said that it is difficult to come up with examples where processing of personal data, which does not lead to profiling, will otherwise lead to an automated decision.[4] As a consequence it infers that such ADM category will never fall into the scope of the GDPR. However, this might not always be the case. More subtle examples might challenge this idea and show that fully automated decisions, processing personal data, can actually be made without profiling. Let's assume that, for Anti-Money Laundering (AML) purposes, each transaction exceeding 5000 € made by a "*politically exposed person*"[5] will be rejected. Or, loan requests will be declined when the customer appears already exposed in the credit bureau system. In both cases, personal data are processed and the automated decisions are made without relying on any profiling activity, but the GDPR provisions will still be applicable.

In any case, the automated process is simple: a firm rule (*deterministic*) is set, and will guide the decision automatically.

---

[3] Article 29 Working Party (2017).

[4] Brkan (2019).

[5] A politically exposed person is someone who has been entrusted with a prominent public function. For most financial institutions, they constitute a potential compliance risk.

2.2 ADM with Profiling

Nevertheless, more complex and sophisticated ADM are usually achievable taking advantage of the profiling activity's results. Since profiling encompasses personal data processing, such ADM category will always fall into the scope of the GDPR. The more precise and accurate the profiling process is, the better the decision will be. It follows that, the added value of this scenario consists of the profiling activity itself which enables such powerful decisions. But, what is meant by profiling?

The Council of Europe, in its Recommendation CM/Rec(2010)132[6] usefully clarifies that profiling may involve three distinct stages.

1. Data collection. This phase is already part of the automated processing[7] and is extremely influential to reach a successful profiling—and decision. More broadly, all the activities concerning the data—collection, cleaning, feature selection, etc.—are essential for a valid ML performance.[8]
2. Automated analysis to identify correlations. This phase is also called *learning* and usually works as follows:
   (a) learning starts with selected information containing patterns or similarities;
   (b) the patterns found in the information are identified. ML algorithms are applied in this identification;
   (c) a model[9] is generated that can recognise the patterns that emerge when fresh data is processed by the model.
3. Applying the model to an individual to identify characteristics of present or future behaviour.[10] This phase includes the following sequential steps:
   (a) the model receives fresh data—never seen before, but similar to that used for learning;
   (b) the model decides which patterns the new data most resembles;
   (c) the model produces an estimated result.[11]

---

[6]Recommendation CM/Rec(2010)13 of the Committee of Ministers to member states on the protection of individuals with regard to automatic processing of personal data in the context of profiling.

[7]"*A comprehensive approach should determine specific legal requirements not only for the usage and further processing of personal data but already for the collection of data for the purpose of profiling and the creation of profiles as such.*" Article 29 Working Party (2013). Advice paper on essential elements of a definition and a provision on profiling within the EU General Data Protection Regulation. Article 29 Working Party (2013).

[8]To understand the role of data in Machine Learning processes see: Singh et al. (2016).

[9]"*Once a machine learning programme has been trained, the method it uses to recognize patterns in new data is called a model. The model is what is used to perform tasks moving forward, with new data.*" Future of Privacy Forum (2018).

[10]Even past behaviours could be foreseen as an "*anomaly detection*" (something I should have done but I did not) or as a "propensity" (something I am going to do in the near future).

[11]Datatilsynet, The Norwegian Data Protection Authority (2018).

For clarification purposes, a fictitious but representative example in the field of insurance relationships will be presented. An insurance company decides to analyse customers' e-payments transaction data to identify certain behaviours that can be linked to road accidents. The analysis identifies that customers that consume alcohol on nights out are more likely to be involved in a car accident (phase 2). The model identified a behavioural pattern that can be used to exploit future customers: if they buy drinks at night in clubs, there are substantial probabilities that they will be more likely exposed to car accidents (phase 3).

More broadly, profiling is a popular data science technique for splitting a population into groups based on similarities between characteristics. An important point to make is that the profiles created are not useful by themselves but can be valuable only when applied. Therefore, some authors argued that the profiling activity itself includes the decision-making process.[12] Whether within the profiling process or as a following step, determinations and conclusions about data subjects based on such profiles will be made—ergo a decision will be taken.

In the case of our example, these profiles may be used by the insurance company to determine insurance premiums or a targeted advertising campaign.

According to the model's outcome it is possible to take a decision that, most of the time, will be fully automated.

2.3 *Why ML?*

Why is Machine Learning technology necessary to make such decisions? This section provides a broad overview of ML technology for the purpose of this analysis and will not go into great technical detail.

First of all, as background, it is taken into consideration the type of ML that is currently—and in a more immediate future—used, called narrow artificial intelligence.[13] Since ML has been defined as "*the science of algorithms that detects*

[12] Kamarinou et al. (2016).

[13] "*Narrow artificial intelligence: An application of artificial intelligence which replicates (or surpasses) human intelligence for a dedicated purpose or application*" Future of Privacy Forum (2018).

patterns in data in order to make accurate predictions for future data"[14] it is certain that ML algorithms appear to be suitable to achieve accurate and satisfactory profiling. More broadly ML is the "*field of study that gives computers the ability to learn without being explicitly programmed*"[15] since it automatically improves with experience. Instead of setting rules that determine the ADM output, ML enables algorithms to set the rules autonomously, and to improve its output—and accuracy—with continued experience. The strength of ML is that it can adapt and improve the quality of its output without direct human intervention and for this reason it is extremely performing.[16]

To sum up, as brilliantly described by Cathy O'Neil "*With machine learning (...) the computer dives into the data, following only basic instructions. The algorithm finds patterns on its own, and then, through time, connects them with outcomes. In a sense, it learns*".[17]

3 ADM in the GDPR

Whether with or without profiling, when personal data are processed to take fully automated decisions, Article 22 of the GDPR applies. This provision was ported to the GDPR from the Article 12(a) and Article 15 of the Data Protection Directive (DPD).[18] In 1995, the intention was to respond to fears in the early days of digitisation, in practice the provision was little known and largely unused.[19] Now, Article 22 has become the subject of much academic attention. In this section it will be analysed, focusing on the circumstances and conditions which allow automated decisions to be made.

Article 22(1) stipulates that: "*The data subject shall have the right not to be subject to a decision based solely on automated processing, including profiling, which produces legal effects concerning him or her or similarly significantly affects him or her*".

The Regulation aims to protect individuals from situations where machines make decisions without any human involvement. In order to lead to the non-application of Article 22, humans should critically intervene, assess the machine's output and substantially influence the decision. Although humans are almost always involved

[14]This definition seems appropriate for the purpose of this paper, however it is narrower than more commonly used ML definitions (clustering, for instance, is not covered by this definition). Session with Ralf Herbrich, Quora. https://www.quora.com/session/Ralf-Herbrich/1.

[15]Samuel (1988).

[16]Kelnar (2016).

[17]O'Neil (2016).

[18]Directive 95/46/EC of the European Parliament and of the Council of 24 October 1995 on the protection of individuals with regard to the processing of personal data and on the free movement of such data, OJ 1995 L 281/31.

[19]Veale and Edwards (2017).

in the design, training and testing of a system incorporating ML, apparently, to avoid the application of this provision, a human decision maker should be present at the stage where the decision is taken.[20]

Moreover, the provision is assessing only the decisions that have certain consequences on individuals, which are not specifically defined in the Regulation. For a decision to produce legal effects it should be binding or create legal obligations or impact the legal interests of the individual. It is definitely more ambiguous to determine when decisions *significantly affect* people. However, this disputed evaluation is out of the scope of this paper.

More broadly, the Article 29 Data Protection Working Party (A29WP) strongly supports Article 22 being read as a general prohibition on solely ADM rather than a right to opt-out.[21] Whether a right or a prohibition, without going through this academic debate, there are three cases which exempt data controllers from this provision: "*when the decision is: 1) necessary for the performance of a contract between the data subject and the controller, 2) permitted by an EU or national law, or 3) based on explicit consent*".[22]

3.1 Contract

Regarding the first exemption which allows fully automated decisions, the example made by the A29WP in its guidelines[23] concerns e-recruiting processes: "*A business advertises an open position. As working for the business in question is popular, the business receives tens of thousands of applications. Due to the exceptionally high volume of applications, the business may find that it is not practically possible to identify fitting candidates without first using fully automated means to sift out irrelevant applications. In this case, ADM may be necessary in order to make a short list of possible candidates with the intention of entering into a contract with a data subject*".

With respect to the example, it is easily arguable that this type of analysis—*a short list of possible candidates*—does not necessarily need to be made by automated means. ADM, in this case, is likely to go far beyond what is required for the performance of a contract, or for pre-contractual relations, more precisely.

In order to clarify the interpretation of the contractual necessity, the A29WP, in its Opinion on Legitimate Interest,[24] states that: "*the provision has to be interpreted strictly and does not cover situations where the processing is not genuinely necessary for the performance of a contract, but rather unilaterally imposed on the data*

[20] Kamarinou et al. (2016).

[21] Article 29 Working Party (2017).

[22] Article 22(2) of the GDPR (n. 1).

[23] Article 29 Working Party (2017).

[24] Article 29 Working Party (2014).

subject by the controller." On the other hand, Maja Brkan[25] reasonably claims that if the meaning of the contractual necessity will be subject to a *very strict textual interpretation,* it will hardly open the door for ADM. Going back to the example, does the recruitment process necessarily need to be sped up with automated means? Or does the ADM just enable and facilitate a faster and more efficient way to enter into a contract? Considering that the margin between *necessity* and *enabling* is particularly subtle, a narrow interpretation of this provision would leave little room for controllers to avail to this exception and thus using ADM.

Moreover, the A29WP explicitly excludes from the scope of the legal ground the cases in which the controller builds "*a profile of the user's tastes and lifestyle choices based on his click-stream on a website and the items purchased: This is because the data controller has not been contracted to carry out profiling, but rather to deliver particular goods and services, for example. Even if these processing activities are specifically mentioned in the small print of the contract, this fact alone does not make them 'necessary' for the performance of the contract.*"[26] This statement suggests that the legal ground cannot be applied to the profiling activity except when the controller has been contracted to carry out profiling, and the contract is unequivocally stipulated for the creation of profiles as such. In other words, profiling can be performed only when it corresponds to the *exact rational of the contract.* If this is true, the contractual necessity is strictly connected to the purpose limitation principle, thus, what is not explicitly covered by the processing purpose cannot find room in this legal ground. It seems that, the profiling activity is interpreted as a processing purpose, even if, as already explained, profiling is just a data processing technique—it requires a purpose and cannot constitute the processing purpose by itself.

Further clarifications on this topic come from the European Data Protection Supervisor,[27] who interpreted that to meet the necessity requirement, the controller must be able to demonstrate that the ADM process is the *least privacy-intrusive* method to perform the contract. But, how can be determined what *privacy-intrusive* actually means? The intrusiveness might be related to several criteria such as the

---

[25] Brkan (2019).

[26] Article 29 Working Party (2014).

[27] "*The measure envisaged should be the least intrusive for the rights at stake: o Alternative measures which are less of a threat to the right of personal data protection and the right for respect of private life should be identified. o An alternative measure can consist of a combination of measures. o Alternatives should be real, sufficiently and comparably effective in terms of the problem to be addressed. o Imposing a limitation to only part of the population/geographical area is less intrusive than an imposition on the entire population/geographical area; a short-term limitation is less intrusive than a long-term; the processing of one category of data is in general less intrusive than the processing of more categories of data. o Savings in resources should not impact on the alternative measures – this aspect should be assessed within the proportionality analysis, as it requires the balancing with other competing objectives of public interest*". European Data Protection Supervisor (2017). Assessing the necessity of measures that limit the fundamental right to the protection of personal data: A toolkit. https://edps.europa.eu/sites/edp/files/publication/17-06-01_necessity_toolkit_final_en_0.pdf. 11 April 2017.

amount of data processed, the data accuracy, the extent of data retention, the access to data within the controller's organization, etc. Let's look at the amount of data—data minimisation—as the main criterion to determine intrusiveness: the more data you process the more invasive the activity will be. If this is true, any computer processing, in general, is more privacy intrusive than manual processing.[28]

Demonstrating that ADM is the least privacy intrusive method may be even more problematic using ML technologies. It is clear that the effectiveness of many ML algorithms depends on the availability of large amounts of data. As quantity does not necessarily affect quality in a negative way, providing the algorithm with more data about data subjects could actually lead to a clearer and more representative picture of them. Therefore, it seems critical to demonstrate that a decision made by a machine is less privacy intrusive than those made by humans. Nevertheless, in some circumstances it might be possible. In this context, particular attention should be given to a crucial phase in data science, called *feature selection*. This technique helps reduce the number of variables that require consideration, choosing only the most relevant and effective ones. In this way, the machine is able to take a decision using a limited number of variables, compared to those that a human would have to process to take the same decision. In other words, the feature selection technique enables a substantial data processing minimization, on condition that the training phase is carried out with anonymous data[29] and thus outside the scope of the Regulation.[30]

Furthermore, most of the data processed by humans to take a decision might reveal itself totally useless. The accuracy of an ADM might be higher than the one made by the human and therefore less likely to lead to adverse consequences to the data subject. Moreover, algorithms can be interrogated—and they will tell the truth, they can be changed, improved, and made better.

For all these reasons—in some circumstances—ADM could not only be more precise, but also less invasive with respect to decisions made by humans.

3.2 Member State or EU Law

The second exception, which provides for the possibility to allow ADM either by Member State (MS) or EU law, has been defined as the most important exception

[28] So this suggests that CVs should be received in envelopes and processed like in the good old days.

[29] Datatilsynet, The Norwegian Data Protection Authority (2018).

[30] "*if profiling does not involve the processing of data relating to identifiable individuals, the protection against decisions based on automated profiling may not apply, even if such decisions may impact upon a person's behaviour or autonomy*". Schreurs et al. (2008).

"*The protection against solely ADM might not apply if the data processed are anonymized*". Savin (2014).

between the three.[31] In Germany, the Federal Data Protection Act,[32] for instance, explicitly authorizes ADM in insurance relationships.[33] This legal ground is able to leave room to manoeuvre for controllers to carry out ADM. Whilst on one hand, taking ADM relying on the contractual necessity is extremely laborious and not very clear, on the other hand, legislators can almost freely determine whether ADM are acceptable or not.

The relevant law must also lay down sufficient suitable measures to safeguard the data subject's rights and freedoms and legitimate interests 22(3). For this legal ground, the GDPR does not specify which minimum measure should be provided, while for the other two scenarios (contract and consent) it is explicitly addressed that the data subject shall have at least (1) *the right to obtain human intervention on the part of the controller,* (2) *to express his or her point of view and* (3) *to contest the decision*. An in-depth analysis of the appropriate safeguards to protect data subject's rights and interests is outside the scope of this paper.

3.3 Consent

The *explicit consent* as a legal ground, which was not included in the DPD, is introduced but not defined in the GDPR.

However, the A29WP guidelines on consent[34] provide guidance on how this should be interpreted. In general, controllers will have to show that data subjects genuinely understand what they are consenting to and they should have enough relevant information about the consequences of the processing to ensure that any consent they provide represents an informed choice. Additionally, when ADM are carried out, the data subject should receive *meaningful information about the logic involved* in the automated processing, including profiling.[35]

It is assumed that, if the controller wants to rely on consent to carry out ADM, all this relevant information should be provided to the data subject. This can be particularly difficult due to technical obstacles, directly related to the complexity of the algorithm. However, in line with the accountability principle, even if this explanation is particularly difficult, controllers should make extra effort to inform individuals to the very best of their ability. Some authors consider almost impossible to explain the logic behind algorithms since sometimes it is infeasible even for the

[31] Brkan (2019).

[32] German Act to Adapt Data Protection Law to Regulation (EU) 2016/679 and to Implement Directive (EU) 2016/680 (DSAnpUG-EU) of 30 June 2017.

[33] "*This provision allows for automated decisions in tortious – and not contractual – relationship between the insurance company of the person who caused damage and the person who suffered damage, under the condition that the latter wins with her claim*". Brkan (2019).

[34] Article 29 Working Party (2018).

[35] Article 29 Working Party (2017) and Veale and Edwards (2017).

developers to understand the reason why an automated decision was taken. However, other authors have proposed technical solutions to achieve a greater algorithmic transparency.[36]

In general, it is difficult to comply with the obligation to inform when the automated processing involves technologies that are non-transparent in nature.[37] ML algorithms may be based on very different computational learning models. Some, like a simple decision tree,[38] are easier for humans to explain and the ADM can be even more traceable than HDM. Other algorithms are so called *black box*[39] because it is nearly impossible to identify the factors that influenced the decision—and especially their weights. Neural networks,[40] for instance, function differently from more simple algorithms, since they are prepared to learn complex patterns. The algorithm tries different combinations of variables—and its weights—until it finds the one that minimizes the prediction error. For this reason, despite the efforts to develop systems that could explain these processes, autonomous decisions taken with help of neural networks are currently inexplicable for most of the data subjects.[41] Thus, some ML algorithms and the models they produce are inexplicable to a typical data subject. As a consequence, this may make it difficult for data controllers to rely on a valid consent and to comply with their duty to inform and with the obligation of transparent processing. In those cases, these difficulties may slow down or prevent the deployments of *black box* ADM, even when their use could potentially lead to less invasive, more accurate and fairer decisions.[42]

In this scenario, is it demonstrable that the given consent represents an informed choice when *black box* technologies are used to take decisions? Depending on the technology involved, what can be explained to a typical data subject to compensate the lack of information regarding the *logic involved*? Which information can be complementary and useful for individuals to understand the decision-making process and thus withdraw a valid consent?

---

[36] Brkan (2019).

[37] "*In algorithmic context, transparency relates to the degree the internals of the algorithms can be seen and understood; for instance, exposing the features of the data the learned model takes into account, the associations and rules that were derived, and the extent to which the model relies on these.*" Singh et al. (2016).

[38] "*Decision tree: a decision support tool that uses a tree-like graph or model of decisions and their possible consequences, including chance event outcomes, resource costs, and utility*". Future of Privacy Forum (2018).

[39] Pasquale (2015).

[40] "*A variation of a machine learning model that, taking inspiration from the human brain, is composed of multiple processing layers*". Future of Privacy Forum (2018).

[41] Singh et al. (2016).

[42] Singh et al. (2016).

4 Conclusions

ML automated decisions are very much used in a wide range of areas which are increasingly improving the efficiency of their activities. This paper analyses the decision-making process and how it can be influenced by profiling individuals. It is argued that is not convenient to interpret the profiling activity as a data processing purpose by itself, since it is just an ordinary data science technique. ML algorithms are playing an important role in the decision-making process, due to their ability of improving the quality of the output by learning automatically, without human intervention or assistance.

Since these decisions might impact individuals' rights and freedoms, the GDPR tolerates them just in three circumstances: (1) when they are necessary for the conclusion of a contract, (2) when authorized by Union or Member State law, and (3) when based on the data subject's explicit consent. Among them, this article aims to explain that, if strictly interpreted, the contractual necessity would never be applicable to make automated decisions, even less when the profiling activity is carried out. The same can be said for the explicit consent, since some ADM are currently inexplicable to a typical data subject. If there is no agreement on how to inform individuals when *black box* technologies are used, the consent will never be valid and then suitable to make automated decisions.

In this scenario, should *black box* ADM development be slowed down or even precluded? Automated decisions, in particular the model behind them, are rigorously tested and analysed, to explore whether they fit for the purpose. Different algorithms are trained and then evaluated in order to select the one that best suits the particular problem area.[43] By analysing the ADM process and functioning the intention is to rise—without an accurate investigation—its potential strengths. The feature selection technique, for instance, enables some sort of data minimization and accuracy, by choosing only relevant and effective variables. For several reasons, ADM could not only be more accurate—and fairer, but also less invasive respect to decisions made by humans. Could these strengths be used to compensate technological limits and threats—such as non-transparency? Is it acceptable that data protection law requirements—like explainability—would impact model assessment, perhaps precluding the use of some approaches, even if more accurate?

The present analysis encourages to go beyond what is merely considered a technological threat. The main aim of the article is to open up a debate as to whether full transparency needs to be sacrificed for more accurate decision-making. The author advocates the need for further research in this area, in particular for a deeper analysis of what concerns the requirements of accuracy.

[43] Singh et al. (2016).

References

Article 29 Working Party. (2013). Advice paper on essential elements of a definition and a provision on profiling within the EU General Data Protection Regulation. Retrieved September 1, 2021, from https://ec.europa.eu/justice/article-29/documentation/other-document/files/2013/20130513_advice-paper-on-profiling_en.pdf

Article 29 Working Party. (2014). Opinion 06/2014 on the notion of legitimate interests of the data controller under Article 7 of Directive 95/46/EC. Retrieved September 1, 2021, from https://ec.europa.eu/justice/article-29/documentation/opinion-recommendation/files/2014/wp217_en.pdf

Article 29 Working Party. (2017). Guidelines on automated individual decision-making and profiling for the purposes of Regulation 2016/679. Retrieved September 1, 2021, from https://ec.europa.eu/newsroom/article29/item-detail.cfm?item_id=612053. Adopted on 3 October 2017 as last Revised and Adopted on 6 February 2018.

Article 29 Working Party. (2018). Guidelines on consent under Regulation 2016/679. Retrieved September 1, 2021, from https://ec.europa.eu/newsroom/article29/item-detail.cfm?item_id=623051. Adopted on 28 November 2017 as last Revised and Adopted on 10 April 2018.

Brkan, M. (2019). Do algorithms rule the world? Algorithmic decision-making in the framework of the GDPR and beyond. *International Journal of Law and Information Technology.* Retrieved September 1, 2021, from https://doi.org/10.2139/ssrn.3124901

Datatilsynet, The Norwegian Data Protection Authority. (2018). Artificial intelligence and privacy. Retrieved September 1, 2021, from https://www.datatilsynet.no/globalassets/global/english/ai-and-privacy.pdf

European Data Protection Supervisor. (2017). Assessing the necessity of measures that limit the fundamental right to the protection of personal data: A toolkit. Retrieved September 1, 2021, from https://edps.europa.eu/sites/edp/files/publication/17-06-01_necessity_toolkit_final_en_0.pdf

Future of Privacy Forum. (2018). The privacy expert's guide to artificial intelligence and machine learning. Retrieved September 1, 2021, from https://fpf.org/wp-content/uploads/2018/10/FPF_Artificial-Intelligence_Digital.pdf

Kamarinou, D., Millard, C., & Singh, J. (2016). Machine learning with personal data. In R. Leenes, R. van Brakel, S. Gutwirth, & P. De Hert (Eds.), *Data protection and privacy: The age of intelligent machines.* Hart Publishing.

Kelnar, D. (2016). The fourth industrial revolution: A primer on Artificial Intelligence (AI). *Medium.* Retrieved September 1, 2021, from https://medium.com/mmc-writes/the-fourth-industrial-revolution-a-primer-on-artificial-intelligence-ai-ff5e7fffcae1

O'Neil, C. (2016). *Weapons of math destruction: How big data increases inequality and threatens democracy.* Crown.

Pasquale, F. (2015). *The black box society: The secret algorithms that control money and information.* Harvard University Press.

Samuel, A. L. (1988). Some studies in Machine Learning using the game of checkers. In D. N. L. Levy (Ed.), *Computer Games I* (pp. 335–365). Springer.

Savin, A. (2014). Profiling and automated decision making in the present and new EU Data Protection Frameworks. Paper presented at 7th International Conference Computers, Privacy & Data Protection, Brussels, Belgium.

Schreurs, W., Hildebrandt, M., Kindt, E., & Vanfleteren, M. (2008). Cogitas, Ergo Sum. The role of data protection law and non-discrimination law in group profiling in the private sector. In M. Hildebrandt & S. Gutwirth (Eds.), *Profiling the European Citizen* (pp. 241–270). Springer.

Singh, J., Walden, I., Crowcroft, J., & Bacon, J. (2016). Responsibility and machine learning: Part of a process. Retrieved September 1, 2021, from https://doi.org/10.2139/ssrn.2860048

Veale, M., & Edwards, L. (2017). Clarity, surprises, and further questions in the Article 29 Working Party draft guidance on automated decision-making and profiling. *Computer Law & Security Review: The International Journal of Technology Law and Practice, 34*(2), 398–404. Retrieved September 1, 2021, from https://doi.org/10.2139/ssrn.3071679

Algorithmic Suspicion in the Era of Predictive Policing

Irmak Erdoğan

Contents

Abstract The rise of new technologies allow massive collection of data and transform the suspicion standards, as well as the old school small-data policing. Our daily routines, social networks, biometrics and thoughts feed private and public databases, while profiling algorithms turn all the noisy data into information. Inferences from digitised data lead to a new type of suspicion that drives not from the observation of individuals' actions, but rather from their interconnected data. This chapter analyses the technologies, which pave the way for a new type of policing and traces the consequences of algorithmic suspicion. It further elaborates the legality and proportionality of the interference with fundamental rights before a citizen does anything overtly criminal. Finally, it questions whether new data protection rules at European Union level meet the challenges of the data-driven predictive policing.

I. Erdoğan (✉)
Istanbul, Turkey
e-mail: ierdogan@gsu.edu.tr

G. Borges, C. Sorge (eds.), *Law and Technology in a Global Digital Society*,
https://doi.org/10.1007/978-3-030-90513-2_5

1 Introduction

The world of small data disappears, as technology allows new models of data collection. The new world with surveillance and analysis technologies promises "a panoramic vision" of bulk data. Old investigation techniques concentrated upon particular suspect(s) at a time, but now the governments tend to "collect-it-all" to find out who is predestined to commit a crime.[1] As biometric pictures, CCTV images, DNAs, and license plates enter into police databases for good, intelligence agencies implement sniffing technologies into internet infrastructures to find out what everyone thinks and shares. Thus, for the sake of not becoming a "potential victim", we all are tracked as potential perpetrators.

On the other hand, the selection of targets, referred as surveillance discretion, is expanded by the use of massive data. It implies that police officers do not have to physically observe the antisocial behaviors, or come up with factual information to target people. Nor do they have to question the informants and associate the clues as traditional human-led investigation suggests (Joh 2016).[2] Thus, ex poste investigation methods are replaced by ex-ante ones, meaning that initial investigation takes place even before the crime is committed.[3]

This article mainly concentrates on the circumstances under which the citizens are targeted for surveillance in the new age of big data. Furthermore, it will explore whether the extension of surveillance discretion is counterbalanced by new rules for accountability. With those objectives in mind, the following Sects. 2.1 and 2.2 will introduce new physical and cyber surveillance methods and programs based on big data. By exploring the new technological methods, these sections will reveal the variety of ways to collect data. Section 2.3 will elaborate how virtually and physically accumulated data are fused and analyzed for the purposes of predictive policing. It will also explain vulnerabilities of predictive modeling, such as data quality, regeneration of bias and correlative assumptions. Finally, Sect. 3 will scrutinize the automated surveillance discretion based on predictive modeling from the data protection perspective at the European level. It will review the lawfulness of such process starting from the data collection until the end of profile formation.

[1] Hu (2015).

[2] Joh (2016).

[3] Kerr and Earle (2013).

2 Watchful Gaze of the New Technologies: 24/7 Surveillance of the Thoughts and Actions

Surveillance transactions spread around public and private spaces to gather traces of citizens. While video surveillance cameras ("CCTVs") collect images, algorithms match them to existing biometric databases. License tracking devices show the whereabouts of registered cars, as "tower dumps," reveal cell phone data and track individuals geographically.[4] Smartphones and Google Maps show whether a person attends religious ceremonies, social meetings or political protests. Intelligence services surveil the telecommunication and electronic messages intensively, analyze social networks and infer a "biographical-behavioral" status.[5] In short, surveillance expands like a rhizomatic plant in different directions for the sake of advertisement, entertainment[6] and above all, for the maintenance of control.

In order to regulate surveillance, one needs to be aware of these methods and programs surrounding the lives of citizens. Thus, the following sections will draw a picture of "physical surveillance" and "cyber surveillance" and briefly introduce how data is collected on a daily basis.

2.1 Physical Surveillance: Disappearance of the Right to Be Left Alone

Flooded by countless images from CCTV and zooming in and out on millions of individuals, the video surveillance promises tracking criminals and maximizing public safety. Camera systems aim to prevent and detect crimes, yet subject everyone to the surveillance with no selection criteria.[7] Although initially CCTV systems were only recording systems and had to be monitored by human watchers, contemporary video surveillance systems utilize object-identification and automated object-tracking, so that a target can be followed through an entire city via networked cameras.[8] It implies subjects who leave their houses cannot remain anonymous. Instead, while walking and driving, they fall under categories such as "law-abiding/suspected", "employed/unemployed", "fast-driver/careful driver". Such classification does not require face-to face police discretion; instead, it is prescribed in a database.[9]

[4]Miller (2014).

[5]Hu (2015).

[6]Timan et al. (2017).

[7]Norris (2003).

[8]Rajpoot and Jensen (2015).

[9]Rajpoot and Jensen (2015).

Surveilling public spaces via cameras may show prima facie more insight than what patrol officers might see on the street, but distant watching has the risk of revealing actually less than what face-to face knowledge might provide.[10] For example, touring and retouring around a shop might be a sign of preparatory acts for robbery, whereas any official at an actual spot might notice the person of interest is waiting for his/her friend who is inside the shop. CCTV operators do not record voices nor can they figure out causality by asking questions on the street. It implies that suspicious behavior will be decided according to a "distanced" common sense.[11] However, common sense might imply assumptions, which rely on biases. As the field research realized by Norris and Armstrong showed, only 24% of people were surveilled due to their explicit crime-related behaviors, whereas teenagers, African Americans and men were systematically over-surveilled, not because of their deviant behavior, but for being under suspicious category.[12] Thus, CCTV's watchful gaze is not concentrated on everyone equally and indiscriminately.

By the help of machine learning, the algorithms, which recognize human emotions, evolve to make surveillance tools even smarter. Such algorithms can distinguish and categorize facial expression and interpret body movements. The recent research carried out by MIT's Media Lab displayed that a facial recognition system might even distinguish between real and faked smiles. It means not only the IDs of persons, but also the emotions such as happiness or "polite disagreement, favor, sarcasm, and empathy"[13] will be categorized soon, and they will give a hint of the suspicious behavior.

Thus, technology itself, will gradually define deviant activity. In other words, algorithms will decide what behavior, image and mimic is suspicious.[14] Even if currently we do not have "robocops" of the science fiction movies, intelligence-led law enforcement systems have become a reality in threat analyses and social control.[15]

2.2 *Cyber Surveillance: Monitoring Digital Personality*

Digital enterprise-oriented surveillance technologies are spreading exponentially and complementing the physical State-Panopticon. In return, surveillance becomes invisible and supposedly "voluntary".[16] In this respect, the Deep Packet Inspection technology ("DPI"), allows for a far-reaching monitoring and helps analyzing each

[10] Norris (2003).

[11] Norris (2003).

[12] Clive and Armstrong (1999).

[13] Hoque et al. (2012).

[14] Lyon (2003).

[15] Norris (2003).

[16] Timan et al. (2017).

data packet sent through Internet. Internet service providers, either voluntarily or upon request of authorities, deploy DPI within their own infrastructures so that the traffic being routed via their servers can be scanned.[17] DPI can manipulate data packets in conformity with a prior instruction, by blocking the movement of detected content, or by regulating the rate of packet flow speed.[18]

As per today, social network surveillance merges the techniques of the DPI network monitoring with data mining. In this way, via filtering algorithms, "uncommon behavioral patterns" can be detected, as well as illegal content. Yet, the effectiveness of such detection remains questionable, since unusual user behavior is not that easy to predefine.[19]

Additionally, remotely controlled Trojan horses are used to integrate backdoor functionalities so that full remote control can be sustained over the infected device.[20] The spywares and viruses combined with the DPI technology surpass the limits for wiretapping for the "crime prevention" or "inspection" purposes. For example, FinFisher spyware can record Skype calls, monitor webcams, listen to microphones, and perform keylogging operations, which allow detecting every move.[21] Malwares such as Crisis, Morcut and Da Vinci Trojan can infect iOS, Android, Blackberry, and Linux systems. They can trace the country of target through GPS and find out cell ID, record incoming/outcoming SMS, and other information such as address book data.[22]

Unless such surveillance methods have a deinstallation function, which removes the software within a time limit, they cannot be deactivated and deleted from the infected personal device. Trojans and DPI have also the risk of exceeding the scope of the imposed measure by capturing everything the user does, and breaching privacy and personal data rights.[23]

Thus, cybersurveillance leads to massive collection of data, which exceed the limits of physical surveillance. On the other hand, such data needs to be further analyzed. Therefore, language analytics as well as sentiment inquiring algorithms are imposed on services like Twitter and Facebook, to detect suspicious activities or "aggressive online comments". Furthermore, social network algorithms are developed to identify personal connections and to map particular groups in the light of mathematical modeling. At the end, police can focus on individuals who may have escaped their attention through traditional surveillance measures.[24]

Based on a pre-emptive model, cyber surveillance invests on data mining and creates a basis for law enforcement action such as preventively arresting someone

---

[17] Schlehahn et al. (2013).

[18] Schlehahn et al. (2013).

[19] Schlehahn et al. (2013).

[20] Marquis-Boire (2013).

[21] Marquis-Boire and Marczak (2013).

[22] Schlehahn et al. (2013).

[23] Schlehahn et al. (2013).

[24] Joh (2016).

based on his/her online behavior.[25] Although such measures aim to prevent crime and maintain public order, their highly intrusive nature risks constituting a disproportionate intervention in private life.

2.3 *Profiling and Scoring Ordinary Suspects via Big Data*

Previously, the article explored the data collection via physical and cyber surveillance measures. These methods led to emergence of private data with public records, thereby integrating public and private data within both government and private databases.[26] As soon as data has been centrally accumulated, it became possible to search through acquired data via algorithms and other electronic modelings with one click.[27]

From crime prevention perspective, the expansion of surveillance and the collection of every kind of data give a "bird's-eye view" for each citizen. From a panoramic vision, algorithms analyse the models and affinities between the datasets. Such an analysis facilitates predicting "potential threats" to society.[28] Police officers do no longer need to observe directly a neighborhood; instead, soon they can rely on alerts created by softwares, which identify patterns of "suspicious" activities derived from available data.[29] In this way, the surveillance discretion has become automated.

The profiling algorithms promise preventing crimes before happening, or at least they aim to render police response just in time. This promise initiated various models of crime prediction softwares such as PredPol, which uses historical crime data and other relevant data to calculate where crimes are to happen.[30] HunchLab is another software, which makes risk terrain analysis, by considering certain features as the following: risky areas like bars, banks or stores, socioeconomic factors, time of the day, weather, and other relevant factors.[31] Thereby, the "environmental backcloth" of criminal opportunities and interwoven factors that influence crime help estimating future crime spots.[32]

On the other hand, technology firms aim to develop new predictive technologies that target "potential criminals" rather than probable crime scenes. New wave of predictive policing made "Minority Report" scenario come true, via profiling individuals based on their past criminal records, economical status, current associations,

---

[25] Timan et al. (2017).

[26] Ferguson (2017b).

[27] Hu (2015).

[28] Hu (2015).

[29] Joh (2016).

[30] O'Neil (2016).

[31] Ferguson (2017a).

[32] Wheeler and Steenbeek (2020).

and combine all relevant factors to foretell their unlawful destiny.[33] Some police departments have already started to use the new model softwares that search through massive amount of records in commercial and public databases. They mine criminal records, recent purchases, online comments and other data to score and to profile internet users.[34] Then again, companies such as Palantir, designed for data collection and analysis, work with governments to spot violent crime and to compose "target lists" or "heat lists" for further surveillance and investigation purposes.[35] All of these data-based softwares encourage more patrolling in certain areas, increase stop, frisk practices, and pave the way for search warrants and raids by police officers.

Especially, low-level crimes and increasing frisks and stops populate the software models with more "flags", and send the law enforcement back to the same neighborhood, creating a vicious circle. The preventive policing boosts new data, and in response justifies more policing. The algorithms behind the system might be "neutral", yet the data is not color blind. Geography is not separable from race and economic status, as the cities are more and more segregated based on poverty and origin.[36] For example, early "involvement" with the police may signal association with crime commitment or recidivism, and people in pour or minority neighborhoods encounter more frequent stops and arrests. As a result, it may cause them to be in a permanent target list. In such cases, automatic systems cannot make adjustment for fairness, nor can hinder the biased data from entering the system.[37] As a result, the toxicated effects of uneven policing enter the system and "impartial" algorithms and "hypothetically scientific analysis" amplify current biases in law enforcement.

3 Legal Assessment of Data-Based Discretion: Surveilling, Profiling and Data-Driven Policing

Digitised information collected by surveillance tools provides a basis for data analytics. Even though this analysis is expected to be objective, it involves discretionary human decisions. It is up to humans to decide on which mathematical model or data to use, and which information to leave out.[38] For example, stop and frisk numbers can be chosen as an input in a model, yet they also involve police discretion. At the end, the model may initiate additional stops or arrests in the same areas, against the same people. Throughout this whole process, from the point of acquisition of data, until profiling and policing stages, there are potential risks to personal data rights, and to the prohibition of discrimination. Undoubtfully,

---

[33] Ferguson (2017a).

[34] Skorup (2014).

[35] Ferguson (2017a).

[36] O'Neil (2016).

[37] O'Neil (2016).

[38] Joh (2016).

data-based policing can lead to due process violations, as well as breaches of further rights such as freedom of speech and association. However, the following part will explore only data-based discretion of law enforcement pertaining to the right to personal data protection.

3.1 Accumulating and Processing Data for the Purposes of Crime Prevention and Detection

Technology-led policing depends on data fusion from different sources. One by one, each data source could have been collected for a particular and a restricted purpose. Their combination, nevertheless, can reveal the identification of an individual.[39] At the European level, once a person is identifiable, the General Data Protection Regulation (GDPR)[40] applies generally to protect the data privacy and ensure the free movement of such data. Furthermore, the Data Protection Directive for Police and Criminal Justice Authorities (Directive 2016/680)[41] specifically standardizes the rules for data protection for the purposes of the prevention, investigation, detection or prosecution of criminal offences.[42]

However, it is not always easy to determine when and how data regulations will apply. Mass data has the potential to avoid data protection rules, primarily because information is not identifiable at each point of collection. Similarly, in predictive policing systems, transfers between databases occur, before the subject is singled out. In that sense, Directive 2016/680 may not be applicable, while collecting and processing unidentifiable data for crime prevention purposes.[43] Nevertheless, identifiability should be interpreted in a broad way. Once the socio-economic, psychological, political stance of a person can be categorised via cookies, web traffic surveillance and by any other methods; it means that the person is singled out. The explicit information regarding the name, the social security number or the address of a person is not required for the person to enjoy data protection rights.[44]

Especially when a person makes it to the "heat list", his identity is revealed to law enforcement or to intelligence services. However, in most of the cases, one may not be aware of being on the list, until a measure is imposed on him/her. In an exemplary

---

[39] President's Council of Advisors on Science and Technology (2014).

[40] Regulation (EU) 2016/679 on the protection of natural persons with regard to the processing of personal data and on the free movement of such data, and repealing Directive 95/46/EC (General Data Protection Regulation), OJ 2016 L 119.

[41] Directive (EU) 2016/680 of the European Parliament and of the Council of 27 April 2016 on the protection of natural persons with regard to the processing of personal data by competent authorities for the purposes of the prevention, investigation, detection or prosecution of criminal offences or the execution of criminal penalties, and on the free movement of such data, OJ L 119, 4 May 2016.

[42] The European Union Agency for Fundamental Rights (2018).

[43] Crawford and Jason (2014).

[44] The European Union Agency for Fundamental Rights (2018).

case, The ECHR has ruled violation of Article 8 of the European Convention, when the applicant could not contest the accuracy of his personal information in secret registers.[45] Nevertheless, data processing performed for policing and criminal justice requires certain flexibility. For example, data quality is not that strict, as in the case of hearsay evidence and information from undercover sources. Similarly, the right to information and access may be restricted to a certain point to render surveillance or investigation possible. However, it is crucial to impose such limits by foreseeable laws and strike a balance between the data protection and security purposes.[46]

As per "lawful data acquisition", the ECHR emphasizes that surveillance regimes have the potential to be abused, and it is vital to have sufficiently clear regulations, which indicate their preconditions and limitations. As secret surveillance measures are not open to public scrutiny, more safeguards are needed against arbitrary interference. In its case law, the ECHR has established the minimum safeguards that should be present in statute law as follows: a catalogue for the offences which may give rise to an interception; a definition for people who fall within target category; specific duration for the measure; an independent examination procedure, rules regarding use and storage of the data; the safeguards for the transfer of the data to other parties; and the rules to erase or destroy the records.[47]

Even though the surveillance measures have a legal basis, their implementation might infringe the principle of proportionality in a democratic society. For example, the collection and retention of cellular samples and DNA profiles may constitute an invasive intervention with right to private life. The ECHR emphasized that storing fingerprints and DNA profiles of all individuals who were suspected but not convicted constitutes "the blanket and indiscriminate nature of the power of retention".[48] Even if the fingerprints and samples were lawfully taken according to domestic law, their retention and use for "matching" cannot be considered necessary in a democratic society. Thus, personal databases shall be reviewed periodically to avoid indiscriminate and open-ended retention regime.

The CJEU has also ruled out that the retention of traffic and location data of all subscribers without "differentiation, limitation or exception according to the objective pursued" is not proportional.[49] In other words, having everyone's data retained, without objectively differentiating persons who are accused of serious criminal

---

[45] ECHR, Cemalettin Canli v. Turkey, No. 22427/04, 18 November 2008, paras. 40–43; ECtHR, Rotaru v. Romania [GC], No. 28341/95, 4 May 2000, para 58–63, available at hudoc.echr.coe.int.

[46] De Hert and Papakonstantinou (2016).

[47] ECHR, Weber and Saravia v. Germany, No. 54934/00, 29 June 2006, paras. 92–95; ECHR, Big Brother Watch and Others v. The United Kingdom, Nos: 58170/13, 62322/14, 24960/15, 13 September 2018, para 306–307, available at hudoc.echr.coe.int.

[48] ECHR, S. and Marper v. The United Kingdom, Nos. 30562/04 and 30566/04, 4 December 2008 para 125, available at hudoc.echr.coe.int.

[49] CJEU, Joined cases C-203/15 and C-698/15, Tele2 Sverige AB v. Post- och telestyrelsen and Secretary of State for the Home Department v. Tom Watson and Others [GC], 21 December 2016, paras 105–106, available at https://eur-lex.europa.eu/legal-content/EN/TXT/?uri=CELEX%3A62015CJ0203.

offences, surpasses the boundaries of what is necessary for fighting against serious crimes.

Directive 2016/180 summarized all the above-mentioned points within Article 4 by pointing out that personal data must based on a legal and a fair ground. Furthermore, data procession should not exceed the purposes for which data is processed. However, it is not crystal clear whether certain data fulfilled their purpose and are no longer necessary in case of crime prevention. Unlike a criminal investigation, data collected for preventive purposes do not have a point of closure. Therefore, in case of preemptive police surveillance, the principle of necessity requires a review within certain periods to decide if concerned data should be kept or deleted. Periodic review must be implemented, especially where personal data are not automatically deleted according to national law.[50]

It is also crucial that legitimate processing rests within its originally specified purpose, because any different purpose of processing will necessitate a separate legal basis.[51] Thus, for example, massive data processing for crime prevention cannot be based on the same legal ground as data processing for a criminal procedure. It also means that interconnecting different public and private databases and mining through them constitutes a violation of purpose limitation. Concurrently, Directive 2016/180 gives no carte blanche for comparing or combining data, which are processed for different purposes.[52]

In short, police discretion should not be based on illegally or disproportionately acquired data. Even if the data is collected for the crime prevention purposes, data subjects must have the opportunity to challenge the acquisition and procession of their personal data, and ask for its rectification, review or erasure.

3.2 Profiling and Predictive Policing for the Purpose of Crime Prevention

Profiling algorithms expose patterns and correlations between data in a way that have not been revealed before. Nevertheless, profiling also carries certain risks like probability and non-accuracy. That's why both the GDPR and Directive 2016/180 have granted data subjects the right to contest profiling and the right not to be subject to solely automated decisions. Unlike the GDPR, the Directive 2016/180 prohibits only automated decisions, which produce "adverse" legal effect on the subject. A

[50] Article 29 Data Protection Working Party, Opinion on some key issues of the Law Enforcement Directive (EU 2016/680), p. 4.

[51] The European Union Agency for Fundamental Rights (2018).

[52] Gutwirth and De Hert (2008).

typical adverse effect would be applying increased security measures or surveillance by the competent authorities based on automated profiling.[53]

With respect to crime prevention, automated decision must be complemented by human intervention, yet the intervenor must have the appropriate authority according to domestic law.[54] However, such intervention must be meaningful. If the human actor routinely follows the software's guidance, such process can still be categorized as solely automated machine profiling.[55] Similarly, for the sake of crime prevention, law officers cannot carry out searches based solely on algorithmic profiling.

Another risk with profiling is that human error can be made during each stage of data acquisition, even at initial data collection for profiling. Therefore, safeguards should be introduced at all stages. For example, the wrong address of the crime scene or a misspelled name may cause a match with the wrong person.[56] Thus, the transparency of data processing must be provided, unless restricted for a period of time to achieve investigatory purposes (in line with Article 15 of the Directive 2016/680).

Even more problematically, the data might have been accumulated as a result of a "biased collection". The more law enforcement dedicates recourses to patrolling in certain neighborhoods, the more arrests are imposed and represented in crime statistics.[57] At the end, the names and data regarding sub-groups based on poverty, race and religion enter databases and profiling confirms such biases formed at the initial stage. To avoid it, the construction of discriminatory profiles by virtue of special categories of data is prohibited within Article 11(3) of Directive 2016/680. On the other hand, biased algorithms may not rely on prohibited data but can still lead to an algorithmic suspicion only upon certain groups. Therefore, an additional legal tool might reinforce more protection against the measures based on the discriminatory suspicion. In such a case, the targeted person should be able to ask for exclusion of any evidence, collected by biased predictive policing and might challenge not only false positives but also the outcomes that are assumed to be "true positives".[58]

Another weakness is that predictive policing based on profiling might inherit the misconception that correlations can determine the risk. Algorithmic correlations indicate only a relation between data, without giving causal explanations. They provide only a prediction (even a hypothesis), based on a past behavior. In that regard, correlations signify a probability that the past will repeat itself in the future,

---

[53] Article 29 Data Protection Working Party, Opinion on some key issues of the Law Enforcement Directive (EU 2016/680), p. 12.

[54] *Id.* at 13.

[55] Hildebrandt (2008).

[56] Ferguson (2017a).

[57] Ferguson (2017a).

[58] Gless (2018).

yet such an assumption is not based on common sense or observation, rather it evolves automatically through data mining.[59] However, the essence of crime prevention or intervention necessitates motivational relationships and facts, which prove causality. In that sense, using correlations to restrict the rights of the subject implies a change in the criminal paradigm, which carries the risk of singling out innocents and investigating them on probabilities rather than facts.

4 Conclusion

Data-based policing raises many questions, and frequently returns to a dystopic scenario, in which people are stopped, searched and arrested before committing a crime. Big data may be visionary, yet it also distorts police discretion. With a quick computer search in databases, any existing information or correlation can be used to justify the police suspicion. However, inaccuracies and unjustifiable correlations in interconnected data may cause the citizens to pay the price.

The most chilling (or, to intelligence agencies, the most welcoming) scenario is to gather all possible data on everyone and everything, without distinguishing the suspects from non-suspects and without really paying attention to data transparency and rectification. A continuum of communications monitoring, visual surveillance and behavioral analysis involves also new actors like ICT service providers, private companies and even the citizens themselves, who share their actions and thoughts without knowing all the consequences. The networks of such different platforms crossbreed data between public and private actors and make it even harder to control the process due to different regimes imposed on private companies and law enforcement authorities. There is a risk of circumventing human rights safeguards, if police can simply access data from private data companies. But still, privacy and data protection laws do not necessarily stop law enforcement from purchasing the stored information.

The data-driven surveillance in pre-crime scene is not actually covered by existing regulations. Yet, it is urgent to address whether data collected by multiple actors for different purposes can be used for the purpose of crime prevention. Furthermore, data mining and profiling technologies should be under the spotlight, as they impact the new automated police discretion. For now, surveillance and big data technologies may not willingly adopt to a moral future. Neither can private companies who sell these technologies to governments. That's why it is essential to set the framework for the use of new surveillance tools and also to surveil more closely the data-driven policing.

[59] Hildebrandt (2008).

References

Clive, N., & Armstrong, G. (1999). CCTV and the social structuring of surveillance. In K. Painter & N. Tilley (Eds.), *Surveillance of public space: CCTV, street lighting and crime prevention* (pp. 157–178). Criminal Justice.

Crawford, K., & Jason, S. (2014). Big data and due process: Toward a framework to redress predictive privacy harms. *Boston College Law Review, 55*(1), 93–128.

De Hert, P., & Papakonstantinou, V. (2016). The new police and criminal justice data protection directive: A first analysis. *New Journal of European Criminal Law, 7*, 7–19. https://doi.org/10.1177/203228441600700102

Ferguson, A. G. (2017a). Policing predictive policing. *Washington University Law Review, 94*(5), 1109–1189.

Ferguson, A. G. (2017b). *The rise of big data policing: Surveillance, race, and the future of law enforcement*. New York University.

Gless, S. (2018). Predictive policing: In defense of 'true positives'. In E. Bayamlıoğlu, I. Baraliuc, L. Janssens, & M. Hildebrandt (Eds.), *Being profiled: Cogitas ergo sum: 10 years of profiling the European citizen* (pp. 76–83). Amsterdam University Press.

Gutwirth, S., & De Hert, P. (2008). Regulating profiling in a democratic constitutional state. In M. Hildebrandt & S. Gutwirth (Eds.), *Profiling the European citizens: Cross-disciplinary perspectives* (pp. 271–302). Springer.

Hildebrandt, M. (2008). Defining profiling: A new type of knowledge. In M. Hildebrandt & S. Gutwirth (Eds.), *Profiling the European citizens: Cross-disciplinary perspectives* (pp. 17–47). Springer.

Hoque, M. E., McDuff, D. J., & Picard, R. W. (2012). Exploring temporal patterns in classifying frustrated and delighted smiles. *IEEE Transactions on Affective Computing, 3*, 323–334. https://doi.org/10.1109/ACII.2015.7344617

Hu, M. (2015). Small data surveillance v. big data cybersurveillance. *Pepperdine Law Review, 42*(773), 773–844.

Joh, E. E. (2016). The new surveillance discretion: Automated suspicion, big data, and policing. *Harvard Law & Policy Review, 10*(1), 15–42.

Kerr, I., & Earle, J. (2013). Prediction, preemption, presumption: How big data threatens big picture privacy. *Stanford Law Review Online, 66*(65), 65–72.

Lyon, D. (2003). Technology vs terrorism: Circuits of city surveillance since September 11th. *International Journal of Urban and Regional Research, 27*(3), 666–678.

Marquis-Boire, M. (2013). Backdoors are forever: Hacking team and the targeting of dissent. *The commercialization of digital spying report*. The Citizen Lab. Retrieved August 26, 2021, from https://citizenlab.ca/storage/finfisher/final/fortheireyesonly.pdf

Marquis-Boire, M., & Marczak B. (2013). From Bahrain with love: finFisher's spy kit exposed? *The commercialization of digital spying report*. The Citizen Lab. Retrieved August 26, 2021, from https://citizenlab.ca/storage/finfisher/final/fortheireyesonly.pdf

Miller, K. (2014). Total surveillance, big data and predictive crime technology: Privacy's perfect storm. *Journal of Technology Law and Policy, 19*(1), 105–146.

Norris, C. (2003). From personal to digital: CCTV, the panopticon, and the technological mediation of suspicion and social control. In D. Lyon (Ed.), *Surveillance as social sorting: Privacy, risk and digital discrimination* (pp. 249–281). Routledge.

O'Neil, C. (2016). *Weapons of math destruction: How big data increases inequality and threatens democracy*. Crown Publishers.

President's Council of Advisors on Science and Technology. (2014). Big data and privacy: A technological perspective report. Retrieved August 26, 2021, from https://obamawhitehouse.archives.gov/sites/default/files/microsites/ostp/PCAST/pcast_big_data_and_privacy_-_may_2014.pdf

Rajpoot, M. Q., & Jensen, C. D. (2015). Video surveillance: Privacy issues and legal compliance. In V. Kumar & J. Swensson (Eds.), *Promoting social change and democracy through information technology* (pp. 1335–1358). IGI Global.

Schlehahn, E., Hansen M., Krieger-Lamina J., & Samaniego, J. S. (2013). Report on surveillance technology and privacy enhancing design. *SurPRISE Project*. Retrieved August 26, 2021, from http://surprise-project.eu/wp-content/uploads/2013/06/SurPRISE-D3.1-Report-on-surveillance-technology-and-privacy-enhancing-design.pdf

Skorup, B. (2014, December 12). Cops scan social media to help assess your 'threat rating'. *Reuters*. Retrieved August 26, 2021, from https://www.reuters.com/article/idUS384038468220141212

The European Union Agency for Fundamental Rights. (2018). *Handbook on European data protection law*. Imprimerie Centrale.

Timan, T., Galič, M., & Koops, B. J. (2017). Surveillance theory and its implications for law. In R. Brownsword, E. Scotford, & K. Yeung (Eds.), *Oxford handbook of the law and regulation of technology* (pp. 731–754). Oxford University Press.

Wheeler, A. P., & Steenbeek, W. (2020). Mapping the risk terrain for crime using machine learning. *Journal of Quantitative Criminology*. https://doi.org/10.1007/s10940-020-09457-7

Part II
Data Protection

Ensuring the Protection of Privacy of Cryptocurrency Users: An Impossible Task?

Loren Jolly

Contents

Abstract The aim of this chapter is to assess the level of privacy available to cryptocurrency users. Even though privacy is often mentioned as a reason to use cryptocurrency, the privacy level is in reality quite low. This should first of all be explained because of the transparency of the transactions' register and second of all because of the difficult applicability of data protection measures. This is being illustrated through the question of the applicability of the General Data Protection Regulation. In this respect, two principal maladjustments problems are addressed: those regarding the addressees of the General Data Protection Regulation's provisions and those resulting from the immutability of the register. These uncertainties have tended to strengthen the development and use of anonymization techniques by the community to compensate the transparency of the register. The chapter finally

L. Jolly (✉)
University of Luxembourg, Luxembourg City, Luxembourg

G. Borges, C. Sorge (eds.), *Law and Technology in a Global Digital Society*,
https://doi.org/10.1007/978-3-030-90513-2_6

addresses these techniques, their potential as a privacy-enhancement tool but also their limits to ensure users' fundamental right to data protection.

1 Introduction

One of the arguments in favour of cryptocurrencies (although far from being the only one) is that it is supposed to be privacy-friendly.[1] It thus provides for a solution to the problem of massive data collection and processing made by companies and authorities in the traditional centralised world. No central authority is controlling the system, storing or processing any data related to the users because the system is decentralised and the ledger distributed to all the nodes of the system.

In most permissionless blockchains, transactions are pseudonymous so that only the public key of the users (which is not linked with the identity of the user since it is delivered in a random way by the system) is disclosed to the other participants to the blockchain. As an example, a transaction is made on the Bitcoin blockchain with the use of a public key that could be considered as a pseudonym. Then, the individual uses his/her private key and the combination of the two makes the transaction feasible. This corresponds to a system of digital signature: the combination of the public and private key is necessary for the user to prove his/her identity[2] and to demonstrate that he/she is the owner of the accounts transferred.

The transactions happen peer-to-peer and are validated by the nodes of the systems by performing a proof-of-work or any similar consensus mechanism. Once a node has validated the transaction it will be included in a block of transactions. Hence, the system is self-sufficient since there is no need for any central authority or man-in-the-middle in order to ensure the functioning of the system or verify the goodwill of the parties to a transaction because the system itself and its nodes are doing it. For all of these reasons, users could suppose that they will have more privacy when using cryptocurrencies than any other kind of services controlled by a central authority. This contribution will prove that this is not that easy and that privacy in the crypto-ecosystem remains an issue. The chapter will first of all argue that the confidentiality on a public blockchain is limited (Sect. 2). It will then prove the inability of the GDPR to ensure users' data protection (Sect. 3) and, as a consequence, the development of anonymization methods in order to overcome this lack of confidentiality will be described (Sect. 4), as well as their limits (Sect. 5). The last part will finally provide concluding remarks (Sect. 6).

[1] See for example Herrera-Joancomartì and Pérez-Solà (2016).

[2] Badev and Chen (2014), p. 8.

2 The Limited Confidentiality on a Public Blockchain

The pseudonymous character of the transactions on a cryptocurrency blockchain could make it a privacy-friendly tool for users. Although this is partly correct, the argumentation is incomplete and it would be incorrect to describe the system as privacy-friendly. This can be explained by the transparency of the transactions' register (Sect. 2.1), and the consequences it has on the categorization of data available on a blockchain (Sect. 2.2).

2.1 A Transparent Register of Transactions

It has to be recalled that in order to create a so-called trustless trust, based on no confidence attributed to a central authority or any third party, permissionless blockchains make available to the participants the record of all the transactions that happened so that the different nodes of the system can verify and accept the transactions before they are included to the blockchain.

Because of that, a blockchain is much more transparent than an ordinary service controlled by a central authority. This transparency should not be problematic because of the public key used as a substitute to users' identities. As we have seen before, the identity of the participants to the blockchain is not disclosed in the transactions. The users only have to use their set of keys so that the transactions can happen and be recorded on the blockchain.

Nevertheless, researches have already showed that the data that one can find on a blockchain, when treated with the adequate technical means, could finally be attributed to natural persons.[3] This can be done for example by linking the data available on the blockchain network (including also the metadata) to off-network data and also by analysing the context of a transaction happening on a blockchain by performing network analysis.[4] With the adequate technical knowledge it has already been proved that it is possible to deduce relationships between different public keys on the Bitcoin blockchain. Also, in some cases, by using off-network information, the users' identities could be inferred.[5] Researchers are also working on clustering all public keys belonging to a same user.[6] Therefore, decentralised systems may end up being more vulnerable in terms of privacy than centralised ones.[7]

---

[3] See for example Kaminstky (2011). On the possibility to link Bitcoin transactions with IP addresses, see also Androulaki et al. (2013) and Reid and Harrigan (2011).

[4] Herrera-Joancomartì and Pérez-Solà (2016).

[5] Reid and Harrigan (2011).

[6] Herrera-Joancomartì and Pérez-Solà (2016).

[7] De Filippi (2017), p. 1.

2.2 Consequences on the Categorization of Data

Financial and banking data is considered by the jurisprudence of the European Court of Human Rights (ECtHR) as personal data falling under the scope of Article 8 of the European Convention on Human Rights (ECHR), irrespective of data being sensitive information or not.[8] Consultation or seizure of banking data constitutes an inference to the data subject's right to a private life.[9] The ECtHR considers that banking data includes bank statements, cheques, fiduciary dispositions and emails.[10]

Even though transactions recorded on a cryptocurrency blockchain are not included in the lists given by the ECtHR of what constitutes banking data, and even though cryptocurrency transactions are made peer-to-peer and therefore do not include the involvement of a bank, cryptocurrency still enables users to buy or sell goods, which is one of the most important functions of money. It could be considered that consultation of stored data on a blockchain could provide as many information on a data subject as the seizure of banking documents would enable. As a consequence, divulgation of the identity of the users should be prevented in order to protect users' right to data protection and this could in theory be made possible by applying the GDPR's provisions to blockchains.

First of all, in order to apply the GDPR to blockchains, it should be assessed whether blockchains and data processed by the system fall under the scope of the GDPR. The GDPR is applicable to different uses of personal data that include the collection, storage or also the fact of making available personal data.[11] As most blockchain use cases are about distributing occurred and occurring transactions to the nodes of the system by delivering to the participants the entire blockchain, it seems clear that it not only collects and stores data relating to the transactions of its users but that it also makes this data available to the nodes of the system. The use of data made in the context of a blockchain should then fall under the definition of processing that is given by the GDPR.

Transactions on a blockchain are often referred to as pseudonymous transactions, because of the fact that most of the time transactions only disclose participants' public keys, which could be compared to a pseudonym to their identity. A question that has then to be answered is whether pseudonymous data on a blockchain—technically speaking—should be considered personal data within the meaning of the GDPR. According to recital 26 of the GDPR, in order to determine whether pseudonymous data should be treated as personal data, it should be assessed whether such data could be attributed to an identified or identifiable person. One should consider all objective factors such as the costs and the amount of time required for identification and the available technology not only at the time of the processing but

[8] ECtHR. 2015. Brito Ferrinho Bexiga Villa-Nova v. Portugal 69436/10: §42.

[9] *Idem*: §44; ECtHR. 2015. M.N. and other v. San Marino 28005/12: §55.

[10] *Idem.*

[11] GDPR, Art. 4(2).

the technological developments should also be taken into consideration.[12] We will assess whether, under these criteria, public keys and transactional data should be considered personal data according to the GDPR's definition.

2.2.1 The Categorisation of Public Keys

Public keys belong to an individual and can potentially be linked to a real identity through the techniques previously discussed. The combination of the public keys with additional data can reveal the identity of the person holding the public key. Furthermore, sometimes, data available on a blockchain can be linked to real identities, without information from the outside, but simply by using various techniques.

The working group on Article 29 specifies that the process of pseudonymization consists of masking an identity. In this case, public keys are used as pseudonyms, but they remain identifiable because pseudonymization is not introduced irreversibly.[13] Therefore, public keys should qualify as pseudonymized data, pursuant to Article 4 (5) of the GDPR. This has been widely recognized: not only among the doctrine,[14] but also among administrations and authorities.[15]

2.2.2 The Categorisation of Transactional Data

Transactional data constitutes personal data if it relates directly or indirectly to an identified or identifiable person. Therefore, if a transaction recording an exchange of a certain amount of cryptocurrency from person A to person B is unlikely to constitute personal data, it may still become personal data. This would for example be the case, if the information is combined with additional information that could allow the user to be identified.

Christian Rueckert illustrates this possibility by the use of information from discussion forums or survey files.[16] The European Parliament considers that this could be the case not only if the additional information relates directly to the data

---

[12] GDPR, Recital 26.

[13] European Parliament, European Parliamentary Research Service. 2019. Blockchain and the General Data Protection Regulation, Can distributed ledgers be squared with European data protection law?, p. 27.

[14] See for example Berberich and Steiner (2016), p. 423.

[15] For example, the CNIL as well as the observatory and forum of the European Union on the blockchain has recognized the fact that public keys constitute personal data according to the definition given by the RGPD. See CNIL. 2018. Blockchain and the GDPR: Solutions for a responsible use of the blockchain in the context of personal data. See also European Blockchain Observatory and Forum. 2018. Blockchain and the GDPR.

[16] Rueckert (2019), p. 7.

subject, but also if it relates to the object of the transaction, if it allows a link to be established with the user.[17]

Moreover, once the identity has been linked to a certain public key, transactional data allows for a particularly detailed picture of the user's activities. Thus, as soon as an actor is able to deduce the identity of any user, he is also able to verify all the transactions related to this user and made from the same public key.[18]

3 The Maladjustment of the GDPR to Ensure Users' Data Protection

Even though the GDPR is only applicable since 2018, it had been initially proposed by the European Commission in 2012,[19] at a time when blockchains, that are now a matter of concern, were at a very early stage of their development. This may be a reason why the GDPR does not mention blockchains and raise a certain number of questions and issues when one tries to apply the provisions of the GDPR to the ecosystem actors.

In terms of data protection, there is a strong need to have a technology-neutral application of the GDPR in order to support an innovative European Union in terms of financial industry.[20] The European Parliament stated that the use of FinTech[21] should not undermine consumer protection and data protection of customers.[22]

However, because of the maladjustment of the GDPR to fit to blockchains' criteria, legal uncertainty remains. This maladjustment is most of all the consequence of the immutability of the system (Sect. 3.1) and its decentralization (Sect. 3.2). Finally, we will see that applying data protection measures through the gatekeepers, even if it seems appealing, is an insufficient solution (Sect. 3.3).

[17] European Parliament, European Parliamentary Research Service. 2019. Blockchain and the General Data Protection Regulation, Can distributed ledgers be squared with European data protection law?

[18] Böhme et al. (2015), p. 221.

[19] European Commission. 2012. Proposal for a regulation of the European Parliament and of the Council on the protection of individuals with regard to the processing of personal data and on the free movement of such data (General Data Protection Regulation).

[20] European Parliament, Committee on Economic and Monetary Affairs. 2017. Report on FinTech: the influence of technology on the future of the financial sector, p. 10.

[21] This may encompass for instance cryptocurrencies or smart contracts, Europol, Financial Intelligence Group. 2017. From suspicion to action—Converting financial intelligence into greater operational contact, p. 38.

[22] European Parliament, Committee on Economic and Monetary Affairs. 2017. Report on FinTech: the influence of technology on the future of the financial sector, pp. 10–14.

3.1 The Maladjustment Resulting from the Immutability of the System

It should also be added that a blockchain is by nature immutable. It is therefore not possible to modify the content of a blockchain or to delete a part of a block that composes it. In terms of data protection, this makes it difficult to imagine the application of certain principles and rights provided for in the GDPR. This is particularly the case for the rights to modify and to be forgotten—essential elements of the GDPR that require the ability to modify or delete the data in question.

It is not impossible to consider the possibility of modifying data contained in a blockchain. For example, this possibility has been studied by Primavera de Filippi[23] as well as directly by the European Parliament in its report on the compatibility of blockchains with the GDPR.[24] It appears from these studies that a modification of the blockchain could possibly be envisaged, under certain specific conditions, but would be expensive and could therefore only be used in very specific cases. This seems rather incompatible with the accessibility of the right to modify or delete data.

3.2 The Maladjustment Resulting from the System's Decentralization

Under the framework of the Directive 95/46/EC on the protection of individuals with regard to the processing of personal data and on the free movement of such data[25] only the controller of the processing was subject to data protection rules. However, the GDPR introduced new rules that not only constrain controllers but also processors.[26] They both share liability for non-respect of the provisions of the GDPR and data subjects may invoke their right to an effective judicial remedy against any of the two.[27]

A controller is defined as the natural or legal person, public authority, agency or other body which, alone or jointly with others, determines the purposes and means of the processing of personal data.[28] A processor is to be understood as a natural or

[23] De Filippi and Reymond (2018), p. 138.

[24] European Parliament, European Parliamentary Research Service. 2019. Blockchain and the General Data Protection Regulation, Can distributed ledgers be squared with European data protection law?, p. 76.

[25] European Parliament, Council. 1995. Directive 95/46/EC on the protection of individuals with regard to the processing of personal data and on the free movement of such data.

[26] *Idem*, Art. 17(3) and 23(1).

[27] GDPR, Art. 79.

[28] *Idem*, Art. 4(7).

legal person, public authority, agency or other body which processes personal data on behalf of the controller.[29] Processing does not only include analysis and treatment of data but also the collection, storage or the fact of making available data.[30]

However, it is difficult to determine to whom on a blockchain; the GDPR's duties should be imposed and who should be responsible in case of a breach. At first sight, we do not find a proper controller and/or processor on a permissionless blockchain. However, there is a number of actors that could bear the responsibility: this could be for instance the creator of the system when it is known, the different nodes of the blockchain or also the miners that contribute to the functioning of the system. This issue has not been solved yet by the legislator in the context of data protection and thus, uncertainties remain on how to effectively apply the GDPR to blockchains.

3.3 The Insufficiency of Data Protection Measures Through the Gatekeepers

The legislator could also decide to regulate the actors around the blockchain that contribute to make the use of blockchains easier for non-technical customers by building a bridge between the online and offline world. This solution has early been chosen in the context of anti-money laundering where it has been decided that cryptocurrency exchange platforms and custodian wallet providers should comply with the Directive on the prevention of the use of the financial system for the purposes of money laundering or terrorist financing provisions.[31] It is true that, even though blockchains are not controlled by any central authority and do not need any third party in order to ensure their functioning, third parties do exist. They render the use of blockchains easier, especially for non-technicians. Also, they make a link between the online world and the offline world.

Exchange platforms for instance, in a way, democratised the use of cryptocurrency, allowing anyone to easily buy cryptocurrency with fiat currency. In order to provide such kind of services, exchange platforms most of the time hold their customers' banking details in order to recognise the transfer of fiat currency made by a customer wishing to be given the equivalent amount of cryptocurrency in exchange.

Shops accepting to be paid in cryptocurrency are also to be mentioned. Because they receive a payment, they are aware of their customers' public keys, but they may also hold additional information. For example, they could ask for the email address

[29] *Idem*, Art. 4(8).

[30] *Idem*, Art. 4(2).

[31] European Parliament, Council of the European Union. 2018. Directive (EU) 2018/843 of the European Parliament and of the Council of 30 May 2018 amending Directive (EU) 2015/849 on the prevention of the use of the financial system for the purposes of money laundering or terrorist financing.

of the user and they may also be aware of the post address where the good that has been bought with cryptocurrency has to be delivered.

The status of permissioned blockchains may also be assessed. Their functioning is technically speaking close to the one of permissionless blockchains but a main difference is that a central authority exercises control on the participants to the blockchain. Anyone cannot enter and participate to the blockchain without having been provided with a right to access.[32] As a consequence, the central authority is generally aware of its customers' identity and this type of blockchain can end up being totally centralised around the authority.[33]

Those third parties, even though they may have pseudonymised their customers' data, are aware of the identity of their users and they could be able to link a digital identity with a real identity and track all transactions involving them.[34] This enhances the idea that pseudonymous data on a blockchain should be considered personal data and not anonymyzed data.[35]

However, targeting such gatekeepers seems insufficient because it will only provide rights and duties regarding data stored and used by these actors and not directly on the blockchain.

4 Anonymization Methods to Overcome This Lack of Confidentiality

The limited level of confidentiality available for cryptocurrency users is well known by the community. The fact that the legislation on data protection is of very little help, encourages the community to consider privacy as a personal duty. To assess this argument, this section will first describe the balanced approach and comprehensive level of data protection wanted by the European legislator (Sect. 4.1). Then, it will be argued that a shift can be seen from a balanced to a strictly individualistic approach of data protection (Sect. 4.2).

[32] Waelbroeck (2017), p. 10.

[33] *Idem*, p. 13.

[34] Böhme et al. (2015), p. 221.

[35] This is for instance the approach that is taken by Matthias Berberick and Malgorzata Steiner in the context of blockchains: "A connection between pseudonymised data and the data subject will usually and necessarily arise in BC transactions effected for offchain goods, e.g. conversion into real money payments, purchase of goods or services, registration data, where the transaction parties must be known. Against that background, there is a strong case for arguing that individual related information on BC is personal data". Berberich and Steiner (2016), p. 423.

4.1 A Balanced Approach and a Comprehensive Level of Data Protection

The provisions of the GDPR include two different ways to implement an adequate data protection framework. It first of all includes measures of individual control. This should be understood as the right for data subjects to keep a certain form of control over their personal data.[36] It is about giving people the possibility to enforce themselves their right to data protection and a way to exercise control over the data relating to them. This theory has been advocated not only by many scholars[37] but has also been addressed by the German Constitutional Court as the right to informational self-determination.[38] Measures of individual control are referred to as the solution to privacy harm, which should be understood as the loss of control of individuals over their data.[39]

Measures based on this theory can be found within the GDPR. For instance the Regulation gives data subjects the possibility to control data by giving them specific rights that they may invoke (e.g. the right to have access to data, to ask for data rectification or deletion). This theory has been seen as a means of empowerment and self-determination of the data subject facing data processing in the era of technology and big data.

Secondly, the GDPR includes measures ensuring the regulation of risks. The development of this theory comes from the growth of technology and is the consequence of the fact that the control of people over personal data is almost infeasible in an era where data is omnipresent.[40] These measures are aimed at imposing duties on data controllers and processors so that they act fairly and in a respectful way of data subjects. It is thus about making sure that even though data subjects do not exercise their right to individual control, they will still get an adequate level of data protection. Such measures are included in the provisions of the GDPR that impose duties on data controllers and processors such as their duty to implement security and organisational measures, data protection by design and by default, or also the obligation to appoint a data protection officer.

---

[36] Van Ooijen and Ursic-Vrabec (2018), p. 92.

[37] See for example Charlies Fried who recognises the importance of choice and control about "facts that most persons in a given society choose not to reveal about themselves", Parent (1983), p. 270. "Privacy is not simply an absence of information about us in the minds of others; rather it is the control we have over information about ourselves" Fried (1968) quoted in Solove et al. (2006), p. 39.

[38] According to the German Constitutional Court, the right to informational self-determination should be understood as the right for individuals to "decide in principle themselves when and within what limits personal matters are disclosed", see German Constitutional Court. 2013. English Translation of essential parts of the German "Volkszählungsurteil" from 15 December 1983, which established in Germany the Basic Right on Informational Self-Determination.

[39] See for instance Rayan Calo who describes privacy harms as "the loss of control over information about oneself or one's attributes" Calo (2011), p. 1134.

[40] Lazaro and Le Métayer (2015), p. 23.

So in general, it can be considered that the GDPR provides a comprehensive level of data protection to data subjects because it not only gives data subjects the possibility to exercise control over the data relating to them, but it also imposes duties on controllers and processors that have to be respected in any case.

4.2 *From a Balanced to a Strictly Individualistic Approach of Data Protection*

The uncertainties that have been examined in this chapter prove that, for the time being, some issues and questions are still to be examined and solved. Legal certainty regarding the fundamental right to data protection for blockchain users is not fully ensured yet. Therefore, it may be argued that data protection as the regulation of risks will not be sufficient as a way to protect data until the uncertainties are solved. The uncertainties may, as a consequence, lead to the need for users to be more involved in the protection of their data and their intervention may go beyond the simple exercise of the individual control measures that are to be found within the GDPR.

To respond to these uncertainties, data subjects could first of all decide to make use of the different rights they get from the GDPR's provisions. They could for instance decide to use their right to access to data, to rectification of data or their right to be forgotten. However, we have seen that the lack of a determined data controller/processor on a blockchain makes this possibility hardly possible.

Therefore, besides the rights provided for by the GDPR, users could also decide to use additional services or engage in additional practices that are meant to provide more security and privacy when using blockchains. Individuals could decide to themselves use techniques meant to ensure privacy such as the encryption of their transactions.

This phenomenon is already happening in the area of cryptocurrencies. First of all, within the community it is largely encouraged to implement basic privacy techniques such as the generation of a new set of keys for any additional transaction made.[41] This was also encouraged by Satoshi Nakamoto in the whitepaper that introduced Bitcoin. In the cryptocurrency ecosystem, it is also encouraged that users repeatedly send fractions of their coins to themselves using a different public key.[42] Also, services have been developed in order to propose more privacy for blockchain transactions. Some services propose to render invisible only the content of the transaction while some others propose to also render invisible the metadata related to those transactions.[43] Another privacy-friendly mechanism is well developed and encouraged: the use of mixing services to provide more privacy to the identities of the parties to a transaction. Mixing services are made to increase anonymity by

---

[41] De Filippi (2017), p. 6.

[42] Reid and Harrigan (2011).

[43] See for example Zerocash, http://zerocash-project.org. Accessed on 10 November 2021.

mixing the coins of multiple users, making it harder to find a relationship between inputs and outputs of a transaction on the blockchain. Mixing services could be compared to shared wallets: the initiator of a transaction would make a transaction to a public key belonging to the mixing service. Then, the mixing service would later transfer the equal amount to the destination address but using a different public key. Thus, the receiver of the transaction will get an output that will not be linked with the input (and hence, with the initiator) of the transaction anymore. As a consequence, the link between the participants to a transaction will be more difficult to establish and more privacy is ensured to the participants.[44]

In addition, in order to answer to the lack of privacy inherent to first generation cryptocurrencies, other more opaque ones have emerged, such as Dash or Monero. They provide a higher level of confidentiality than pseudonymous cryptocurrencies. The addresses of the parties and/or the amount of the transactions are hidden. While they all aim to achieve a higher level of confidentiality, they use different means in order to achieve this. Some prevent the traceability of transactions, make it more difficult to link the transaction to the virtual identity, or even hide the amount of the transaction as well as the public key of the parties. These systems offer all the advantages of cryptocurrencies without the transparency counterpart.

5 Limits to Anonymisation Techniques

These anonymisation techniques are used quite often within the community as a way to compensate not only the transparency of the transactions register but also the lack of practically applicable data protection measures. However, these techniques also have limits and limitations. First of all, relying mostly on anonymisation measures taken by users would lead to a situation where technical knowledge would be a prerequisite for an adequate level of data protection (Sect. 5.1). Second of all, these anonymization technique do not meet all the interests at stake and may progressively be subject to forced limitation (Sect. 5.2).

5.1 Technical Knowledge as a Prerequisite for an Adequate Level of Data Protection

If we consider that data protection is only a matter of individual control, the level of protection that data subjects get should depend on their involvement in protecting their data. First of all, data subjects should be aware enough of the issues regarding data protection and the rights they get from the GDPR in order to use them. Secondly, they should have a sufficient understanding of the technology that will

[44]Möser (2013).

allow them to implement other techniques in order to ensure their privacy. Data processing is becoming nowadays very complex and is hence difficult to be fully understood by non-technical experts.[45] However, data subjects should not be expected to become privacy experts in order to get an adequate level of protection.[46]

The shift we examined from a balanced to a strictly individualistic approach of data protection is contrary to the will of the European legislator. It should be recalled that the GDPR aimed, among other things, at implementing data protection by design/by default so that by default, data subjects could get an adequate level of data protection. The processing of data should be respectful of data subjects' fundamental right to data protection even if data subjects are not involve in the protection of their personal data.

Some authors argued that a right to ignorance should exist as part of our fundamental rights.[47] The idea behind this is that the ignorance of an individual, or in this specific case his/her lack of technical knowledge, should not result in weakening the protection of his/her fundamental rights. It seems that this view is shared at the European level: even though the European institutions encourage control over personal data,[48] it is clear for the European legislator that individual control measures should not exist without a privacy-friendly environment and adequate measures that the data subject would benefit from.[49] Actually, the protection of personal data cannot be fully ensured by the individuals themselves if they are not helped by measures imposed on data controllers/processors meant to ensure their privacy. Data protection based on individual control only could lead to different levels of data protection and also to a new hierarchy where technology specialists will get an adequate level of protection of their fundamental right to data protection through the use of additional services aimed at ensuring privacy, when others will not.

Of course, an adequate level of data protection should however not be subject to technological and scientific knowledge. This dual level of data protection that result from the uncertainties regarding the application of the GDPR to blockchains is counterproductive for the development of innovation and technology. On the contrary, a strong protection of fundamental rights for users would benefit the improvement and development of blockchains because it will improve the image such systems present to potential users and to the society as a whole.

---

[45] Moerel and Prins (2016), p. 62.

[46] Solove (2013), p. 1901.

[47] Moerel and Prins (2016), p. 84 and the reference they make to Arnon Grunberg who argued for the introduction of the right to ignorance as a human right.

[48] See for example Kroll (2012).

[49] Lazaro and Le Métayer (2015), p. 21.

5.2 *Limitation to Anonymization Techniques?*

Generally, among politicians and lawyers, technological means to enhance the confidentiality of transactions are viewed very negatively. Even basic techniques used by ordinary users to ensure confidentiality, such as the generation of an additional public key for each new transaction, is often seen as a means of preventing judicial authorities from carrying out their investigative work.[50] In the context of cryptocurrency, there is currently much discussion about the possible prohibition of this type of tools and there is a tendency to demonize these various means, by saying that the objective could not be other than criminal.[51] It is true that, while these various technological means aimed at greater confidentiality can be used legitimately, they are also used for criminal purposes, in order to make judicial investigations more difficult. This is why, in some communications of various authorities, but also in doctrinal works, one regularly finds the argument of the necessary prohibition of anonymization mechanisms used in the ecosystem.[52] For example, at a conference sponsored by Europol, Interpol and the Basel Institute, it was mentioned that the existence of these services should not be tolerated.[53] In addition, the text of the fifth Directive, which states that virtual currencies should not be anonymous, invites us to think that this could be a legal basis for the prohibition of anonymization services.

6 Conclusion

This article has shown that the inherent privacy-level on a blockchain is far from being optimal. This is mostly because of the transparency of the transactions register.

The GDPR is supposed to be adapted to the recent issues that society faces nowadays. However, the recent developments in the financial sector are not addressed by the GDPR and no reference is made to blockchains or cryptocurrencies. Nothing is said on how the differences with traditional financial actors should be addressed so as to ensure that customers get an equivalent level of data protection.

---

[50] "Therefore, it is possible in principle for someone to create, for ex, 100k bitcoin transactions between addresses they control before converting the bitcoins into another form. Reassembling such a chain of transactions, particularly if it needs to be done manually, would be at least very time consuming, if not impossible. Such a technique may form part of a complex laundering technique involving multiple individuals, VC and so on." UNODC. 2014. Basic Manual on the Detection And Investigation of the Laundering of Crime Proceeds Using Virtual Currencies, p. 44.

[51] Feuvrier (2019).

[52] See for example Houben and Snyers (2018), p. 14.

[53] See for example the proposal made during a global conference on money laundering and digital currencies organised by the Basel Institute in 2017.

This article aimed at highlighting the inadaptability of the measures provided for by the GDPR to cryptocurrencies. This can be explained by the decentralization of the system and by its immutability. Targeting third parties evolving around the blockchain, even if to some extent useful, can in no case constitute an optimum solution.

The legal uncertainties that have been addressed in this chapter have consequences on the right to data protection of users. As a consequence, users may decide to act themselves to protect their privacy. They could decide to implement their own privacy practices by using additional privacy-friendly techniques. This is already incentivised within the ecosystem.

This shift in the way data protection is seen and ensured may have negative consequences for users. One could fear that technical knowledge would become a prerequisite to get an adequate level of privacy through the use of additional techniques. In this case, an adequate level of protection of fundamental rights will not be ensured for all data subjects anymore.

On top of that, anonymization techniques are seen in a very negative way by legislators in general, who consider that they are mostly used for criminal purposes. Thus, we can see that there is a tendency towards more severity regarding the use and acceptance of anonymization techniques that could even become a strict prohibition in the future.

Finally, if there is no doubt on the fact that there is a strong need to ensure an adequate level of privacy to cryptocurrency users, this should be ensured by default, for all users, and not by anonymization techniques put in place by some individuals themselves. In order for this to be possible, and because of the inadaptability of the system to data protection measures in general, research should be conducted on the legal and technical means available in order to reconcile two interests that are largely opposed: the right to privacy of legitimate users on one hand, and the need to prevent and/or detect behaviours constituting criminal offenses on the other hand.

References

Androulaki, E., Karame, G. O., Roeschlin, M., Scherer, T., Capkun, S. (2013). Evaluating user privacy in Bitcoin. In A. R. Sadeghi (Eds.), *Financial cryptography and data security*. FC 2013. Lecture Notes in Computer Science (vol. 7859). Springer.

Badev, A., & Chen, M. (2014). *Bitcoin: Technical background and data analysis*. Finance and Economics Discussion Series. Division of Research & Statistics and Monetary Affairs Federal Reserve Board 2014-104. Board of Governors of the Federal Reserve System (U.S.).

Berberich, M., & Steiner, M. (2016). Blockchain technology and the GPDR – How to reconcile privacy and distributed ledgers. *European Data Protection Law Review, 2*(3).

Böhme, R., Christin, N., Edelman, B., & Moore, T. (2015). Bitcoin: Economics, technology and governance. *Journal of Economic Perspectives*, 29.

Calo, R. (2011). The boundaries of privacy harm. *Indiana Law Journal*, 86.

De Filippi, P. (2017). The interplay between decentralization and privacy: The case of blockchain technologies. *Journal of Peer Production*, (7).

De Filippi, P., & Reymond, M. (2018). Blockchain et droit à l'oubli. In T. Nitot & N. Cercy (Eds.), *Numérique: reprendre le contrôle*. https://hal.archives-ouvertes.fr/hal-01676888/document
Feuvrier, L. (2019). Bitcoin et lutte anti-blanchiment: la course à la réglementation. *Journal du Coin*. https://journalducoin.com/analyses/bitcoin-lutte-anti-blanchiment-course-reglementation/
Herrera-Joancomartì, J., & Pérez-Solà, C. (2016). Privacy in Bitcoin transactions: New challenges from blockchain scalability solutions. In V. Torra, Y. Narukawa, G. Navarro-Arribas, & C. Yañez (Eds.), *Modeling decisions for artificial intelligence*. 13th International Conference, MDAI 2016 Sant Julià de Lòria, Andorra, September 19–21, 2016 Proceedings, Springer.
Houben, R., & Snyers, A. (2018). *Cryptocurrencies and blockchains*. European Parliament, Policy Department for Economic, Scientific, and Quality of Life Policies.
Kaminstky, D. (2011). Black Ops of TCP/IP. https://www.blackhat.com/presentations/bh-asia-02/Kaminsky/bh-asia-02-kaminsky.pdf
Kroll, P. (2012). *Take control of your personal data*. European Commission – Directorate-General for Justice Publications Office of the European Union.
Lazaro, C., & Le Métayer, D. (2015). The control over personal data: True remedy or fairy tale? *Project-Teams Privatics Research Report* N° 8681. https://arxiv.org/ftp/arxiv/papers/1504/1504.03877.pdf
Moerel, L., & Prins, C. (2016). Privacy for the *Homo Digitalis*, proposal for a New Regulatory Framework for Data Protection in the Light of Big Data and the Internet of Things. Tilburg University. https://doi.org/10.2139/ssrn.2784123
Möser, M. (2013). *Anonymity of Bitcoin transactions. An analysis of mixing services*. Münster Bitcoin Conference. University of Münster. https://www.wi.uni-muenster.de/sites/wi/files/public/department/itsecurity/mbc13/mbc13-moeser-paper.pdf
Parent, W. (1983). Privacy, morality and the law. *Philosophy and Public Affairs, 12*.
Reid, F., & Harrigan, M. (2011). An analysis of anonymity in the Bitcoin system. In Y. Altshuler, Y. Elovici, A. Cremers, N. Aharony, & A. Pentland (Eds.), *Security and privacy in social networks*. Springer.
Rueckert, C. (2019). Cryptocurrencies and fundamental rights. *Journal of Cybersecurity, 5*(1).
Solove, D. (2013). Privacy self-management and the consent dilemma. *Harvard Law Review*, 126.
Solove, D., Rotenberg, M., & Schwartz, P. (Eds.). (2006). *Privacy, information and technology*. ASPEN Publisher.
Van Ooijen, I., & Ursic-Vrabec, H. (2018). Does the GDPR enhance consumers' control over personal data? An analysis from a behavioural perspective. *Journal of Consumer Policy, 42*(1).
Waelbroeck, P. (2017). Les enjeux économiques de la blockchain. *Annales des Mines – Réalités industrielles*, 3.

Are We Obligated to Employ Algorithmic Decision-Making?

Jonas Pfister

Contents

Abstract While algorithmic decision-making has been used in the private sector for years, we have only recently started to apply it in mainstream fields of the public sector. The key concept is to increase efficiency through automated processes. In many legal frameworks, efficiency is also a guiding legal principle for public administration. Hence, the argument could be made, that public bodies must employ algorithmic decision-making, insofar as it is more efficient than conventional procedures. In the article, the merits of this argument shall be evaluated at the example of the Austrian legal framework. Economic efficiency is described as a legal principle and public interest and its relationship to other interests such as privacy is explored.

J. Pfister (✉)
Centre for Computers and Law, University of Vienna, Vienna, Austria
e-mail: jonas.pfister@univie.ac.at

G. Borges, C. Sorge (eds.), *Law and Technology in a Global Digital Society*,
https://doi.org/10.1007/978-3-030-90513-2_7

1 Introduction

Algorithmic decision-making systems take full advantage of the four V's of Big Data: "volume", "variety", "velocity" and "value". These four terms imply the processing of enormous amounts of data at high speed from various sources to gain additional value.[1] When a person is evaluated by means of an algorithm,[2] a decision is often made based on predictive models with little to no direct human influence. The reduction of the human component is justified by efficiency gains and, from the point of view of profit-maximising industries, appears to be desired. Still, the technology remains widely criticised by the media and ethicists.[3]

While predictive modelling has been used for analysis in the private sector for years, we have only recently started to apply it in mainstream fields of the public sector, where it can serve as a basis for algorithmic decision-making. Even though the stakeholders in the public sector are different, the justification remains the same. Economic efficiency is not only a decisive factor in the private but also in the public sector—and for good reason:

Economic efficiency is a legal principle, of which traces can be found in various legal frameworks of member states of the European Union. The German Data Ethics Commission argues that algorithmic decision-making is not just a method we should but rather must employ in certain legal fields, since the public administration has a legal obligation to use its resources sparingly.[4] Some Austrian scholars even argue that economic efficiency of the administration is a constitutional principle.[5] Therefore, the argument could be made, that the public administration is obligated to employ technologies such as algorithmic decision-making, insofar as it is more efficient than conventional procedures.

In the following article, the merits of this argument shall be evaluated at the example of the Austrian legal framework. Economic efficiency shall be described as a legal principle and public interest and its relationship to other interests such as privacy shall be explored.

[1] Ehrnberger and Kröpfl (2016), p. 149.

[2] https://www.svr-verbraucherfragen.de/wp-content/uploads/GI_Studie_Algorithmenregulierung.pdf.

[3] Zerilli et al. (2019).

[4] www.bmi.bund.de/SharedDocs/downloads/DE/publikationen/themen/it-digitalpolitik/gutachtendatenethikkommission.pdf;jsessionid=24408A25564E43E84F826E5C6157DC1F.2_cid287?__blob=publicationFile&v=6.

[5] Lachmayer (2004), pp. 135–160.

2 Defining Efficiency

For dogmatic reasons, it seems appropriate to attempt a defintion of the term "efficiency". The expression is described differently not only in economics and law, but also within the legal doctrine and jurisdiction of the highest courts. The discipline which is mainly concerned with efficiency within the legal sciences is called "Economic Analysis of Law". In essence, this discipline is based on the principles of neoclassical economics and is not without controversy. In their fundamental discussion paper on the economic analysis of law, *Kirstein* and *Schmidtchen* argue that economic efficiency is achieved, when the generation of wealth from resources is at its maximum.[6] Provided that produced goods are used in such a way that maximum benefit is created for the members of society, one speaks of "allocative efficiency".[7] However, various models of such allocative efficiency exist. "Pareto efficiency" is reached when marginal utility and marginal cost are equal. In accordance with this criterion, a situation is inefficient if it is possible to change the use of resources for the production of goods and the distribution of the produced goods in such a way that the benefit of a member of society increases while nobody suffers a loss of benefit.

This criterion is, however, often viewed as insufficient when defining efficiency from a legal perspective, since it doesn't weigh the specific gains against the losses. Therefore, the aforementioned authors state, that one must rather apply other paradigms. *Stringham* for example argues, that the so called Kaldor-Hicks-Efficiency is rapidly gaining traction in the legal field.[8] *Kirstein* and *Schmidtchen* define said principle as follows: "*[...] An allocation is efficient if there is no more possibility to generate a benefit increase for one part of the society through re-allocation, which is greater than the losses generated by this re-allocation for the other part of the society.*"[9] In contrast to our legal system and common notions of justice, all these definitions are unbiased.[10]

For the Austrian legal framework, an attempt to define efficiency as a legal principle has been made by *Potacs*.[11] He circumscribes efficiency as "achieving effectiveness while avoiding effort at the same time". Similarly, other authors interpret efficiency in the Austrian legal framework as optimal relationship between the use of resources (input) and the result (output).[12]

---

[6] Kirstein and Schmidtchen (2003), p. 4.

[7] Kirstein and Schmidtchen (2003), p. 5.

[8] Stringham (2001), p. 41.

[9] Kirstein and Schmidtchen (2003), p. 6.

[10] Kirstein and Schmidtchen (2003), p. 7.

[11] Potacs (2008), mn. 598.

[12] Kroneder-Partisch (2001), mn 34 f.

3 Sources

To determine whether the public administration is obligated to employ algorithmic decision-making, it is necessary to identify a general efficiency principle. It is apparent that the distinction of potential legal sources plays a central role in answering the question posed above. The sources differ significantly in their origin and scope. The separation is therefore not only useful for dogmatic reasons. Rather, both the origin and the formulation have significance for the specific application. This in turn has direct consequences for the extent to which these principles are to be considered in the context of the potential use of algorithmic decision-making in the administration.

3.1 Constitutional Law

The most central source of Austrian public law is the Federal Constitutional Law (B-VG).[13] The B-VG contains various efficiency principles with different addressees.

3.1.1 Efficiency and Self-Governing Bodies

The constitutional efficiency principle for self-governing bodies such as municipalities or social insurance agencies is defined explicitly. A typical characteristic of a self-governing body is budgetary sovereignty and thus financial independence.[14] Despite their explicit autonomy, however, self-governing bodies need to adhere to certain principles. For example, Art. 120 c para. 2 B-VG dictates that the tasks of the self-governing bodies must be fulfilled in an economical and economic fashion and be financed through contributions from their members or by other means. *Berka* et alia interpret this provision as a constitutional requirement for efficiency.[15] However, while this commandment is expressly mentioned in the constitution, the scope of application is limited to self-governing bodies. Even though this concerns a large subset of decisions made within the field of public law, the principle cannot simply be applied to all administrative procedures.

[13] Bundes-Verfassungsgesetz (B-VG) StF: BGBl I 1930/1 idF BGBl. I 2020/24.

[14] Kahl and Weber (2015), p. 94; Öhlinger (2002), mn. 191 (192).

[15] Berka (2019), p. 63 (67).

3.1.2 Auditing Efficiency

A more general principle can be found in Art 126b (5) B-VG: "The audit of the audit office must cover the correctness of figures, the compliance with the existing regulations, as well as economy, efficiency and expediency." *Berka* derives a general principle of economic efficiency from Art. 126b (5) B-VG, which is not limited to self-governing bodies. Some authors even go so far as to call this interpretation the prevailing doctrine and name Art 126b (5) B-VG as the central provision.[16] This opinion has already found expression in standard literature.[17]

It could, however, be criticized that Art 126b (5) of the Federal Constitutional Law merely specifies the competence of the audit office to verify the accuracy of figures, compliance with existing regulations, economy, efficiency and expediency. In addition, the consequence of a negative audit result would only be politically relevant. A strict interpretation of the words would lead to the result that this standard is merely an instruction for action or a definition of competence for the audit office. The provision was, however, further developed by literature and courts. This led to a broadening of the field of application.[18] Indirectly, the provision results in an instruction to the executive branch to orientate its own actions according to this norm.

3.1.3 Principle of Equality and Rationality

In addition to the constitutional principle of economic efficiency within the meaning of Article 126b (5) of the Federal Constitution, the idea has been expressed in literature that the principle of equality and rationality in its formulation within the framework of Article 7 of the Federal Constitution can be seen as the basis for the constitutional principle of efficiency. According to *Lachmayer* the principle of efficiency is also anchored in the general requirement of objectivity and in the fundamental proportionality test. He justifies this idea with the more recent case law of the Constitutional Court, which will be discussed in the corresponding section.[19]

[16] Sumper (2020), mn. 17.

[17] Kahl and Weber (2015), p. 94.

[18] See case law below.

[19] Fuchs et al. (2004), mn. 454; Lachmayer (2004), pp. 135–160.

3.2 Ordinary Laws

Besides constitutional law, ordinary law can also be seen as a source of efficiency principles. These principles shall only be mentioned for the sake of completeness. § 18 of the Administrative Procedure Act (AVG)[20] stipulates, that the authority must carry out procedures as expediently, quickly, simply and cost-effectively as possible. According to *Hengstschläger* and *Leeb,*[21] this provision constitutes an efficiency maxim. Other examples such as § 28 of the Administrative Court Procedure Act[22] allow for more freedom in the procedure if it is in the interest of a speedy procedure or is associated with considerable cost savings.[23] Even though these laws are not on the same hierarchy level as the constitutional principles, they are a good indicator, that efficiency is a public interest.

3.3 Case Law

Very early on, the Constitutional Court also recognized economic efficiency as a public interest. In a decision from 1970/71,[24] it stated, with reference to its preliminary ruling, that the avoidance of "an uneconomic administrative burden" can constitute an objective justification for differentiation and encroachment in a fundamental right. The case furthermore highlights an important aspect of the principle: If there are two variants of interpretation of a norm and the second variant leads to much higher costs and no special interests lean in favour of the second option, the first variant of interpretation must be chosen. In further decisions, the Constitutional Court even showed a tendency to reject inefficient interpretations of the law.[25]

Later, in the well-known Austro-Control-GmbH decision, the Constitutional Court dealt with outsourcing. In its ruling the court states:

> As with every legislative act, the granting of loans to outsourced legal entities must comply with federal constitutional requirements, such as the requirement of objectivity resulting from the principle of equality (cf. VfSlg 8457/1978, 11369/1987, 11639/1988) or the **constitutional requirement of efficiency** [...].[26]

---

[20] Allgemeines Verwaltungsverfahrensgesetz 1991 (AVG) StF: BGBl I 1991/51 idF BGBl I 2018/58.

[21] Hengstschläger and Leeb (2014), mn. 117.

[22] Bundesgesetz über das Verfahren der Verwaltungsgerichte (Verwaltungsgerichtsverfahrensgesetz—VwGVG) StF: BGBl I 2013/33 idF BGBl I 2018/57.

[23] For more examples see § 39 AVG, § 3 Verwaltungsvollstreckungsgesetz 1991 (VVG) StF: BGBl I 53/1991 idF BGBl I 2013/33 or § 79 Abs 3 u 356 Abs 1 Gewerbeordnung 1994—GewO 1994, BGBl. I 1994/194 idF BGBl. I 2016/120.

[24] VfGH B 381/70 VfSlg 6509/1971 and VfGH B 319/91 VfSlg 13.639/1993; for a detailed analysis see Potacs (2008), mn. 602f.

[25] VfGH G 220/88, G 221/88, G 237/88 VfSlg 12.009/1989.

[26] VfSlg 14.473/1996; VfSlg 17023/2003.

Therefore, the Constitutional Court established, that the general principle of efficiency even binds the legislator. This doctrine was later solidified in further decisions:

> Finally, in its more recent case law in connection with outsourcing, the Constitutional Court has ascribed the meaning of a **general principle of efficiency**, which also binds legislation, to the requirement for economical, economic and expedient budget management derived from the Federal Constitution.[27]

In the opinion of the Constitutional Court, efficiency is a motive that can objectively justify a differentiation, which constitutes an interference in the equality principle, provided that proportionality is maintained. From these statements, scholars developed the principle of "graduated efficiency": The more economically efficient a regulation (compared to alternatives) is, the more likely it is to justify an encroachment in or interference with fundamental rights.[28]

3.4 Interim Conclusion

As an interim result, it can be stated that principles of efficiency are explicitly defined for self-governing bodies at constitutional level and for administrative authorities in ordinary law. Furthermore, constitutional principles of efficiency were developed by doctrine based on Article 126b para. 5 B-VG and on Article 7 B-VG. The classification as a legal principle leads to complex interpretation. Most aptly, this problem was broken down by *Sumper*[29] on the relevance of the efficiency principle in the area of tax law, which can be summarized as follows: The author points out the distinction between legal rules and legal principles. In line with the teachings of legal theory, a strict distinction must be made between these types of norms. While rules are simply to be fulfilled, principles are to be considered only to a certain extent and must be weighted.[30] Legal principles are also qualified as "optimization imperatives".[31] Thus they demand that an objective is realized as far as possible within the legal framework. If two or more principles collide, the more important principle must prevail. The less important principle, however, does not become invalid. Efficiency principles, especially those under Art. 126b (5) of the Federal Constitution or Art. 7 B-VG, are in any case aimed at optimizing administrative action. But the administration must optimize its actions only to the extent that it is legally possible and compatible with other principles—such as human rights. The efficiency principle therefore requires granular coordination with other principles.

---

[27] Translated from VfGH G 222/02 VfSlg 17023/2003.

[28] Potacs (2008), mn. 598 (604).

[29] Sumper (2020), mn. 19.

[30] Alexy (1986), mn.71 f.

[31] Sumper (2020), mn. 19.

4 The Crux with Trade-Offs

Ott and *Schäfer* argue that, as a rule of thumb, efficiency losses must be justified by gains in equity.[32] The efficiency principle cannot be interpreted on its own but must rather be weighed against other principles. This in turn results in a complex system of trade-offs. If the remarks of the German Data Ethics Commission mentioned at the beginning of this article are taken into account, it must be concluded that the argument can only be applied to the Austrian legal framework in certain cases. An economically inefficient measure can be unconstitutional if the inefficiency is not justified by an improvement in the effectivity of human rights enforcement or equity. Likewise, a measure can be efficient, but not proportionate from a human rights perspective. Therefore, one can argue, that the administration could have a legal obligation to employ technologies such as algorithmic decision-making. However, this obligation could only apply, insofar as the technology leads to an increase in efficiency and does not disproportionately infringe on human rights. Hence, the question arises, which system or procedure leads to the highest amount of efficiency, while not disproportionately infringing on human rights.

5 Criticism

Before going more into detail, the criticism of legal efficiency principles, as set out in the Economic Analysis of the Law, should also be briefly mentioned. Critics fear that efficiency will be elevated to the status of a sole principle and that efficiency will be interpreted in provisions that do not actually aim at it.[33] On the one hand, the criticism appears justified, since the case law indicates that the Constitutional Court does not focus solely on the subjective will of the legislator but rather on the objective wording.[34] There is thus, of course, the danger of locating efficiency principles in norms that did not intentionally pursue such a goal. On the other hand, however, one can clearly see that the Constitutional Court makes very sensitive and precise considerations and always defines efficiency as only one of several principles and objectives. Furthermore, the administrative courts have already ruled that not every measure can be justified by efficiency gains.[35]

---

[32] Ott and Schäfer (1998), mn. 213 (218).

[33] Huber (1997), mn. 594.

[34] VfGH G 220/88, G 221/88, G 237/88 VfSlg 12.009/1989.

[35] VwGH 93/04/0196 VwSlg 14.062 A/1994.

6 Efficient Implementation and Data Protection

What exactly constitutes an efficient and proportionate measure? Answering this question already seems difficult when deciding between conventional procedures and algorithmic decision-making. Gains in efficiency when transitioning between these options could result from the reduction of the need of human resources and a significantly faster procedure. On the other hand, one must consider additional R&D costs, hard- and software costs, organisational restructuring and the acquisition of expertise needed for an effective implementation. As stated above, one must also consider the possible trade-off, since the measure must be proportionate in addition to being efficient. Concerns typically raised are potential discrimination and privacy issues.[36]

Both of these issues were key to the creation of recent legislation by the European Union. Art 22 of the General Data Protection Regulation (GDPR)[37] deals with automated decision-making. The article restricts automated decision-making and implements safeguards for data subjects. It is, however, limited in scope. In essence, this leaves a controller with two possible ways of introducing algorithms in the decision-making process:

On the one hand, an algorithm can create a "pre-evaluation" and the actual decision-making can be left to agents of the controller. When utilizing this form of implementation, it is usually the aim of the controller to not fall within the provision, since it only applies to "fully" automated decision making.[38] On the other hand, a controller can fully automate the process and create a final output through the algorithmic decision-making system. In this case, Art 22 GDPR shall apply and a data subject may profit from the safeguards. The data subject therefore has the right to express their own opinion, to obtain human intervention on the part of the controller and to contest the decision.[39]

Deciding between these two options again is a compromise: The first option leaves the possibility of correction of a discriminatory pre-evaluation through the human agent. This, however, results in higher needs of human resources. Agents

---

[36] COM (2020) 65 final; https://www.svr-verbraucherfragen.de/wp-content/uploads/GI_Studie_Algorithmenregulierung.pdf.

[37] Regulation (EU) 2016/679 of the European Parliament and of the Council of 27 April 2016 on the protection of natural persons with regard to the processing of personal data and on the free movement of such data, and repealing Directive 95/46/EC (General Data Protection Regulation) OJ L 119, 4.5.2016, pp. 1–88.

[38] Art 22 par 1 GDPR. For more detail see: Schulz (2018).

[39] Art 22 par 3 GDPR; Note: These safeguards are only explicitly defined in case the legal basis for the processing is a contract between data subject and controller or in case the data subject has given consent. According to the opinion of the Article 29 Data Protection Working Group, however, similar safeguards need to be implemented in case the processing is authorised by Union or Member State law, see Art 29 Data Protection Working Party (2018). Guidelines on Automated individual decision-making and Profiling for the purposes of Regulation 2016/679. 17/EN, WP251rev.01, p. 30.

must not only be employed, but also trained and sensitized to the possible issues. This obligation is not taken lightly by the data protection authorities, as the recent case of the Austrian data protection authority demonstrates.[40] In essence, the Austrian labour market service aimed to introduce an algorithmic decision-making system to evaluate an unemployed person's chances of reintroduction into the labour market. The output of the algorithm was supposed to serve as a basis for the decision-making of the agents of the labour market service. Ultimately the decision should fully lie within the discretion of the individual agent. The data protection authority, however, ruled, that since in some parts of the country an agent on average only spent 10 min in a consultation session with an unemployed person,[41] the risk of the agent simply applying the algorithmic pre-evaluation as final decision was too high. Therefore, said data processing could fall within the scope of Art 22 GDPR. Hence the necessary safeguards were not provided for by the controller.[42] The Data Protection Authority did not go into detail as to what would constitute an appropriate amount of time. Therefore, a controller choosing this option runs the risk of not being in compliance. Additionally, if more time was invested by the individual agent, efficiency gains could become marginal.

The second option would be to fully automate the decision-making process. This certainly has the potential to reduce the costs of human resources. It is, however, not as widely accepted[43] and seems to raise additional concerns of increased human rights violations, since there is little room for correction.[44] Furthermore the aforementioned safeguards apply in this case. Compliance therefore still requires investing in a certain amount of human resources to deal with providing information to data subjects, complaints and fulfilling other requirements. An increased chance of human rights violations could also lead to additional lengthy court proceedings as well as rendering the whole measure disproportionate.

7 Solving the Dilemma

The data protection legislation in combination with the efficiency principles force a public sector controller to make difficult ex-ante estimations on the potential efficiency increase of introducing algorithmic decision-making. One could therefore ask whether such a complex assessment with potential direct consequences for data subjects should be the sole responsibility of a public controller. If one follows the opinion in the literature and the Constitutional Court, it becomes clear that the executive branch must optimize its own actions. Therefore, it can also be argued

---

[40] DPA (Austria), 19.08.2020, 213/1020 2020-0.513.605 (not published).

[41] DPA (Austria), 19.08.2020, 213/1020 2020-0.513.605, p. 5.

[42] DPA (Austria), 19.08.2020, 213/1020 2020-0.513.605, p. 11.

[43] Spiekermann (2019).

[44] COM (2020) 65 final.

that the administration must automate its procedures if this leads to an obvious gain in efficiency. However, this inevitably leads to a dilemma under the rule of law. After all, the administration is bound by the legal framework and administrative action must at least be provided for by law.[45] If an administrative procedure is pre-determined in detail, there is little room for optimization. Of course, the administration often has a certain amount of leeway within the scope of discretion, which leaves at least some room for optimization. But the administration is largely dependent on the legislator.

Lachmayer proposed an innovative approach to solving this dilemma. Like the Constitutional Court, he also derives the principle of efficiency from the powers of the audit office and the principle of rationality. In essence, he proposed to incorporate efficiency checks in the legislative procedure. Efficiency could then already be considered ex ante in a formal process. The Constitutional Court would then only have to examine compliance with the procedural steps.

8 Digital-Check

Lachmayer's idea could be transposed into a "Digital Check". This check would force the legislator to examine whether a procedure could be efficiently automated through algorithmic decision-making. Considering that—according to the Constitutional Court—the legislator is bound by the constitutional efficiency principle anyway, the argument could be made that the legislator already has to consider efficiency as a principle in the legislative process. This is reinforced by the fact that the Constitutional Court is limited to a mere control of justifiability in matters of efficiency, and the essential balancing must be done by the legislator.[46]

In addition, without the necessary legislation, the administration cannot substantially change and optimize any procedure. Therefore, it does not seem unreasonable to place the responsibility for implementing the principle of efficiency on the legislator. For reasons of constitutionality and the rule of law it would also be desirable to ultimately trace back the decision on whether algorithmic decision-making should be used to the legislator. The idea also seems to be in line with current political goals in Austria: The following passage can be found on page 319 of the current governmental programme: "Digital check of new laws: Wherever possible, administrative processes arising from new laws should be able to be carried out digitally - if possible, automatically. Existing administrative processes should be screened for the possibilities of digitization."[47] Therefore, it is also a political concern to optimize administrative processes through automation and to make them more efficient.

[45] Art 18 B-VG.

[46] Kahl and Weber (2015), p. 94.

[47] www.dieneuevolkspartei.at/Download/Regierungsprogramm_2020.pdf.

9 Conclusion

In summary, the considerations of the German Data Ethics Commission regarding the obligation to use algorithmic decision-making systems in administration can be applied to the Austrian legal framework to a certain extent. In Austria, the administration and the legislator are bound by the efficiency principle. It is questionable, however, how far this obligation extends for the respective branches and how it should be fulfilled in practice. Interferences in fundamental rights based on "efficiency" are possible, but the specific considerations within the framework of the graduated efficiency principle are highly complex. The question of whether an algorithmic decision-making system can or must be used depends largely on the efficiency gains associated with the implementation. A "digital check" in the legislative process could serve as a possible solution to the problem.

References

Alexy, R. (1986). *Theorie der Grundrechte* (mn. 71 f). Suhrkamp.

Art 29 Data Protection Working Party. (2018). Guidelines on Automated individual decision-making and Profiling for the purposes of Regulation 2016/679. 17/EN, WP251rev.01, (p. 30).

Berka, W. (2019). Verfassungsfragen der Neuorganisation der Sozialversicherung in Österreich. In: G. Baumgartner (Ed.), *Jahrbuch Öffentliches Recht 2019* ((p. 63) (67)). NWV Verlag Gmbh.

COM (EU). White Paper on Artificial Intelligence - A European approach to excellence and trust Brussels, 19.2.2020 COM (2020) 65 final.

Datenethikkommission. (2019). Gutachten der Datenethikkommission (p. 214). Retrieved August 25, 2021, from www.bmi.bund.de/SharedDocs/downloads/DE/publikationen/themen/it-digitalpolitik/gutachtendatenethikkommission.pdf;jsessionid=24408A25564E43E84F826E5C6157DC1F.2_cid287?__blob=publicationFile&v=6

Ehrnberger, V., & Kröpfl, M. (2016). Big Data - zwischen Recht und Ethik. In D. Jahnel (Ed.), *Jahrbuch Datenschutzrecht 2016* (p. 149). NWV Verlag GmbH.

Fuchs, C., Gruber, M., & Weinhandl, M. (2004). Recht und Ökonomik -Tagungsbericht zur 44. Assistententagung Öffentliches Recht. *ZfV, 2004*(88), 454.

Gesellschaft für Informatik. (2018). *Technische und rechtliche Betrachtungen algorithmischer Entscheidungsverfahren. Studien und Gutachten im Auftrag des Sachverständigenrats für Verbraucherfragen*. Sachverständigenrat für Verbraucherfragen. Retrieved August 25, 2021, from https://www.svr-verbraucherfragen.de/wp-content/uploads/GI_Studie_Algorithmenregulierung.pdf

Hengstschläger, J., & Leeb, D. (2014). *Verfwaltungsverfahrensrecht* (p. 117). Facultas.

Huber, P. (1997). Effizienz als Rechtsprinzip. Möglichkeiten und Grenzen der ökonomischen Analyse des Rechts. *ZfV, 1997*(5), 594.

Kahl, A., & Weber, K. (2015). *Allgemeines Verwaltungsrecht* (p. 94). Facultas.

Kirstein, R., & Schmidtchen, D. (2003). Ökonomische Analyse des Rechts, Rechts, CSLE Discussion Paper, No. 2003-04, Universität des Saarlandes, Center for the Study of Law and Economics (CSLE), p. 4. Retrieved August 25, 2021, from www.econstor.eu/bitstream/10419/23035/1/2003-04_oear.pdf

Kroneder-Partisch, C. (2001). Art 126b B-VG. In: K. Korinek, M. Holoubek, et al (Eds.), *Österreichisches Bundesverfassungsrecht* (mn 34 f). Verlag Österreich.

Lachmayer, K. (2004). Effizienz als Verfassungsprinzip – eine Maxime für staatliches Handeln in Österreich? In: M. Bundenberg et al. (Eds.), *Recht und Ökonomik, 444. Assistententagung des öffentlichen Rechts Jena 2004* (pp. 135–160). C.H. Beck.
Öhlinger, T. (2002). Die Bedeutung der Selbstverwaltung in der Sozialversicherung. *DRdA, 2002*(3), 191 (192).
Ott, C., & Schäfer, H.-B. (1998). Die ökonomische Analyse des Rechts. Irrweg oder Chance wissenschaftlicher Rechtserkenntnis? *Juristenzeitung* 1988, 213 (218).
Potacs, M. (2008). Ökonomische Effizienz als Interpretationsmaxime? *ZfV, 2008*(5), 598.
Schulz, S. (2018). Art 22 GDPR. In P. Gola (Ed.), *DS-GVO*. C.H. Beck.
Spiekermann, S., (2019, September 23). Warum das AMS keine KI auf österreichische Bürger loslassen sollte. *DerStandard*. Retrieved August 25, 2021, from https://www.derstandard.at/story/2000108890110/warum-das-ams-keine-ki-auf-oesterreichische-buerger-loslassen-sollte
Stringham, E. P. (2001). Kaldor-Hicks efficiency and the problem of central planning. *The Quarterly Journal of Austrian Economics, 4*(2), 41.
Sumper, M. (2020). Effizienz und Leistungsfähigkeit im Spannungsfeld eines demokratischen Steuersystems. *ZfV, 2020*(2), 17.
Zerilli, J., Knott, A., Maclaurin, J., et al. (2019). Algorithmic decision-making and the control problem. *Minds & Machines, 29*, 555–578. https://doi.org/10.1007/s11023-019-09513-7

Digital Personal Assistants with AI and Data Protection GDPR & E-Privacy-Reg

Christian Jaksch

Contents

Abstract Digital assistants become more relevant and create some data protection related questions. This paper gives an overview over this new data protection challenges.

1 Overview Digital Assistant

1.1 Definition

A digital assistant is an application that can perform tasks or services for an individual. They let people ask questions by speaking in their natural language and listening to verbal answers. Digital assistants make it possible for people, that they can ask about anything that is public knowledge (public in the internet) or where

C. Jaksch (✉)
Vienna, Austria

G. Borges, C. Sorge (eds.), *Law and Technology in a Global Digital Society*,
https://doi.org/10.1007/978-3-030-90513-2_8

the digital assistant has access to and they get immediately an answer. A digital assistant works only "online" via internet because the whole processing and the analytics of the user requests take place in the cloud infrastructure of the provider. Digital assistants are also called "personal assistant," "virtual assistant," "intelligent agent" or "voice assistant." From a roughly technical point of view, a digital assistant is an artificial intelligence system that emulates human interaction to perform particular tasks. Technical backgrounds are a well-functioning cloud infrastructure, deep neural networks, and other advancements in AI technologies. The artificial intelligence behind the technology indicates the capacity for learning, reasoning, and understanding—three key elements required to emulate a customer service agent's ability to solve problems. A digital assistant performs standard tasks automatically and provides support for complex knowledge work in the form of proactive information processing and the submission of recommendations. The digital assistant learns from the user's decisions, behavior and communication.[1] So a digital assistant can be defined as a tool that allows the user to interact with different systems and processes in natural language, learns from the user's habits and supports him or her with suggestions.

1.2 *"Siri", "Alexa" and "Google Assistant" as Today's Examples of Digital Assistants*

"Siri" ("**S**peech **I**nterpretation and **R**ecognition **I**nterface") from Apple Inc. is one of the first well-functioning and for it one of the most famous digital assistants. The origins of "Siri" go back to a military research project nearly twenty years ago: In 2003 the US Defense Department initiated through DARPA (*The Defense Advanced Research Projects Agency*) the so called PAL-Program (***P**ersonal **A**ssistant that **L**earns - Program*). The PAL-Program was expected to spawn innovative ideas that bring new science, fundamental approaches to current problems, and algorithms and tools and yield technology of significant value to the military.[2] One project of the PAL-Program was called CALO (***C**ognitive **A**ssistant that **L**earns and **O**rganizes*) and focused on improving the way that computers support humans through the use of "cognitive systems".[3] "*PAL focused on creating a framework for the use of cognitive systems that work alongside humans, learning from experience and applying logic and reason to commands from human operators.*"[4] In 2007 some of the lead researchers of the military CALO-project started to figure out how to use

[1]Hoy et al. (2018), pp. 81–88; Conrad (2017), pp. 740 ff; Gray (2016); pcmag.com, encyclopedia definition of virtual assistant; techtarget.com, definition virtual assistant (AI assistant).

[2]Mazzucato (2014), pp. 136 ff; SRI International (2020); Bosker (2013); Trillo (2013); DARPA (2011); Roush (2010).

[3]Bosker (2013); ai.sri.com (2020) (web.archive.org/); pal.sri.com (2020) (web.archive.org/).

[4]SRI International (2020).

the best ideas from CALO for the commercial sector. As the result of these considerations the start-up company Siri Inc. was founded. The start-up company Siri Inc. was allowed to license key software from the military CALO project in exchange for giving SRI (Stanford Research Institute) a stake in the company. So the idea of a digital assistant in the form of a voice-controlled iPhone app (the first iPhone was released on June 29th 2007) seemed the ideal business model and a perfect way to help mobile users complete all kinds of tasks.[5] In 2010 Apple Inc. acquired the start-up Siri Inc. for about $ 250 Million. The application Siri was then integrated into the iPhone by Steve Jobs as Apple's digital assistant and released for the first time in October 2011 with the iPhone 4S.[6]

Google Assistant is the digital assistant from google Inc. The Google Assistant helps to trigger apps and devices, user can ask all kinds of questions, make plans, and carry out a variety of useful tasks.[7] The Google Assistant monitors its users and learns about their behavior and can effectively support the user out of this collected knowledge (favorite music, favorite app, etc.). A successful utilization of Google Assistant depends on three factors:

- an excellent core AI.
- how much information Google Assistant is able to collect about its users and build up detailed knowledge about its users in the cloud (profiling).
- on which external data and digital services and public information Google Assistant has access and can process data to fulfill user requests or to match information.[8]

Microsoft Cortana is the digital assistant from Microsoft Inc.[9] Microsoft describes Cortana in this way: *"Cortana is Microsoft's personal productivity assistant that helps you save time and focus attention on what matters most.* (...) *Here are some things Cortana can do for you:*

- *Manage your calendar and keep your schedule up to date*
- *Join a meeting in Microsoft Teams or find out who your next meeting is with*
- *Create and manage lists*
- *Set reminders and alarms*
- *Find facts, definitions, and info*
- *Open apps on your computer."*[10]

---

[5] Mazzucato (2014), pp. 136 ff; Bosker (2013); Nusca (2013); Naone (2007).

[6] Golson (2011); Roush (2010); Schonfeld (2010); Damaschke (2016), p. 9.

[7] Lacoma (2017).

[8] Schmidt et al. (2017).

[9] Warren (2014).

[10] microsoft.com, Cortana: https://support.microsoft.com/en-us/topic/what-is-cortana-953e648d-5668-e017-1341-7f26f7d0f825 (last visit August 31st 2021).

Amazon Alexa is the digital assistant from amazon Inc. It is famous for its Alexa voice service "echo".[11] "*Alexa is able to play music, provide information, deliver news and sports scores, tell you the weather, control your smart home and even allow Prime members to order products from Amazon. Alexa sits in the cloud, expanding the information offered all the time and refining the responses to give you more accurate information. No matter what Alexa device you ask, all can return these sorts of answers, be that on your soundbar or in your car.*"[12]

2 Data Protection Related Questions

2.1 Data Flows and Data Processing Activities

In a roughly technical point of view, the (personal) data flow from the user to the provider of the digital assistant works in the following way.

I. The digital assistant is activated by a voice command such as "computer", "Siri", "Alexa", "Echo" or "Jarvis", whereby the assistant understands the voice command with help of artificial intelligence (*1st neural network*[13]);
II. After activation via the activation command or by pressing a button, the digital assistant records the user's subsequent voice command, compresses the recorded voice file and transfers this compressed voice file to the cloud. With the help of artificial intelligence (*2nd neural network*), the spoken word of the user (compressed voice file) is converted into text form in the cloud. This is followed by storage in the cloud.
III. The voice command converted into text form is analyzed with help of artificial intelligence (*3rd neural network*) regarding its content and context, with the goal of finding an answer for the user. The finally found answer is transmitted from the cloud back to the digital assistant, which then communicates it to the user. A digital assistant can also make autonomous suggestions and support the user in the respective situation in which he or she finds him-/herself. This depends on how much information the digital assistant has about the user and which applications are connected to the digital assistant.[14]

The three main relevant data protection processing activities of a digital assistant are:

---

[11] Kühl (2016) and Emmer and Dirscherl (2017).

[12] O'Boyle (2020).

[13] Neural networks are computing systems inspired by the biological neural networks that constitute animal brains. Such systems learn (progressively improve performance on) tasks by considering examples, generally without task-specific programming.

[14] Damaschke (2016), pp. 13 ff; Public Discussion: "Licht und Schatten der Digitalisierung" with Y. Hofstetter, U. Schäfer and B. Baginski (Published on Jun 23rd 2017): https://www.youtube.com/watch?v=fBYpp2AyLRs (last visit August 31st 2021).

- voice recording, transmitting in the cloud and conversion to text;
- processing and storing in a cloud computing infrastructure;
- profiling of the user and (automated) individual decisions.

2.2 *Application of the GDPR and Roles*

Before analysing data flows and processing activities, it is necessary to have a look on the application area of the GDPR to the *user* and to the *provider* of the digital assistant and in which role in the GDPR context they act. In the context of digital assistants it is prima facie not so clear to define correctly the application of the GDPR and, if applicable, who are all the "controllers" and/or "processors".

2.2.1 User Perspective

GDPR recital 18 (to Art. 2 Par. 2 lit.c GDPR) says clearly: *"This Regulation does not apply to the processing of personal data by a natural person in the course of a purely personal or household activity and thus with no connection to a professional or commercial activity. Personal or household activities could include correspondence and the holding of addresses, or social networking and online activity undertaken within the context of such activities."*[15]

This means that if the data processing of a digital assistant by a user is only *"a purely personal or household activity"*, the user falls out of the scope of the GDPR and only the provider of the service is bound to the GDPR requirements (recital 18 and Art. 2 Par. 2 lit. c GDPR). A simple purely private user does not fall under the GDPR when using a digital assistant at home.[16]

Because of its large impact, it is necessary to analyse Art. 2 Par. 2 lit. c GDPR (ex Art. 3 Par. 2 directive 95/46/EC) more detailed: The exception for purely personal or household activities exists in a different form since the first draft of the EC-Data Protection Directive in the year 1990. In the document COM(90) 314 final, EC-Commission explains in clear examples the use of personal data for only private purposes:

- *"personal electronic diary"*,
- *"the files are registers of members of an association whose consent to appear therein can be presumed from their very membership and the information contained in the register is not transmitted to third parties."*[17]

[15] Recital 18 and Art 2 Par. 2 lit. c GDPR.
[16] Fercher and Riedl (2016), pp. 18 ff.
[17] COM(90) 314 final., p. 21.

In COM(92) 422 final the EC-Commission again uses the example of an *"electronic diary"* to describe *"a purely private activity"* in the 1992 draft of a European Data Protection Directive.[18]

In the opinion of *Dammann/Simitis* (1997) the following activities are covered from the "household exception": *"holding of records of telephone numbers, addresses and birthdays of relatives, friends"*, *"private correspondence"*, *"the usage of personal data for the individual consumption (leisure)"*, *"correspondence with a catalog company"*, *"private account management"*, etc.[19]

The ECJ had already to decide, if a data processing for a legally purely personal or household activity takes place or not. In the case *Lindqvist* the ECJ stated: *"That exception must therefore be interpreted as relating only to activities which are carried out in the course of private or family life of individuals, which is clearly not the case with the processing of personal data consisting in publication on the internet so that those data are made accessible to an indefinite number of people."*[20] In the case *Satamedia* the ECJ spoke out: *"It follows that the latter exception must be interpreted as relating only to activities which are carried out in the course of private or family life of individuals (see Lindqvist, paragraph 47). That clearly does not apply to the activities of Markkinapörssi and Satamedia, the purpose of which is to make the data collected accessible to an unrestricted number of people."*[21]

So the crucial question is, if the personal data, which are used for purely private purposes, are made accessible to an indefinite/an unrestricted number of people (for example: publication on the internet)[22] or are transmitted to third parties (e.g. cloud provider and its support companies?),[23] or not. GDPR legal comments also deal with this important question:

The GDPR household exception (recital 18 GDPR) can be seen as problematic in the perspective of new technological advances like smart homes and internet of things, because in most of the smart home and internet of things business models, the personal data are no longer processed only locally at the private home as the originally intention of the household exemption was for, instead the data processing for only purely personal or household activities today takes place in a worldwide cloud infrastructure of the cloud provider with classical commercial interests. From this point of view, the GDPR extends the historical scope[24] of the household exemption (local storage of private personal data at home—for example electronic diary at the private personal computer hard disk at home) also to the use of social networking and online activities (recital 18 GDPR), where this personal data are not

[18]COM(92) 422 final., pp. 12, 67.

[19]Dammann and Simitis (1997) Art. 3 mn. 8.

[20]ECJ Judgement from November 6th 2003, C-101/01 recital 47, ECLI:EU:C:2003:596.

[21]ECJ Judgement from December 16th 2008, C-73/07 recital 44, ECLI:EU:C:2008:727.

[22]ECJ Judgement from November 6th 2003, C-101/01 Recital 47, ECLI:EU:C:2003:596; ECJ Judgement from December 16th 2008, C-73/07 recital 44, ECLI:EU:C:2008:727.

[23]COM(90) 314 final., p. 21

[24]COM(90) 314 final. and COM(92) 422 final.

only stored locally at home, instead these purely personal or household data are stored in worldwide clouds and there is no strict prohibition of a further processing of this only purely personal or household personal data by the provider of the services.[25] In the context of smart homes and digital assistants problems arise, when other data subjects than the user (e.g. guests, postmen, craftsmen, etc.) enter a private flat and it is not clear who has to fulfill and how to fulfill the GDPR requirements (transparency obligations, consent, etc.).[26] Recital 18 says to this situation: *"However, this Regulation applies to controllers or processors which provide the means for processing personal data for such personal or household activities."*[27] These legal problems, mentioned by *Geminn*[28] (in the context of smart homes), are not solved by the GDPR. This was one of the reasons, why Austria (and originally also Slovenia) voted against[29] the GDPR. On April 14th 2016 Austria explained in the European Council in Brussels, why it still voted against the GDPR: *"Private activities on social media are not included in the scope of the protection afforded by the Regulation ('household exemption'; Recital 18 and Article 2(2)(c). It has not been possible to resolve in a satisfactory manner the fundamental problem that the private use of data can also encroach on and violate the fundamental rights of others."*[30]

Before the GDPR with its recital 18 came into force, a well discussed problem[31] dealt with the messenger WhatsApp: A WhatsApp user transmits automatically his/her private telephone directory of his/her private smartphone (= purely personal or household purpose locally stored on smartphone) to WhatsApp Inc./facebook Inc. (= commercial purpose) during the installation process of the app. Many data protection experts (in Austria) see in this data transfer from the user's private smartphone to WhatsApp Inc. a relevant processing of personal data, which could not be covered by the old "household exception" at least of the former § 45 Austrian Data Protection Act 2000[32] (compare the same discussion about WhatsApp and the

[25] Fercher and Riedl (2016), pp. 18 ff.

[26] Geminn (2016), pp. 575 ff. (576 f; 579).

[27] Recital 18 General Data Protection Regulation (EU) 2016/679.

[28] Geminn (2016), pp. 575 ff. (576 f; 579).

[29] EU-Datenschutzreform steht, Österreich stimmt dagegen: https://futurezone.at/netzpolitik/eu-datenschutzreform-steht-oesterreich-stimmt-dagegen/136.195.665 (last visit August 31st 2021).

[30] Council of Europe (2016), Dok. 7920/16.

[31] Keßler (2016).

[32] Former Austrian § 45 DSG 2000 *"(1) Natural persons shall be permitted to process data for purely personal or family matters that have been disclosed to them by the data subject himself or that they have received in a lawful manner, in particular in accordance with § 7 para. 2. (2) Data that are processed by a natural person for purely personal or family matters shall be transmitted for another purpose only with the consent of the data subject, unless expressly provided for otherwise by law."*

former § 1 Par. 2 No. 3 BDSG 2003[33] in Germany[34]). The basis of the old national law "household exemptions" was Art. 3 Par. 2 EC-Data Protection Directive 95/46/EC, which is now Art. 2 Par 2 lit c GDPR. But, if this strict point of view would be followed also under the GDPR (by ignoring or reducing the interpretation of recital 18), the whole GDPR would be applicable also for private (WhatsApp) users too, because by the use of tools, that transmit the original purely personal data to a commercial controller, who uses this data later for himself for commercial purposes, this transfer and processing of this data by the user is not covered by the old househould exemption (historical interpretation):[35] The Austrian Data Protection Authority follows recital 18 GDPR and basically accepts the usage of the messenger WhatsApp and says: *"The exclusively private use of services such as WhatsApp is covered by the scope of the "household exception", provided that it is not accompanied by unrestricted publication of personal data on the internet"*[36].

If personal data are made accessible to an indefinite/an unrestricted number of people[37] or are transmitted to third parties, then—out of the historical law perspective—European Data Protection Laws are applicable for the private users too.[38] *Kühling/Raab* have the opinion that the ECJ has to decide about recital 18 GDRP and this topic.[39]

As already mentioned, the GDPR tries to solve this problem for the users through recital 18 by extending the "household exemption" for the private use of social networking and online activities.[40] If the clear statement of recital 18 is accepted by all data protection authorities and courts, there should be no data protection problem anymore for private users, who use their cloud based digital assistant for only purely personal or household activities in a worldwide cloud infrastructure. The classical local stored personal electronic diary and private correspondence at home from the 1990s as mentioned in COM(90) 314 final. and COM(92) 422 final. is no longer the

---

[33] Former German § 1 Par. 2 Number 3 BDSG 2003: *"This Act shall apply to the collection, processing and use of personal data by (...) private bodies (...) except where the collection, processing or use of such data is effected solely for personal or family activities."*

[34] *Solmecke,* 2017; Gericht: Nutzer verantwortlich Kontakte-Weitergabe an Whatsapp illegal?: https://www.n-tv.de/technik/Kontakte-Weitergabe-an-Whatsapp-illegal-article19908476.html (last visit August 31st 2021).

[35] *Keßler,* 2016; Gericht: Nutzer verantwortlichKontakte-Weitergabe an Whatsapp illegal?: https://www.n-tv.de/technik/Kontakte-Weitergabe-an-Whatsapp-illegal-article19908476.html (last visit August 31st 2021).

[36] Austrian Data Protection Authority 03.05.2021, 2021-0.285.169.

[37] ECJ Judgement from November 6th 2003, C-101/01 Recital 47, ECLI:EU:C:2003:596; ECJ Judgement from December 16th 2008, C-73/07 recital 44, ECLI:EU:C:2008:727.

[38] COM(90) 314 final., page 21.

[39] Kühling and Raab (2020), Art. 2 mn. 23 ff.

[40] Keßler (2016).

only relevant point of view to find out what is seen "purely personal and household".[41]

But the extension of the household exemption through recital 18 GDPR did not solve the open legal data protection problems behind it: Other data subjects like guests, postmen, craftsmen, etc. enter the private smart home of a user and a purely personal and household digital assistant is activated and collects personal data of these natural persons and transfers these data into a cloud, without consent and knowledge and chance to prevent this data processing activities for the affected data subjects (because a purely private and household user is not responsible under GDPR for his/her digital assistant in a smart home environment).[42] Nevertheless in one of the handbooks for Amazon Echo, the author of the handbook addresses clear words on the users: *"Part of a responsibly use of Amazon Echo is to clarify others - family members, friends, guests - in a basic form about the features of Amazon Echo"*[43]

So for these questions of only purely private activities, the (old) national civil law provisions (for example § 823 BGB in Germany; or the §§ 16, 17a, 1328a ABGB in Austria), dealing with (general) personality rights, have to be taken.

2.2.2 Provider Perspective

For the provider of the digital assistant the whole GDPR is applicable even if the user of the digital assistant falls under the "household exemption". Recital 18 GDPR says clear: "(...) *However, this Regulation applies to controllers or processors which provide the means for processing personal data for such personal or household activities."*[44]

The GDPR contains two different roles, it is important to find the correct role for the provider:

- A controller is defined as a person, who, *"alone or jointly with others, determines the purposes and means of the processing of personal data"* (Art. 4 Par. 7 GDPR).
- A processor is defined as *"a natural or legal person, which processes personal data on behalf of the controller;"* (Art. 4 Par. 8 GDPR).

Again the same question appears, as it emerged for search engines some years ago, where sometimes the argument was used to say, that a provider of a search engine or a provider of persons rating portal should only be seen as "processors"

---

[41] Fercher and Riedl (2016), pp. 18 ff; COM(90) 314 final., p. 21; COM(92) 422 final., pp. 12, 67.

[42] Fercher and Riedl (2016), pp. 18 ff; Council of Europe (2016), Dok 7920/16; Schillerhof (2017), p. 26.

[43] Schillerhof (2017), p. 26 (translated by the author).

[44] Recital 18 and Art 2 Par. 2 lit. c GDPR.

(Art. 4 Par. 8 GDPR) for its users, who themselves underlie the "household exemption", because these providers fulfil only their search requests in the area of only purely personal and household activities. *Kamp* and *Apostel* show that a provider of a person's rating portal and a person's search engine are the "men of the data", because these providers have a big interest on the personal data of their users and the providers demand all copyrights to user-generated content for themselves and formulate this demands strictly in their general terms and conditions. So it is obvious, that providers of online services do not and cannot act only as "processors" (Art. 4 Par. 8 GDPR) of their users ("household exemption").[45] So a provider of a digital assistant is without doubt "controller" (Art. 4 Par. 7 GDPR).

2.3 Legal Bases

For the whole data processing process (*collection, recording, organisation, storage, retrieval, consultation, making available, alignment, combination*, etc.), a controller (Art. 4 Par. 7 GDPR) needs a legitimation through a legal base in Art. 6 Par. 1 GDPR and for special categories of personal data also a legitimation under Art. 9 Par. 2 GDPR. So it is necessary to analyze the three main processing activities of a digital assistant and the relevant legal bases in this context (see Sect. 2.1)

- voice recording and conversion to text;
- user profiling and (automated) individual decisions;
- cloud computing (transferring, processing, storage).

In March 2021 (adopted in July 2021) the European Data Protection Board (EDPB) published new guidelines on virtual voice assistants. The EDPB has the opinion that a virtual voice assistant as an internet of things device (e.g. smart speaker as "IoT device") has to be seen as "terminal equipment" (Art. 1 directive 2008/63/EC) and underlies—additionally to GDPR—also Art. 5(3) ePrivacy-directive 2002/58/EC in the version of Art. 2(5) Citizens' Rights directive 2009/136/EC. This means that the gaining of access to information stored in the terminal equipment (e.g. "IoT device") is only allowed:

- informed consent of the subscriber or user; or
- it is necessary for the sole purpose of carrying out the transmission of a communication over an electronic communications network; or
- it is strictly necessary for the provider of an information society service (e.g. provider of the digital assistant) to fulfill a service explicitly requested by the user.[46]

---

[45] Kamp (2011), pp. 20 ff; Apostel (2013), pp. 169 ff.

[46] EDPB (2021), pp. 12 ff.

Art. 5(3) ePrivacy directive has been implemented in national law in Austria in § 96(3) TKG 2003[47] and in Germany it has been implemented in § 25 TTDSG.[48]

2.3.1 Voice Recording and Conversion to Text

In Germany as well as in Austria the recording of voice is not only a data protection question, because human voice is additionally protected by national criminal law (§ 201 StGB[49] Germany, § 120 StGB[50] Austria) with the obligation of at least "implied consent" before recording the voice of a human.

Out of this national criminal law obligations it seems obvious in the first step, that the informed consent under GDPR is also the suitable data protection legal base for the users' voice recordings (voice commands). Additionally it has to be clarified, if this users' voice data (voice commands) have to be seen as special categories of personal data (Art. 9 Par. 1 GDPR), namely "biometric data" (Art. 4 No. 14 GDPR)? This is important with regard to the legal basis (Art. 6 Par. 1 lit. a GDPR and/or Art. 9 Par. 2 lit. a GDPR).[51] In my opinion it has to be distinguished: If the voice data are only collected for the purpose to interact with the digital assistant, then no special categories of personal data ("biometric data") are processed, with the application of Art. 6 Par. 1 lit. a GDPR as correct legal base. If these voice data are also processed to identify the user (speaker recognition), then this voice data have to be seen as special categories of personal data (Art. 9 Par. 2 lit. a GDPR). *Feiler/Forgó* specify with reference to recital 51 sentence 3 GDPR, that it depends on the intended use of the data: Personal data like voice commands have only to be seen as "biometric data" in the sense of Art. 4 No. 14 in conjunction with Art. 9 Para. 1 GDPR if they are processed specifically for the unambiguous identification of a natural person. It therefore depends on the purpose of processing whether the voice data are to be regarded as special categories of personal data (Art. 9 Par. 1 DSGVO) or not. If the recorded voice commands of the users are not used to identify the speaker, no special categories of personal data are present in the case of voice commands, despite the existing identifiability, but due to the fact that they are not processed for this purpose "speaker identification".[52] If the voice commands are used for the purpose to identify the user (speaker recognition), explicit consent of the user according to Art. 9 Par. 2 lit. a GDPR in conjunction with consent under Art. 5(3) ePrivacy-directive seems the only possible legal base for collecting the voice. Because of the transparency obligations, consent this will only work with separate accounts for each specific subscriber or user.

---

[47] BGBl. I Nr. 70/2003; BGBl. I Nr. 102/2011; BGBl. I Nr. 78/2018.

[48] BT-Drs 19/27441, pp. 25 f.

[49] Kargl (2017) § 201 mn. 23 f; Heuchemer, as of 01.05.2019, § 201 mn. 19.

[50] Lewisch and Reindl-Krauskopf, as of 17.10.2017, § 120 mn. 14; Tipold (2017), § 120 mn. 13 ff.

[51] Weichert (2020), Art 4 No 14 mn. 3 f.

[52] Feiler and Forgó (2017), Art 9 mn. 3; Art 4 mn. 32 f.

When using (explicit) consent as the legal base, it must be assured that the digital assistant can be activated, controlled and used by pressing buttons (remote control) as alternative.

If the voice commands are not used for the purpose to identify the user, only Art. 6 Par. 1 GDPR is applicable. Besides the informed consent other legal bases of Art. 6 Par. 1 GDPR could be relevant as legitimation too: The legal base "contract fulfillment" according to Art. 6 Par. 1 lit. b GDPR seems an additional applicable legal base for the collection of the user's voice commands. The contract fulfillment as legal base allows it for the controller to argue, that "implied consent" (e.g. user's contract approval for the requested service "virtual voice assistant")—according to the national criminal law obligations—of the data subject as end-user for the collection of the voice commands is available. However, the collection of voice commands alone under the legal base "legitimate interests" (Art. 6 Par. 1 lit. f GDPR) seems actually to be a legal risk. On the one hand because out of the sensitivity of voice data from the perspective of the balance of interest test (risks for identity thefts, etc.) in conjunction with the "implied consent" national criminal law obligations (§ 201 StGB Germany/§ 120 StGB Austria) and on the other hand because of Art. 5(3) ePrivacy-directive, that the gaining of access to information stored in the terminal equipment (e.g. "IoT device"[53]) is only allowed under informed consent, or—without consent—for the sole purpose of carrying out the transmission of a communication over an electronic communications network, or for the service fulfillment with an information society service (e.g. service/contract fulfillment). This means, that a controller can get access to terminal equipment to process voice data under this GDPR/ePrivacy legal bases without subscribers' or users' consent only to execute requests from subscribers or users and so no additional advertisement and marketing purposes (profiling, tracking, etc.) are possible (because only service/contract fulfillment as purpose is allowed). Because of Art. 5 (3) ePrivacy-directive and national criminal law obligations for voice data, the legal base Art. 6 Par. 1 lit. f GDPR alone is not suitable to justify the collection of voice commands from a digital assistant (e.g. smart speaker as "IoT device").

2.3.2 User Profiling and Automated Individual Decisions

The user profile is required for a digital Assistant to function properly. The profiling takes place in the cloud and not in the smart speaker, smartphone or laptop. When "profiling", personal data are processed (Art. 4 No. 4 GDPR) in order to evaluate certain personal aspects relating to the user with the aid of the digital assistant, e.g. to analyze or predict aspects such as personal preferences, general interests, reliability, behavior, whereabouts or change of location. A digital assistant creates a user profile and on the basis of this profile the digital assistant independently makes decisions for the user, e.g. in the form of proactive information provision (interest and preference-

[53] EDPB (2021), pp. 12 ff.

based provision to the user) or automated appointment scheduling (prioritization of appointments according to different parameters).

The *Art 29 Working Party* makes a clear distinction between the following profiling stages:

- profiling as pure personal data processing (recital 72 in conjunction with Art. 6 Par. 1 GDPR);
- decisions (with final human decision) based on profiling (recital 72 in connection with Art. 6 Par. 1 GDPR);
- only automated decisions—e.g. based on profiling—which have legal effect for data subjects or which affect data subjects in a similar way (recital 72 in conjunction with Art. 6 Par. 1 in conjunction with Art. 22 GDPR).[54]

Profiling, which does not result in an exclusively automated decision, is subject only to the general requirements for data processing in accordance with Art. 5 ff GDPR.[55] As legal bases for the profiling as data processing activity the (opt-in) consent of the user (Art. 6 Par. 1 lit. a GDPR), the contract fulfillment with the user (Art. 6 Par. 1 lit. b GDPR) or the legitimate interest of the controller (Art. 6 Par. 1 lit. f GDPR) can be argued. In my opinion the "best" legal base in this case would be the contract fulfillment, but this demands that the profiling as data processing is absolutely necessary for the controller to fulfill the contract "digital assistant" with the user. If this necessity cannot be argued, (opt-in) consent or legitimate interest are possible alternatives.

If special categories of personal data (Art. 9 Par. 1 GDPR) are part of the profiling process, the only possible legal base for the controller is the explicit consent of the data subject (Art. 9 Par. 2 lit. a GDPR).

Profiling with subsequent (automated) decision making involves three stages, the last stage being the application of Art. 22 GDPR:

a) Collecting the user data for a profile (Art. 5 ff GDPR);
b) Model development (Art. 5 ff GDPR);
c) (Automated) single decision (Art. 5 ff and Art. 22 GDPR).[56]

The "*collection of user data to form a profile*" is legally "profiling" according to Art. 4 No. 4 GDPR as a type of automated processing of personal data, which consists of using personal data to evaluate certain personal aspects relating to a natural person. Pure profiling is not covered by the prohibition rule of Art. 22 DSGVO.[57]

The "*model development*" takes place with the help of machine learning algorithms. It is possible to find out correlations and personal user attributes. The data is

[54] Art 29 Working Party, WP 251rev.01, 2018, pp. 8 f.

[55] Recital 72 GDPR; Art 29 Working Party, WP 251rev.01, 2018, pp. 5 ff; Martini (2018), Art 22 mn. 21 f.

[56] Kamarinou et al. (2016), pp. 8 ff.

[57] Recital 72 GDPR; Kamarinou et al. (2016), pp. 8 ff: Buchner (2020) Art. 22 mn. 14 ff; mn. 20 ff.

not only processed to obtain a descriptive profile of the user, but also compared with predefined patterns of the average behavior of numerous other persons to determine further information, e.g. whether the profile corresponds to the normal behavior or deviates from it and draw appropriate conclusions.[58]

On this information basis "*(automated) individual decisions*" are then made. The final decision is made either by a human being—supported by suggestions from his digital assistant—or by the digital assistant on its own. If the final decision is made by a human being, the general standards of data protection law remain applicable (Art. 5 ff GDPR). However, if the decision is made exclusively by a computer (digital assistant), i.e. a machine, the additional applicability of Art. 22 GDPR must be examined: Art. 22 Par. 1 GDPR contains a prohibition of automated individual decisions (= a machine makes the final decision on human beings) and is formulated in the GDPR as the "right" of the person concerned. In case of violation, the person concerned has a right to injunction against the person responsible.[59] Art. 22 GDPR would like to prevent and abolish the unchecked subjection of individual under the sole decision of machines. The procedure of automated individual decisions is regulated, i.e. it is assumed that a decision is made by a machine. A creative act with final effect must therefore be present.[60] Automated preliminary decisions (preselection) before the actual final decision are not covered by Art. 22 GDPR.[61] Like its predecessor provision Art. 15 EC-Data Protection Directive 95/46/EC and its national transpositions in § 6a BDSG-old version in Germany and § 49 DSG 2000-old version in Austria, Art. 22 Par. 1 GDPR does not deal with the question of whether personal data that may contribute to the assessment of a person may be processed at all (legal basis; recital 72), but only with automated decisions themselves, which—based on such processing already classified as lawful—are to be supported (use of certain results).[62] Art. 22 GDPR imposes a series of conditions and restrictions on the use of purely mechanical results of automated data processing, in the form of additional conditions of legality for decision-making, which are additional to the general rules on the legality of data processing.[63]

The prerequisites for Art. 22 Par. 1 GDPR to apply are:

- existence of an exclusively automated decision based on the processing of personal data—including profiling;
- this exclusively automated decision has legal effect on the person concerned; or
- this exclusively automated decision significantly affects the person concerned in a similar way as a decision with legal effect.

[58] Kamarinou et al. (2016), pp. 8 ff.

[59] Taeger (2017), pp. 3 ff.

[60] von Lewinski Stand: 01.02.2021, Art 22 mn. 1 ff.

[61] BT-Drs 14/4329, p. 37 (German Bundestag printed matter).

[62] von Lewinski, Stand: 01.02.2018, § 6a BDSG 2003, mn. 1 ff; Scholz (2014), § 6a mn. 4; Jahnel (2010) mn. 8/66 ff; Art 29 Working Party, WP 251rev.01, 2018, pp. 6 ff; Schulz (2018), Art 22 mn. 3 f.

[63] Buchner (2020), Art 22 mn. 11 ff; mn. 23 ff.

Permission for such a purely automated individual decision is only granted if this automated individual decision (Art. 22 Par. 2 GDPR):

- is necessary for the conclusion or performance of a contract between the data subject and the controller;
- is provided for by Union or national legislation offering appropriate guarantees; or
- with the explicit consent of the person concerned.

An automated individual decision within the meaning of Art. 22 Par. 1 DSGVO does not exist if it is not an exclusively automated decision. Mere preliminary decisions, pre-selections, etc., where a person makes the final decision and can thus deviate from the automated proposal, are not covered by Art. 22 Par. 1 GDPR. However, this requires a person who is the final human decision-maker, who is equipped and instructed with the appropriate decision-making authority and the intervention of the natural person takes place before the decision is made.[64] A precondition is, however, that the natural person can actually influence the final decision. If there is no competence to influence a computer decision by human intervention of the natural person, an automated decision in the sense of the GDPR is still possible despite the involvement of a human being.[65]

In the actual cases dealing with digital assistants, Art. 22 GDPR will not apply. Digital assistants like "Alexa", "Siri" or "Cortana" do not decide exclusively, they just make suggestions and give advice to users but do not generally formulate "automated individual decisions".

2.3.3 Cloud Computing

The digital assistant itself "lives" in the cloud, this means that the voice commands, the user profile and the software of the digital assistant are stored in the cloud. This cloud can be the controller's own cloud ("Private Cloud") or the controller of the digital assistant uses the cloud of a third party ("Public Cloud") to run his digital assistant.[66]

If the controller of the digital assistant uses the cloud of a third party ("Public Cloud"), this has to be done strict in accordance with GDPR. In most of the use cases the cloud provider can be seen as a processor (Art. 4 No. 8 GDPR). This means that a cloud provider as processor with processing the data inside the European Union has especially to fulfill the obligations out of Art. 28, 29 and Art. 32 GDPR including that the processor has to make a data processing agreement (Art. 28 Par. 3 GDPR) with the controller where these obligations are settled in detail.[67]

[64] Schulz (2018), Art 22 mn. 12 ff; BT-Drs 14/4329, p. 37.

[65] Art 29 Working Party, WP 251rev.01, 2018, pp. 20 f.

[66] Art 29 Working Party, WP 196, 2012, p. 30.

[67] Art 28 Par 3; Art 29; Art 32 GDPR; Art 29 Working Party, WP 196, 2012, pp. 15 ff.

If the data of the digital assistant should be transferred outside the European Union, special legal obligations arise. An international data transfer under GDPR is only possible if:

- there is an adequate level of data protection in the third country (Art. 45 GDPR). The following countries have an adequate level of data protection: Andorra, Argentina, Canada (only commercial organizations), Faroe Islands, Guernsey, Israel, Isle of Man, Japan, Jersey, New Zealand, Switzerland and Uruguay (adequacy talks were concluded with South Korea);[68] or
- there are appropriate safeguards like standard contractual clauses (SCCs), binding corporate rules (BCRs), etc. (Art. 46, Art. 47 GDPR).

If there is neither an "adequate level of data protection in the third country" nor exist "appropriate safeguards for the data transfer", a legal base according to Art. 49 Par. 1 GDPR must allow the transfer. The problem is, that Art. 49 GDPR is not suitable for such transfers, that only take place out of business reasons (cost reduction, usability, etc.) and not out of an objective necessity (contract fulfillment, etc.). So it is not a good idea to build an international cloud computing business model on the legal base according to Art. 49 Par. 1 GDPR, although two permissions for a transfer can be seen as relevant for such cases as:

- the data subject has explicitly consented the proposed transfer (Art. 49 Par. 1 lit. a GDPR);
- the transfer is necessary for the performance of a contract between the data subject and the controller (Art. 49 Par. 1 lit. b GDPR). This means that the transfer of user data is absolutely necessary for the contract fulfillment of the digital assistant, which seems to be difficult to argue.

These two legal bases are not as legal secure as an "adequate level" or "appropriate safeguards".

Actually for European companies, who use US cloud providers many data protection problems arise since the United States do not have an "adequate level of data protection" and also SCCs and BCRs cannot bring "appropriate safeguards" for EU-US data transfers: The reason ist, that US IT-companies underlie special US legal obligations such as FISA 702 and EO. 12.333, which allow US secret services access to personal data from Non-US-Citizen.[69] FISA 702 is addressed to "*electronic communication service providers*" which includes "*telecommunications carriers*", "*providers of electronic communication services*", "*provider of a remote computing service*" and "*any other communication service provider who has access*

[68] EU-Commission Adequacy decisions: https://ec.europa.eu/info/law/law-topic/data-protection/international-dimension-data-protection/adequacy-decisions_en (last visit August 31st 2021).

[69] ECJ Judgement from October 6th 2015, C-362/14 ("Schrems I"), ECLI:EU:C:2015:650; ECJ Judgement from July 16th 2020, C-311/18 ("Schrems II"), ECLI:EU:C:2020:559.

to wire or electronic communications either as such communications are transmitted or as such communications are stored".[70] So US IT-companies, which offer cloud computing services fall under these special US legal obligations (FISA 702, EO. 12.333). Since the two ECJ decisions ("Schrems I" and "Schrems II") it is generally not possible for US companies, who underlie these US laws, to achieve an adequate level of data protection under GDPR without supplementary measures.[71] The *European Data Protection Board* (ex *Art 29 Working Party*) demands effective encryption as an example for additional safeguards, when using SCCs or BCRs in countries like the US. This means for example that the data have to be sufficient encrypted as well during the data transfer as during the whole processing and storage phases in the US, that the US cloud provider does not have access to clear data and cannot give access to US public authorities.[72]

The actual problem out of my perspective is, that the legal obligations out of GDPR and Art. 8 Charter of Fundamental Rights of the European Union and today's IT-business reality are so completely different. It is very complex to bring GDPR legal obligations and necessary commercial IT-business demands effectively together.

2.4 Data Protection Principles

The next step is the compliance with the principles relating to processing of personal data:

Personal data shall be processed lawfully, fairly and in a transparent manner in relation to the data subject (Art. 5 Par. 1 lit. a GDPR). "Lawfulness" means that there has to be a legal base (see Sect. 2.3), the data subject has to be informed about the data processing ("fairness"; exception: Art. 14 Par. 5 GDPR) and has to give this information in a transparent manner ("transparency"). This includes the information to data subjects that a profiling takes place (Art. 13 Par. 2 lit f, Art. 14 Par. 2 lit. f GDPR) and/or that the data are transferred outside the European Union. It means also transparency for data subjects through privacy by design and privacy by default settings.[73]

The data protection principle of "purpose limitation" (Art. 5 Par. 1 lit. b GDPR) protects data subjects by setting limits on how data controllers are able to use their data while also offering some degree of flexibility for data controllers. Personal data

[70] 50 U.S. Code § 1881 – Definitions: https://www.law.cornell.edu/uscode/text/50/1881 (last visit August 31st 2021).

[71] ECJ Judgement from October 6th 2015, C-362/14 ("Schrems I"), ECLI:EU:C:2015:650; ECJ Judgement from July 16th 2020, C-311/18 ("Schrems II"), ECLI:EU:C:2020:559.

[72] EDPB (2020), pp. 26 f (Use Case 6).

[73] *Herbst* 2020, Art. 5 mn. 13 ff; mn. 19; Feiler and Forgó (2017), Art. 5 mn. 2 f; Dammann and Simitis (1997), Art. 6 mn. 3; Recital 38 EC-Data Protection Directive 95/46/EC; Recital 39 General Data Protection Regulation (EU) 2016/679.

must be collected for specified, explicit and legitimate purposes (purpose specification) and may not be further processed in an incompatible way (Art. 6 Par. 4 GDPR) with those purposes (compatible use) or if Union or Member State law, which constitutes a necessary and proportionate measure in a democratic society to safeguard the objectives referred to Art 23 Par 1 GDPR exists (incompatible use).[74] This means for a provider of a digital assistant, that the provider is only allowed to process personal data of the user (data subject) as declared in the privacy policy of the provider (for example: user consent for specific purposes; purpose contract fulfilment with user, purpose direct marketing, etc.).[75]

The data protection principle of "data minimization" (Art. 5 Par. 1 lit. c GDPR) demands that the processing of personal data has to be adequate, relevant and limited to what is necessary in relation to the purposes for which they are processed. In October 2017 it was discovered by media that Google Mini (a variant of a Google Home speaker, which enables users to speak voice commands in their homes to interact with services through Google Assistant) was listening to users even when the users did not activate the system via pressing the device's button or the voice command by saying "OK Google." Some users had found out that sound files had been uploaded to Google's servers from the Google Mini without their consent. This case would be an informative example of a(n) (accidentally) collecting of personal data (users' private voices in their homes) against the data protection principle of "data minimization", because it was not adequate, relevant and not limited to what is necessary in relation to the purpose(s) of a smart home of a user. A digital assistant should collect personal data (voice files) only after being activated by its users and not more.[76] A similar critic is addressed to Amazon Alexa from the *Consumer Advice Center of North Rhine-Westphalia*. If switched on, Amazon Alexa is listening to the conversations that take place in the user's home. The reason is, because Amazon Alexa has to react immediately if the call word "Alexa" or "Echo" or "Amazon" or "Computer" is spoken out by the users, therefore "Alexa" has to listen the whole time. The real data collection and the transmission of the recorded audio files to the amazon servers should only exclusively start after one of the call words is said by the users. So before the outspoken call words, amazon Alexa should only listen but should not transmit any audio files to the amazon servers.[77] Amazon is now criticized by the *Consumer Advice Center of North Rhine-Westphalia*, because Alexa starts to collect and transmit audio files of user not only when the real call word "Alexa" or "Echo" or "Amazon" or "Computer" is outspoken, it is also activated and starts transmitting audio files after it listened a word, which sounds only similar. So in the opinion of the *Consumer Advice Center of North Rhine-Westphalia*, there is a high data protection risk, that normal private sensible

[74]Recital 50; Art 6 Par 4 GDPR; Jaksch (2019), pp. 141 ff; Art. 29 Working Party, WP 203 - Opinion 03/2013 on purpose limitation (2013), pp. 1 ff.

[75]Conrad (2017), pp. 740 ff (743).

[76]Wilkens (2017) and Kling (2017).

[77]Schillerhof (2017), p. 26.

conversations are collected and transmitted to the cloud servers without knowledge of the users, because they do not expect that amazon Alexa had been activated by a word during a private conversations that maybe sounds a little bit similar to the call words ("Alexander" instead of "Alexa"; "holidays at Amazonas" instead of "Amazon" or in German sentence "Komm Peter" because it sounds similar to call word "Computer").[78] All this data collections without knowledge of the user could violate Art. 5 Par. 1 lit. c GDPR.

The data protection principle of "accuracy" (Art. 5 Par. 1 lit. d GDPR) requires, that personal data shall be accurate and, where necessary, kept up to date. It is a data protection obligation that demands a certain form of data quality and means that stored personal data should always be conform with the actual reality where necessary. So this obligation of "accuracy" has to be seen in regard to the different purposes of processing activities. This means that the GDPR does not state that personal data always have to be kept up to date, it says in all cases *"where necessary"*.[79] This would mean for a provider of a digital assistant, that for personal data, which should be used especially for user profiling purposes and for automated decision making, significant efforts to a good data quality should be made to be compliant with data protection principle of "accuracy". Old personal data, which became inaccurate, principally have to be erased or at least a concept has to be developed how old or inaccurate data can be rectified without delay, to be able to use this data again.

The data protection principle of "storage limitation" (Art. 5 Par. 1 lit. e GDPR) requires, that personal data are no longer stored than it is necessary for the purposes for which the personal data are processed. After purpose achievement, personal data have to be deleted or made anonymous. This means for a provider of a digital assistant, that the provider has to develop a concept for data deletion and/or data anonymization. For this reason the provider of a digital assistant has to know exactly, how long which (categories of) personal data are needed to achieve the defined purposes and how to correctly delete or anonymize personal data, which are no longer needed for the defined purposes. Recital 39 GDPR says: *"In order to ensure that the personal data are not kept longer than necessary, time limits should be established by the controller for erasure or for a periodic review."*[80] A famous provider of a digital assistant apparently stores the audio files from the users for 6 months with an identifier and then after 6 months the identifiers are removed and the audio files without the identifiers are stored for about further two years.[81] For *Müller-Peltzer/Franck* the storage of the audio files and some temporary analyses of the completed user transactions are in compliance with the European Data Protection law, but a permanent collection of the audio files of user spoken words is not allowed

---

[78] Verbraucherzentrale NRW (2017).

[79] Herbst (2020), Art. 5 mn. 60 ff; Dammann and Simitis (1997), Art.6 mn. 13 f.

[80] Recital 39 General Data Protection Regulation (EU) 2016/679.

[81] Damaschke (2016), pp. 16 f.

under European Data Protection law.[82] The *Federal Commissioner for Data Protection and Freedom of Information (BfDI)* in Germany sees a high data protection risk, because all words, which the user speaks with his/her digital assistant, are stored in the cloud of the provider, although this storage would be—in the opinion of the *BfDI*—not necessary. The problem for the *BfDI* is now, that the hours of outspoken words of each user, stored in the international cloud infrastructure of the providers, could be effectively utilized for identity thefts, if cloud providers are hacked.[83]

The data protection principle of "integrity and confidentiality" (Art. 5 Par. 1 lit. f GDPR) requires for a provider of digital assistant to implement appropriate technical and organizational measures to ensure data security. In the opinion of *Feiler/Forgó* this data protection principle has two aims: On the one hand it is addressed to Art. 32 GDPR (Security of processing—confidentiality, integrity, availability and resilience of processing systems and services), so that enough technical and organizational measures for the data security (IT-security) are being implemented and on the other hand, it is addressed to achieve lawfulness (Art. 24 GDPR: all technical and organizational measures to ensure and to document Compliance).[84] Because of the reason that many providers of digital assistants are cloud providers too, a second law obliges them to IT-Security. The Directive on security of network and information systems NIS Directive (EU) 2016/1148 is addressed to digital service providers (*cloud computing service*[85]) and these providers have to identify and they have to take appropriate and proportionate technical and organizational measures to ensure a level of security of network and information systems (Art. 16 NIS Directive). To document the compliance with these legal demands (Art. 32 GDPR, Art. 16 NIS Directive), the following standards can be considered:

- "ISO/IEC 27001:2013" is the international standard that describes best practice for an information security management system (ISMS).
- "ISO/IEC 27002:2013" is a popular, internationally-recognized standard of good practice for information security.
- "Critical Security Controls for Effective Cyber Defense" of the Center for Internet Security (CIS) is a publication of best practice guidelines for computer security.
- "IT baseline protection" ("IT-Grundschutz") of the German Federal Office for Information Security (FSI), in German "Bundesamtes für Sicherheit in der Informationstechnik (BSI)", is a methodology to identify and implement computer security measures in an organization.[86]

For a provider of a digital assistant, the fulfillment of a special IT-Security Standard (*ISO/IEC 27001; Critical Security Controls for Effective Cyber Defense*

[82] Müller-Peltzer and Franck (2017), p. 251.

[83] Federal Commissioner for Data Protection and Freedom of Information (BfDI), Sprachassistenten, 2017, p. 2.

[84] Feiler and Forgó (2017) Art. 5 mn. 11 ff; Art. 24 mn. 4.

[85] Annex III i.c.w. Art. 4 Par. 19 i.c.w. Art. 16 Par. 1 NIS Directive (EU) 2016/1148.

[86] Feiler and Forgó (2017), pp. 27–28; Donnelly (2017).

or/and *IT baseline protection*) will be an obligation to document the compliance with the data protection principle of "integrity and confidentiality" (Art. 5 Par. 1 lit. f; Art. 32 GDPR) and also with the requirement form Art. 16 NIS Directive (EU) 2016/1148.

2.5 International Application

A provider of a digital assistant that operates only from a third country (non EU/non EEA) is also in the scope of the GDPR (Art. 3 Par. 2 GDPR) if:

- the provider is offering services (digital assistant) to data subjects in the Union, and
- the provider monitors the behavior of data subjects in the Union, which takes within the Union (e.g. usage of the digital assistant).[87]

2.6 Additional GDPR Requirements

2.6.1 Data Protection Officer

The provider of a digital assistant will have to designate a data protection officer because a digital assistant requires regular and systematic monitoring of data subjects on a large scale (Art. 37 Par. 1 lit. b GDPR). If the digital assistant is also used to support elder or handicapped people (Ambient Assisted Living—AAL), it will process also a large scale of special categories of data (Art. 37 Par. 1 lit. c GDPR).[88]

2.6.2 Data Protection Impact Assessment

Basically there has to be made a data protection impact assessment because a systematic and extensive evaluation of personal aspects relating to natural persons which is based on automated processing, including profiling, takes place.[89]

---

[87] Art 3 Par 2 GDPR (EU) 2016/679.

[88] Jaksch (2017), pp. 140 ff.

[89] Art. 35 Par. 3 lit. a GDPR.

2.7 *ePrivacy-Regulation Draft*

2.7.1 General

The ePrivacy Regulation draft defines special rules for the lawful collection, processing and storage of data in connection with public electronic communication. Large parts of this ePrivacy Regulation draft will address only to "electronic communications networks" (Art. 2 No. 8 directive (EU) 2018/1972 European Electronic Communications Code) as transmission systems (telecom infrastructure) and "electronic communications services" (Art. 2 No. 4 directive (EU) 2018/1972 European Electronic Communications Code) and will have no direct impact for the legal relationship between the controllers of digital assistants and its users (Art. 5—Art. 7 ePrivacy-Regulation draft).[90] The ePrivacy Regulation draft contains three relevant provisions for controllers of digital assistants that will definitely have impact on digital assistants (Art. 8, Art. 10 and Art. 16 ePrivacy Regulation draft).

2.7.2 Art. 8 ePrivacy Regulation Draft

The use of processing and storage capabilities of terminal equipment and the collection of information from end-users' terminal equipment, including about its software and hardware, is only lawful if (version draft EC-Commission):

- necessary for the sole purpose of carrying out the transmission, or
- an informed consent of the end-user under GDPR conditions exists, or
- necessary for providing an information society service requested by the end-user, or
- necessary for web audience measuring, provided that such measurement is carried out by the provider of the information society service requested by the end-user.

The collection of information emitted by terminal equipment to enable it to connect to another device is only lawful:

- for the purpose of establishing a connection,
- notice under the conditions as in Art. 13 GDPR defined (standardized icons) as well as any measure the end-user of the terminal equipment can take to stop or minimize the collection.[91]

Terminal equipment is defined in Art 1 directive 2008/63/EC as *"equipment directly or indirectly connected to the interface of a public telecommunications*

[90] COM(2017) 10 final.

[91] COM(2017) 10 final.

network to send, process or receive information; in either case (direct or indirect), the connection may be made by wire, optical fibre or electromagnetically".[92] This means that laptops, smartphones and smart speakers have to be seen as "terminal equipment" in accordance with Art. 1 directive 2008/63/EC and Art. 8 ePrivacy Regulation draft.[93]

In the scope of Art. 8 ePrivacy Regulation draft a digital assistant can only interact with its cloud infrastructure legally as a "collection of information from end-users' terminal equipment", if the end-user gives his consent (Art. 8 ePrivacy Regulation draft/Art. 6 Par. 1 lit. a GDPR) or if the collection is necessary for the fulfillment of a contract with the end-user (Art. 8 ePrivacy Regulation draft/Art. 6 Par. 1 lit. b GDPR). The GDPR legal base "legitimate interest" (Art. 6 Par. 1 lit. f GDPR) cannot alone justify a data collection from the users' terminal equipment (e.g. smart speaker) in the scope of application of the ePrivacy Regulation draft.[94]

2.7.3 Art. 10 ePrivacy Regulation Draft

This provision gives special rules for privacy by design and privacy by default for software placed on the market permitting electronic communications. Art. 10 ePrivacy Regulation draft will regulate, that such a software shall offer the option to prevent third parties from storing information on the terminal equipment of an end-user or processing information already stored on that equipment. Upon installation, the software shall inform the end-user about the privacy settings options and, to continue with the installation, require the end-user to consent to a setting. The aim is, that the ePrivacy Regulation draft should provide for the possibility to express consent by using the appropriate settings of a browser or other application. The choices in the privacy settings of an end-users should be binding on, and enforceable against, any third parties.[95]

So Art. 10 ePrivacy Regulation draft can be seen as a specification for software permitting electronic communication to the general obligations in Art. 25 GDPR. Controllers of a digital assistant will have to respect Art 10 ePrivacy Regulation in addition to Art. 25 GDPR.

2.7.4 Art 16 ePrivacy Regulation Draft

Art. 16 ePrivacy Regulation draft regulates the direct marketing communication based on electronic communications services (email, messenger, phone calls, etc.). It

---

[92] DIRECTIVE 2008/63/EC on competition in the markets in telecommunications terminal equipment.

[93] EDPB (2021), pp. 12 ff.

[94] COM(2017) 10 final.

[95] COM(2017) 10 final.

is very similar to today's Art. 13 ePrivacy directive 2002/58/EC (version directive 2009/136/EC). In general consent is necessary before starting electronic direct marketing communication. One exception of the consent requirements exists if:

- electronic contact details for electronic mail are collected from customers in accordance with GDPR in the context of the sale of a product or a service;
- the electronic contact details for direct marketing are only used to advertise own similar products or services; and
- customers are clearly and distinctly given the opportunity to object;
- this right to object is given each time a message is sent.[96]

A digital assistant with commercial purposes has to respect Art. 16 ePrivacy Regulation draft.

2.8 Conclusion

For controllers of a digital assistant the whole GDPR is applicable. Because of recital 18 GDPR the usage of a digital assistant for purely private and household activities causes no GDPR data protection questions for its purely private users at home.

GDPR and at least Art. 8, Art. 10 and Art. 16 of the ePrivacy Regulation draft will be applicable on digital assistants for its controllers. In EDPB's opinion from March 2021 (adopted in July 2021) the "old" Art 5(3) ePrivacy directive in the version Art 2(5) Citizens' Rights directive 2009/136/EC should be applicable on digital assistants (as "IoT devices") too, that makes the situation more complex:

The collection of voice commands (voice data) and as well as the collection of information from terminal equipment (e.g. "IoT devices") can be justified with "informed consent" and "contract fulfilment" (e.g. registered subscribers or users). This conclusion results out of an analysis of Art 8 ePrivacy Regulation draft and Art 5(3) ePrivacy directive in conjunction with national criminal law (§ 201 StGB Germany/§ 120 StGB Austria) and Art 6 Par 1 lit a and lit b GDPR. The legal base Art 6 Par 1 lit f GDPR alone cannot justify—the collection of voice data (§ 201 StGB Germany/§ 120 StGB Austria: at least "implied consent" necessary = GDPR "consent" or "contract approval").

Out of a general position it can be finally said: If the access to information stored in the terminal equipment does not contain any voice data, this conclusion above could be seen differently. The access and processing of non-voice data as information stored in the terminal equipment, could be additionally possible in accordance with Art. 5(3) ePrivacy directive 2002/58/EC (Art 8 ePrivacy Regulation draft in conjunction with Art 6 Par 1 lit f GDPR, if this access and processing is only strictly necessary for the sole purpose to execute non-verbal requests from non-registered

[96]COM(2017) 10 final.

subscribers or users (= no informed consent and no contract available) as a service explicitly requested by (non-registered) subscribers or users.

The legal base "legitimate interest" (Art. 6 Par. 1 lit. f GDPR) can justify the profiling process and other processing activities for transparent purposes that take place in the cloud backend systems.

According with the two ECJ "Schrems"-decisions the transfer of personal data to US cloud providers causes legal, technical and business model questions in lack of comparable European IT-companies. In the opinion of the EDPB sufficient encryption can bring additional technical and organizational measures to justify data transfers under SCCs or BCRs (Art 46, Art 47 GDPR).

References

ai.sri.com. (July 14th 2020 - web.archive.org/), Cognitive Assistant that Learns and Organizes: http://web.archive.org/web/20200714113445/http://www.ai.sri.com/project/CALO (last visit August 31st 2021).

Apostel. (2013). Personensuchmaschinen, pp. 169 ff.

Austrian Data Protection Authority 03.05.2021, 2021-0.285.169.

Bosker. (2013). In huffingtonpost.com (January 22nd 2013 - Updated December 06th 2017), SIRI RISING: The Inside Story Of Siri's Origins -- And Why She Could Overshadow The iPhone: https://www.huffingtonpost.com/2013/01/22/siri-do-engine-apple-iphone_n_2499165.html (last visit August 31st 2021).

Buchner. (2020). In Kühling & Buchner (Eds.), DSGVO BDSG[3]. Art 22.

Conrad. (2017). Künstliche Intelligenz — Die Risiken für den Datenschutz, DuD 12/2017, pp. 740 ff.

Council of Europe. (2016). Brussels April 14th 2016, Dok 7920/16: DE: http://data.consilium.europa.eu/doc/document/ST-7920-2016-INIT/de/pdf and EN: http://data.consilium.europa.eu/doc/document/ST-7920-2016-INIT/en/pdf (last visit August 31st 2021).

Damaschke. (2016). Siri Handbuch.

Dammann, & Simitis. (1997). EG-Datenschutzrichtlinie.

DARPA. (2011). In darpa.mil (August 5th 2011 - web.archive.org/) Overview Personalized Assistant that Learns (PAL) Project: https://web.archive.org/web/20110805162949/http://www.darpa.mil/Our_Work/I2O/Programs/Personalized_Assistant_that_Learns_%28PAL%29.aspx (last visit August 31st 2021).

Donnelly. (2017). itgovernance.eu (July 5th 2017), How ISO 27001 can help with your GDPR project: https://www.itgovernance.eu/blog/en/how-iso-27001-can-help-with-your-gdpr-project/ (last visit August 31st 2021);

Emmer, & Dirscherl. (2017). In pcwelt.de (June 1st 2017), Alexa Smart Home: Die besten Geräte für Amazon Echo: www.pcwelt.de/a/alexa-smart-home-die-besten-geraete-fuer-amazon-echo,3391001 (last visit August 31st 2021).

European Data Protection Board - EDPB. (November 10th 2020), Recommendations 01/2020 on measures that supplement transfer tools to ensure compliance with the EU level of protection of personal data, pp. 26 f.

European Data Protection Board - EDPB. (2021). Guidelines 02/2021 on Virtual Voice Assistants, pp. 12 ff.

Feiler, & Forgó. (2017). EU-DSGVO.

Fercher, & Riedl. (2016). DSGVO: Entstehungsgeschichte und Problemstellungen aus österreichischer Sicht. In Knyrim (Ed.), Datenschutz-Grundverordnung, pp. 18 ff.

Geminn. (2016). Das Smart Home als Herausforderung für das Datenschutzrecht, DuD 9/2016, pp. 575 ff.
Golson. (2011). In macrumors.com (October 4th 2011), Siri Voice Recognition Arrives On the iPhone 4S: https://www.macrumors.com/2011/10/04/siri-voice-recognition-arrives-on-the-iphone-4s/ (last visit August 31st 2021).
Gray. (2016). In interactions.com (June 1st 2016), The Rise of Intelligent Virtual Assistants: https://www.interactions.com/rise-intelligent-virtual-assistants/ (last visit August 31st 2021).
Herbst. (2020). In Kühling & Buchner (Eds.), DS-GVO BDSG[3]. Art. 5.
Heuchemer. (2019). In v. Heintschel-Heinegg (Ed.), BeckOK StGB[42. Edition] (as of 01.05.2019) § 201.
Hoy, A., Siri, C., et al. (2018). An introduction to voice assistants. *Siri in Medical Reference Services Quarterly, 37*(1), 81–88. https://www.researchgate.net/publication/322456429_Alexa_Siri_Cortana_and_More_An_Introduction_to_Voice_Assistants (last visit August 31st 2021).
ISO/IEC 27002:2013 Information technology — Security techniques: http://www.iso27001security.com/html/27002.html (last visit August 31st 2021).
Jahnel. (2010). Handbuch Datenschutzrecht.
Jaksch. (2017). Die Bestellungspflichten eines Datenschutzbeauftragten gemäß DSGVO, ZIIR 2017/2, pp. 140 ff.
Jaksch. (2019). Der Grundsatz der Zweckbindung und Zweckvereinbarkeit im Rahmen von Weiterverarbeitungen personenbezogener Daten, *Jahnel* (Hrsg) Jahrbuch Datenschutzrecht 2019, pp. 141 ff.
Kamarinou, Millard, & Singh. (2016). Machine Learning with Personal Data, 8 ff: https://www.researchgate.net/publication/316595297_Machine_learning_with_personal_data_is_data_protection_law_smart_enough_to_meet_the_challenge (last visit August 31st 2021).
Kamp. (2011). Personenbewertungsportal, pp. 20 ff.
Kargl. (2017). In Kindhäuser, Neumann, & Paeffgen (Eds.), Strafgesetzbuch[5]. § 201.
Keßler. (2016). In futurezone.at (March 31st 2016), "Legale WhatsApp-Verwendung ist praktisch unmöglich": https://futurezone.at/netzpolitik/legale-whatsapp-verwendung-ist-praktisch-unmoeglich/189.971.340 (last visit August 31st 2021).
Kling. (2017). In zdnet.de (October 11th 2017), Datenschutz-Panne: Google Home Mini hört dauernd mit: http://www.zdnet.de/88315221/datenschutz-panne-google-home-mini-hoert-dauernd-mit/?inf_by=5a390395681db884538b494e (last visit August 31st 2021).
Kühl. (2016). In zeit.de (October 28th 2016), "Alexa. Ich bin dein Vater": http://www.zeit.de/digital/mobil/2016-10/amazon-echo-alexa-test-deutschland (last visit August 31st 2021).
Kühling, & Raab. (2020). In Kühling & Buchner (Eds.), DS-GVO BDSG[3]. Art. 2.
Lacoma. (2017). In digitaltrends.com (September 30th 2017), What is Google Assistant? Here's the guide you need to get started: https://www.digitaltrends.com/mobile/what-is-google-assistant/ (last visit August 31st 2021).
Lewisch, & Reindl-Krauskopf. (2017). In Höpfl & Ratz (Eds.), WK StGB[2] (as of 17.10.2017) § 120.
Martini. (2018). In Paal & Pauly (Eds.), DS-GVO BDSG[2]. Art 22.
Mazzucato. (2014). Das Kapitel des Staates – Eine andere Geschichte von Innovation und Wachstum, pp. 136 ff.
Müller-Peltzer, & Franck. (2017). Grüß Bot! Aktuelle Rechtsfragen zum Einsatz von Chatbots. In Taeger (Ed.), Recht 4.0, pp. 241 ff.
Naone. (2007). In technologyreview.com (November 30th 2007), Intelligent Machines Software That Learns from Users, https://www.technologyreview.com/s/409120/software-that-learns-from-users/ (last visit August 31st 2021).
Nusca. (2013). In zdnet.com (January 23rd 2013), The 'largest AI program in history' led to Apple's Siri: http://www.zdnet.com/article/the-largest-ai-program-in-history-led-to-apples-siri/ (last visit August 31st 2021).

O'Boyle. (2020). In pocket-lint.com (November 18th 2020), What is Alexa and what can Amazon Echo do? https://www.pocket-lint.com/smart-home/news/amazon/138846-what-is-alexa-how-does-it-work-and-what-can-amazons-alexa-do (last visit August 31st 2021).
pal.sri.com. (October 21st 2020 - web.archive.org/), Overview The PAL Framework: http://web.archive.org/web/20201021151932/https://pal.sri.com/ (last visit August 31st 2021).
pcmag.com. Encyclopedia definition of virtual assistant: https://www.pcmag.com/encyclopedia/term/61228/virtual-assistant (last visit August 31st 2021).
Roush. (2010). In xconomy.com (June 14th 2010), The Story of Siri, from Birth at SRI to Acquisition by Apple—Virtual Personal Assistants Go Mobile: https://www.xconomy.com/san-francisco/2010/06/14/the-story-of-siri-from-birth-at-sri-to-acquisition-by-apple-virtual-personal-assistants-go-mobile/?single_page=true (last visit August 31st 2021).
Schillerhof. (2017). Amazon Echo, p. 26 (translated by the author).
Schmidt, Gerdt, & Skoruppa. (2017). In computerbild.de (May 17th 2017), Google Assistant: Neue Funktionen ... aber vorerst nur in den USA!: http://www.computerbild.de/artikel/cb-News-App-Check-Google-Now-Google-Assistant-16367107.html (last visit August 31st 2021).
Scholz. (2014). In Simitis (Ed.), Bundesdatenschutzgesetz[8]. § 6a BDSG 2003.
Schonfeld. (2010). In echcrunch.com (April 28th 2010), Silicon Valley Buzz: Apple Paid More Than $200 Million For Siri To Get Into Mobile Search: https://techcrunch.com/2010/04/28/apple-siri-200-million/ (last visit August 31st 2021).
Schulz. (2018). In Gola (Ed.), DS-GVO[2]. Art 22.
Solmecke. (2017). In wbs-law.de (June 26th 2017), Urteil: Weitergabe der WhatsApp-Kontaktdaten ist illegal: https://www.wbs-law.de/it-und-internet-recht/datenschutzrecht/urteil-weitergabe-der-whatsapp-kontaktdaten-ist-illegal-droht-eine-abmahnwelle-22712/ (last visit August 31st 2021).
SRI International. (2020). In medium.com/ (July 30th 2020), 75 Years of Innovation: CALO (Cognitive Assistant that Learns and Organizes) A research project that developed an enduring personalized cognitive assistant: https://medium.com/dish/75-years-of-innovation-calo-cognitive-assistant-that-learns-and-organizes-9f176f2de05d (last visit August 31st 2021).
Taeger. (2017). Verbot des Profiling nach Art. 22 DS-GVO und die Regulierung des Scoring ab Mai 2018, RDV 2017/1, pp. 3 ff.
techtarget.com. Definition virtual assistant (AI assistant): http://searchcrm.techtarget.com/definition/virtual-assistant (last visit August 31st 2021).
Tipold. (2017). In Leukauf & Steininger (Eds.), StGB[4]. § 120.
Trillo. (2013). In giga.de (January 23rd 2013), Siri: Ein Militär-Projekt, das fast bei Android gelandet wäre, https://www.giga.de/apps/siri/news/siri-ein-militar-projekt-das-fast-bei-android-gelandet-ware/ (last visit August 31st 2021).
Verbraucherzentrale NRW. (December 20th 2017), Digitaler Sprachassistent: Alexa reagiert auch ungefragt: https://www.verbraucherzentrale.de/aktuelle-meldungen/digitale-welt/digitaler-sprachassistent-alexa-reagiert-auch-ungefragt-21363 (last visit August 31st 2021).
von Lewinski. (2018). In Wolff & Brink (Eds.), BeckOK Datenschutzrecht[23. Edition] (Stand: 01.02.2018) § 6a.
von Lewinski. (2021). In Wolff & Brink (Eds.), BeckOK Datenschutzrecht[35. Edition] (Stand: 01.02.2021) Art 22.
Warren. (2014). In theverge.com (April 2nd 2014), The story of Cortana, Microsoft's Siri killer: https://www.theverge.com/2014/4/2/5570866/cortana-windows-phone-8-1-digital-assistant (last visit August 31st 2021).
Weichert. (2020). In Kühling & Buchner (Eds.), DS-GVO BDSG[3]. Art 4 No 14.
Wilkens. (2017). heise.de (October 12th 2017), Datenschutzpanne mit Google Home Mini: Einschaltknopf wird deaktiviert: https://www.heise.de/newsticker/meldung/Datenschutzpanne-mit-Google-Home-Mini-Einschaltknopf-wird-deaktiviert-3858861.html (last visit August 31st 2021).

GDPR Compliance for East European Non-EU Companies

Oleksandra Malichenko

Contents

Abstract GDPR compliance for non-EU companies face a significant legislation and implementation gaps. One of the most common myths about GDPR that non-EU companies can easily avoid the implementation of the regulation and penalties. But GDPR is built in such a way that not only to force a non-EU company to fully comply with GDPR, but also force the company from the EU territory to carefully select its counterparties for the processing of personal data from outside the territory of the European Union. This study aims to determine how GDPR compliance for East European non-EU companies can be better targeted in order to increase privacy of personal data floating out of European Union. Specifically, it investigates on example of Ukraine, as non-EU country and other neighboring countries whether the fundamental privacy rights should be granted a higher level of protection and non-compliance with the requirements of the GDPR can have a negative impact for non-EU organizations and companies for non-compliance.

Eastern Europe is one of the most popular directions for IT business and IT startups. The conclusions are based on the data provided by European Startup Monitor.[1] Most start-up's outside the European Union (EU) are confused about the EU General Data Protection Regulation (GDPR).[2] One of the most common myths about GDPR, is that non-EU companies can easily avoid the implementation of the regulation and penalties.

[1] http://europeanstartupmonitor.com.

[2] EU Regulation (EU) 2016/679, 2016 (hereinafter GDPR in footnote and GDPR or "the Regulation" in the text).

O. Malichenko (✉)
Kyiv, Ukraine

G. Borges, C. Sorge (eds.), *Law and Technology in a Global Digital Society*,
https://doi.org/10.1007/978-3-030-90513-2_9

The reaction about GDPR activated a highly public and ongoing debates, both from domestic privacy specialists and from international organizations concerned about the privacy of personal data floating out of European Union.

Nonetheless, the Regulation is built in such a way that not only to force a non-EU company to fully comply with GDPR, but also force the company from the EU territory to carefully select its counterparties for the processing of personal data from outside the territory of the European Union. To make this connection explicit, companies trying to increase this choice how personal information is used and visibility of this information to mitigate some risks or run business in most apart way from GDPR. The trend is started to go exponentially because of GDPR, that's why non-EU companies also started to be actively involved in better interactions with consumers and users. Although, from the non-EU customer point of view arises a tricky question, how to control the data that was collected from you by some EU and non-EU companies. Acquisti and Brandimarte et al. examined request for data privacy settings on social networks and access of private information and how easily sensitive information is collected.[3] Thus, it matters that the data protection policies have been broadened and extended over time.

Generally, GDPR established how personal data of EU citizens should be collected, processed and stored. Any company that operates within EU borders as well as by those companies that provide services for EU citizens and access their personal data in the process shall apply an internal route map to the GDPR requirements as soon as possible. EU citizens now have more rights and policies in place to ensure their data is kept securely. If an organization that stores data does not have the correct processes in place to manage it in a secure manner, then citizens have the lawful right to request that their information is securely deleted.

The fact of extending the extraterritorial application entails the obligation to appoint a representative on the territory of the EU, which will monitor compliance with the obligations of the non-EU companies. The regulation contains a clear paragraph that states that the representative should be subject to enforcement procedures. Companies are required to appoint a representative who will be responsible for their breach of personal data.

Since many transnational corporations' process HR data at the cluster level, where a non-EU subsidiary is often the center of such clusters in Eastern Europe, along with EU member states (such as Poland, Slovakia, Hungary, Czech Republic, etc.). The importance of meeting GDPR requirements in the process of transferring HR data to non-EU company in the case of violation the process, only the exporter will be liable for noncompliance as the European personal data owner and as a direct employer of EU residents, whose data will be processed in non-EU company in Eastern Europe.[4]

Ukraine has been a member of the Council of Europe since 1995, thus Ukrainian national legislation takes into consideration the developments and standards

---

[3] Brandimarte et al. (2012).

[4] Tarasyk (2018), p. 28.

developed by the council. Specifically, in 1997, ratified legal the European Convention on Human Rights. The Convention 108 was ratified in 2010. Following the ratification, Ukraine adopted the Law of Ukraine On Personal data Protection, that was mostly supported the provisions of Convention 108 with the EU's data Protection Directive of 1995 and mirrors most of their provisions. Indeed, Ukraine is not a part of EU, but is a country in Eastern Europe very close to European Union to not take into account GDPR features in the processing of HR data (data on employees, their background and monitoring of work activity). Though, the issue of cross-border data transmission (in precise, HR data), which shall regulate with the GDPR during the export of personal data of EU resident workers to their processing in Ukraine, some implications of EU practices remain unclear.

But, bringing the Ukrainian legislation in line with the Regulation is part of the implementation of the National Human Rights Strategy for the period up to 2020[5] and shows the importance of the European document for the protection of human rights in Ukraine. Ukraine would need to establish a solid legal framework covering data protection issues within the GDPR extent and local peculiarities, represent efficient enforcement within the real case-law.

In this way of implication of GDPR any company outside the EU should be aware of extraterritorial GDPR applicability, since it applies to companies anywhere in the world if they come into contact with EU citizens data.

GDPR established that even non-EU companies if they are data operators and/or processors of personal data of the customers of personal data from the EU or whose activities are related to:

- the provision of goods or services to customers in the EU, regardless of whether payment of the goods and services offered or not (for example, using the EU website in the language of the EU member country or the currency of the EU member country to provide goods or services);
- monitoring the behavior of customers within the EU (profiling them, for example, to make a decision regarding them or to analyze, predict their personal preferences, behavior and attitudes).

With respect to organizations located outside the EU territory and processing data from Europeans, the indirect influence of legislation implicated through their business partners in the EU, who will have to take into account the risks of cooperation with organizations located outside the European Union, taking into account possible GDPR fines. Consequently, it seems that most authors agree that fundamental privacy rights should be granted a higher level of protection and non-compliance with the requirements of the GDPR can lead to the negative consequences for non-EU organizations and companies.[6]

[5]Order of President of Ukraine (2015).

[6]Goddard (2017), p. 704.

Data Protection Regulation does not explicitly comment nor require companies to conduct a dataflow analysis. But it is stated to "maintain a record" of relevant "processing activities". In the beginning of analyzing a data source and complete map of the data moving through entire infrastructure and highlighting all data flows between all systems, shall be defined with a complex project and possible deadlines. Otherwise, this flow analysis can turn as never-ending project process.

After basic analysis, the company will know what personal data is collected, why they need to collect and use it, and where it is being processed and stored. In order to overcome these shortcomings, Laybats et al. and Davies et al. (2018) through an interview with John Davies, Managing Director of digital agency Reading Room they propose that company shall focus on implementing regulations, considering the effect on organizations coming under its jurisdiction, and then goes on to consider the implications for organizations currently out of the geographical area the GDPR controls.[7]

The second aspect for a company to take actions on personal data, it should specify the aim of collecting data and gain individuals' consent to do so, like agenda of processing undertakings. The processing of consent must be freely given, specific, informed and unambiguous. The consent request must be delivered before gathering of any personal data—basically, during the first visit to a non-EU company or organization website. Before obtaining consent, non-EU company should inform individuals (customers) of forthcoming collecting, processing and storing personal data. This argument is even more understandable with regard to the second paragraph of Article 8 of the Charter of fundamental rights of the European Union[8] specifying that personal data must be processed fairly for specified purpose.

The most important change is that GDPR requires the following guidelines for the consent received from the data subject:

- the controller shall be able to demonstrate that the data subject has agreed to the processing of the data;
- the consent shall be presented in a method which is clearly distinguishable from the other matters (for example, the consent might be included in the employment contract or in an addendum to the employment contract);
- the controller shall stipulate the right of the data subject to withdraw consent at any time and such withdrawal should be received an easy-going way.

Non-EU countries legislation also contains detailed clauses related to the obligatory consent obtaining, namely:

(1) Law of Ukraine "On Protection of personal data" dated from June, 6th, 2010, article 10: "the use of personal data includes any actions by the owner concerning the processing of these data, actions for their protection, as well as actions to grant partial or complete right to process personal data to other

[7]Laybats and Davies (2018), p. 82.

[8]https://eur-lex.europa.eu/LexUriServ/LexUriServ.do?uri=OJ:C:2010:083:0389:0403:en:PDF.

subjects of the relationship related to the personal data carried out with the consent".[9]

(2) Law Republic of Moldova "On personal data protection" dated from July, 8th, 2011, article 3: "consent of the subject of personal data - any expression of will, free, specific and unconditional, given in written or electronic form - in compliance with the requirements for electronic documents - the form by which the subject of personal data agrees to processing of his personal data".[10]

(3) Federal Law of the Russian Federation dated from July 27, 2006, "On Personal Data", article 9: "The subject of personal data decides on the provision of his/her personal data and gives consent to their processing freely, by his/her own will and interest. Consent to the processing of personal data must be specific, informed and conscious. Consent to the processing of personal data may be given by the subject of personal data or his/her representative in any form allowing to confirm the fact of receipt, unless otherwise established by federal law. In case of obtaining consent for the processing of personal data from a representative of the subject of personal data, the authority of the representative to give consent on behalf of the subject of personal data is checked by the operator".[11]

Moreover, the user of the any service can ask for information with what kind of data operates the service, what happens to the data, why the company processes it and shared processors, how long this data is stored and what will happen next to them.

One of the most impressive provisions of the GDPR that that is implemented in laws of Ukraine, Moldova and Russian Federation, is the right of the user to withdraw his consent to data processing at any time. The rules require that it should be as easy to do as to give consent for the processing of information.

This right has received the name right to oblivion (right to erasure, right to be forgotten). This rule operated before, but only concerned search engines, from where EU residents could demand the removal of individual search results items.

Although this is quite remarkable logic, the information should be given in an easily understandable way and cover such main accents as:

- in what purposes personal data will be collected, processed and stored;
- what kind of data to be collected and processed further;
- for how long personal data will be stored and where;
- how individual can access personal data, discover who was using it, in what purpose and how to delete all personal data that was gathered;
- contact information regarding personal data protection, details and instructions how to contact.

[9] http://zakon2.rada.gov.ua/laws/show/2297-17.

[10] http://lex.justice.md/ru/340495/.

[11] http://www.consultant.ru/document/cons_doc_LAW_61801.

To sum this up, after conducted analysis of ongoing processes and checking basic requirements—data flow map shall be created, which can then be assessed using a process called gap analysis.

Decisions based on automatic processing is one of the important subjects that is not regulated by non-EU Eastern European countries legislation. Consequently, the right of the subject of personal data to prohibit the controller from accepting, for such a subject, decisions based exclusively on automatic processing, including profiling, if such decisions shall have legal consequences. As a possible example of such an automatic solution is the decision of a financial institution to grant or not to grant a loan to a person on-line, based on the data of the questionnaire, which was filled by it or other types of services, in which there is a so-called "scoring" or customer assessment as potential vendor with the help of automatic systems. Another example may be the online recruiting, which may include automatic decisions about the possibility of offering jobs or rejecting a particular subject of personal data. EU data protection becomes progressively more important for third-country controllers.[12]

Although shall European individuals care what types of personal data non-EU Eastern European companies managing and how they protect? After our general basic analyses non-EU Eastern European companies' websites, we may sum up that they typically hold a vast amount of personal data, from employee data to that of suppliers and consumers from EU, all of which must be protected.

This includes data held in consumer/management system, customers database, employee HR files such as addresses (including email addresses), banking/payment card data, dates of birth, records/screening forms, questionnaires, consent forms, consumer contact records, supplier personnel data and some others. Furthermore, individuals should be protected by the right to restrict data processing, the right to object or curtail the collection of certain types of data and the right to data portability.

To start with, a result of full scope data flow analysis of the company shall be Data Protection Impact Assessment (DPIA) with detailed description of data processing purposes, assessment of the need and adequacy of processing, a list of persons who have access to data, requirements for data protection, identification and analysis of risks, processes that ensure the protection of personal data, future measures to protect information.

For instance, if some non-EU IT company operates for customers in EU, they may provide service for booking tickets on concerts and theaters. Thus, their services may involve partners from all Europe. Therefore, this company got acquainted with GDPR and should choose: to leave the European market or start preparing for the GDPR or leave everything as it is.

Consequently, it seems that maybe leaving the European market or waiting for a company to be fined is not the best choice, because it was decided to prepare for GDPR. Company shall start to work on road map of their personal data flow. The "road map" shall indicate where the personal data comes from (user account,

[12] Brkan (2016), p. 817.

registration form on the site, registration through a mobile application, etc.). So, if our IT tickets booking company collects personal data on its website while registering users in the system. We receive the following information: name and surname, contact details, payment details, visitor names, concert and event dates. The company also collects information from the user's computer: IP address, browser type, language, location. If the user has his own account, then the reservation history, maybe photos, reviews are stored in addition.

Further, after registering, we see that personal data is stored on the servers of the company, access to personal data is available to all employees of the company.

Personal data may be transferred to third parties in several cases:

- data is transferred to partner companies where event is booked;
- data is transferred to the payment service;
- the data is transferred to third parties for marketing purposes.

Let's assume, that consent to the processing of personal data to the company were given by individuals, but a mark of such consent is affixed by default, that is obvious violation of the GDPR. At the same time, the privacy policy of the IT tickets booking company did not comply with the GDPR: is was very long and confusing, according to the policy, data could be transferred to third parties without restrictions.

There are also other factors that will indicate that our IT tickets booking company identified the following high risks, as access to personal data have all employees of the company; third parties to whom they transfer personal data for marketing purposes do not protect this data at the proper level; consent to the processing of personal data is given by default and that privacy policy did not fully comply with GDPR.

Furthermore, after the risks are identified and assessed by the company, a set of measures shall be made to protect the data. Organizational measures shall be takes, as writing a new privacy policy for employees, drawing up instructions, training employees, appointing responsible persons for working with personal data and much more, entering contracts with third parties with data protection obligations. Not only the organizational scope, but also the provisions in the instruments have extended quite significantly to technical measures. Technical measures include encryption, storing data in protected locations, creating backup copies, etc.[13]

Likewise, for a company is worth observing own employees' attitude to the data flow, while they process data, using questionnaires is better to check their opinions and how in-depth their current knowledge is, as well as arranging meetings to discuss the data mapping process with departments.

Meanwhile, facilitation workshops may be useful for doing this—creating freeform diagrams on the whiteboard or mind maps and other methods, to get the relevant information down. This will allow a company to identify and analyze gaps, identify any risks and in turn allow to focus on the most crucial areas needed to improve.

[13] Laybats and Davies (2018), p. 82.

Many East European non-EU companies think that they have to take very sophisticated actions in an order to comply with GDPR. Due to non-EU countries shared security priorities, area countries will remain interested in cooperating with the European Union and each other on issues such as defense and technology diversification.

Regardless of whether a company is based in the EU or not, the GDPR hold the main principle of extraterritoriality, so if the personal data of EU residents is collected by the company that is out of EU, these personal data are anyway a subject to the new GDPR. For companies and organizations in EU or not in EU, adopting a GDPR primarily means money spending.

Indeed, each organization that conducts activities (or potentially can conduct it) with EU citizens must optimize the processes of collecting, processing and storing personal data and designate the employees responsible for these processes. This includes data protection, as the regulations clearly state that companies must avoid any kind of leaks and report on hacking or theft of user bases as soon as possible.[14] So, cyber security costs will increase.

In non-EU countries, the conformity of the GDPR is worth thinking of online stores operating in Europe and receiving revenues in euros. Online stores and websites require personal data from customers and visitors for analysis, content marketing, and effective lead generation. If a company does not comply with the GDPR, it does not have the right to work in the EU and with EU citizens.

If the non-EU company business involves interaction with the EU counterparty, then it should be audited for compliance with the standards of the GDPR. And if there is no match with GDPR requirements, better to refuse to cooperate with such a company. This also works in the opposite direction when non-EU company act as a counterparty: companies may refuse from services due to non-compliance with the GDPR.

Taking into account that it is quite difficult to unequivocally divide users into residents of the EU and all the rest, most of them will have to adapt to the requirements of the GDPR, first of all, in order to avoid huge fines. It is quite natural that many users who had never even thought about how much and what kind of information they left on Internet resources will now simply refuse to provide it if they do not plan to make a purchase right now.

Accordingly, over time, the data in the user database will become obsolete and deleted, since it can be stored for a fairly limited time. And this reduces the effectiveness of targeted advertising and increases the cost of contacts and advertising campaigns. In addition, for small sites that previously could have used photos copied from the Network without the express permission of the persons depicted on them, there is either the risk of getting a fine, or additional costs for purchasing such materials on commercial sites with stock photos.

Very likely that a customer from the EU is going to tell a non-EU company what to do if he wants to ensure that his personal data is properly processed.

[14] Dewar (2017), p. 96.

The most convenient way to ensure that data is properly processed by your partner non-EU company, is a standard data processing agreement (DPA) listing out the requirements for technical and organizational measures that a processor from non-EU company must implement.

There are other ways to ensure data transfer protection. For instance, a controller may require a processor to complete certification or work out certain corporate rules that have to be approved by EU regulatory agencies.

To sum up, first of all, the company shall assess and classify the information they collect and process and form a methodology of processing.

Secondly, to evaluate internal processes of information flow and conduct an audit of the processing of data in the organization or company for compliance with the requirements of the GDPR and regulations of the countries where the organization's offices are located in the field of data protection.

Thirdly, to develop an action plan and a roadmap for achieving compliance with the GDPR and determine the necessary organizational measures, processes and controls in terms of data protection (confidentiality bases).

Also, to implement processes and organizational and administrative documentation. For example, non-EU organizations and companies are required to ensure the integrity and availability of the data environment, as a part of the compliance process.

For instance, the Code of conduct inside of the company might be a central document with which all employees and contractors must be acquainted and defines actions and prohibitions that all users must follow. The policy shall provide guidelines for managers, developers, legal personnel, and any other employees within the companies to use as a reference for best data privacy practices within the corporate infrastructure. Also, the policy requirements and restrictions shall apply to network infrastructures, encryption, hardcopy reports, digital or physical media, wireless networks, databases and any other methods used to convey knowledge and ideas across the interaction with personal data of customers. Company shall not forget about instruments to conduct continuous monitoring of the functioning of the implemented data protection controls, as privacy impact assessment (PIA) and data mapping.

Because not full understanding of the internal processes of the data flow will make difficult to ensure that companies data processing activities shall comply with the new obligations set out in the GDPR. GDPR also requires an on-going monitoring mechanism to ensure alignment with the internal documents of the companies or organizations.[15]

As has been stressed, organizations and companies in non-EU area have the unique opportunity to interpret and concretize GDPR for their specific sector and needs. This does not only trigger a practical implementation of Europe's new data privacy rules overall but also enhances trust and strengthens the protection of

[15] Goddard (2017), p. 705.

personal data while safeguarding innovation-friendly solutions for non-EU businesses and organizations.

Likewise, EU-oriented East European companies shall establish job positions of a data protection officers to work closely with EU authorities, because even these functions are usually under responsibilities of legal councils or compliancy officers and there is no separated position for such responsibilities. In any case, such an adaptation cannot be avoided on the way to coherent European integration.

Considering the latest private data leakage case in Ukraine with company Nova Poshta LLC[16] (a courier company based all over Ukraine, EU and US) on leakage its customers' personal data, as passport data and payment details due to simple human factor, is another proof that the GDPR is going to become more critical for East European Area.

Though it remains to be seen exactly how the GDPR updates will affect a number of non-EU companies in Eastern Europe, however—this is a step in the right direction in the bid to protect users and the huge grasp of data sharing.

Thus, natural and legal persons of non-EU countries, especially in geographically close Eastern Europe that oriented on the market of the European Union Member States should analyze in detail whether they are subject to the provisions of the Regulation and, if so, to adapt their policies and processes of processing and collect personal data in accordance with GDPR.

And if today it is difficult to establish and investigate exactly how they (natural and legal persons of Ukraine) can be held liable under the Regulations, this does not mean that it is worth ignoring its provisions, since in any case, the negative consequences will come, as assumed by Tarasyk.[17] Even if it is not a fine, provided by Art. 83 of the Regulation, the option, as a consequence, would be the inability to open a bank account in the EU, the weakening of positions in the EU market, the cancellation of value contracts with EU entities (since the latter, after May 25, 2018, already elect contractors without violations in the field protection of personal data).[18]

The question arises: if natural and legal persons of non-EU countries may fall under the extraterritorial rule of the Regulation, can such an offender be prosecuted and sanctioned under the Regulations? Obviously, as of now, there is no clear mechanism for applying fines in non-EU countries according to EU legislation. And for now, it is not entirely clear how the perpetrators, physical and legal persons out of EU can be brought to the above liability, this argument is also supported by O'Brien et al (2016).[19] In the context of this, it will mean the termination of the contract between the two companies: the company that administers this site will not be able to administer it, because it will not have access to personal data. A company that develops software will lose sight of cooperation with a company in the EU, because it will not be able to properly program, test, and administer the software.

---

[16] Nova Poshta LLC (2018).

[17] Tarasyk (2018), p. 29.

[18] https://eur-lex.europa.eu/legal-content/EN/TXT/?uri=celex%3A32016R0679.

[19] O'Brien (2016), p. 82.

Thus, the company may care only if they are going to the European market. If the Ukrainian company is going to cooperate with a company from the EU, then it is bound to adhere to these provisions. So, if some Ukrainian company violates the provisions of the regulation and waives its liability, the question will arise how to penalize such company and stop the process of leakage.

The minimum of outcomes that may be that a company will ignore request from a supervisory authority or court, the supervisor will instruct a company located in the territory of the EU to suspend a flow or transfer of data to Ukraine.

To be clear, fines for companies failing to comply with the GDPR requirements are way higher than any average fines in non-EU countries. For now, this whole GDPR thing for non-EU companies is a good opportunity to expand expertise and show to everyone how they are perfectly able to ensure compliance. Failure to comply with GDPR can result in a fine ranging from 10 million Euro to 4% of the company's annual global turnover. Fines will depend on the severity of the breach and on whether the company is deemed to have taken compliance and regulations around security in a serious enough manner.

Consequently, the violations of GDPR provisions by controllers/operators of non-EU countries may lead to the following scenarios:

(1) if non-EU companies opening the account in EU bank for conducting business in the EU and if in the database there will be information on the violation of the GDPR regulation by the entity, then, most likely, the account will not be opened;
(2) the supervisory authority in EU can prohibit non-EU company operates in EU if will be leakage of personal EU citizens data or if data flow or transfer will be out of EU without monitoring this issue and informing EU citizens.

There are certain aspects that can be seen as fundamental and arguably, but more akin basic and priority steps of adaptation can be assumed, as follows:

(1) approval and development by the Regulation and EU standards by governments in non-EU countries an effective state policy on the protection of personal data and permanently control in this area;
(2) the introduction and consolidation of the institutions on personal data protection, which will provide supervision and control and improve the legal framework for the protection of personal data, as well as interaction with the competent authorities of the EU and countries—EU Member States for data protection;
(3) the assignment to the central executive authorities of functions aimed at the protection of personal data, the creation of special units that could act as part of the non-EU Ministry of Justice, for example;
(4) functions of organizing the provision of technical protection of personal data in non-EU countries shall be also assigned to state authority and conducting investigations concerning the protection of personal data;
(5) as regards the judiciary, it is possible to take into account the experience of the United Kingdom where the data protection court operates and set up specialized court for the protection of information (data), where it is possible to consider not

only the protection of personal data, but also violations of privacy, circulation of public information, media activities, etc.;

(6) establishment internal positions of the Data protection specialists in the non-EU organizations and companies, but also hiring on a contractual basis, as they shall stay independent in their functions.

1 Conclusion

Summarizing and analyzing the above, we can again emphasize the fact that the rapid pace of technological development allows individuals around the world use new information and services communication technologies to improve their lives. At the same time, technology strengthens the ability of governments, companies and individuals to supervise, interception and collection of data that may violate or abuse rights, in particular, the right to data privacy.

Principally, the main purpose of integration into the data privacy policy and preparing for a GDPR is a need for each company to work with European countries in the future. Sooner or later, all companies working with the European market will comply with the new rules for the protection of personal data. Non-EU companies need to start preparing now. Since the national means of influence are already inadequate, requires the development of security mechanisms at a more global level, therefore a breakthrough in this area is GDPR.

Furthermore, every Non-EU company shall successfully make all the steps to comply with the GDPR. After the entry into force of the GDPR, users have become very picky about the protection of personal data. Company shall retain all of its customers and purchase new ones as well since it prepared itself on time and managed to outrun the competition.

However, the broader cycle, is bringing non-EU company powered into compliance with GDPR requires a number of steps. Gaining individuals' consent on data gathering, processing, storing and, if necessary, delivering to third-parties. This consent should be taken after individual will receive all necessary information on the subject matter, stated in a simple and understandable manner.

Implementing technical measures that will ensure the safety of personal data as well the ability to delete personal data, trace it transfers, processing stages and any other data-related actions. Non-EU company specialists shall integrate GDPR-consent module on any pages of a website that gather personal data.

The third is involving the EU authorities into the audit of provision the adequate (required) level of personal data protection outside of EU. EU laws and the GDPR prohibits to transfer data outside the European Economic Area unless the recipient's jurisdiction is considered as safe as possible. The European Commission is responsible for listing out such jurisdictions. For now, the European commission okayed the level of personal data protection in Andorra, Argentina, Canada, Faroe Islands, Guernsey, Israel, Isle of Man, New Zealand, and Uruguay, but still no clear answer on non-EU East European countries. However, IT company from Ukraine or other

Eastern European country or simply non-European country that works with the personal data of European citizens should consequently comply with the rules of GDPR applying all possible measures.

Indeed, non-compliance with the requirements is a automatic rejection to provide relied and safe services to the European customers and may have a bad impact on future IT business of such non-EU company. For disclosure of personal data or leakage, penalties may be applied to such non-EU company and this is a second critical argument of losing reputation on the market. At the same time, legitimate interests are not clearly defined, which gives a lot of space for searching loopholes. To overcome these shortcomings, in the digital era, data protection legislation shall be defined as crucial for the protection of human rights, but, indeed, in many countries they are absent or practically absent.

References

Brandimarte, L., Acquisti, A., & Loewenstein, G. (2012). Misplaced confidences: Privacy and the control paradox. *Social Psychological and Personality Science, 4*(3), 340–347. https://doi.org/10.1177/1948550612455931

Brkan, M. (2016). The unstoppable expansion of the EU fundamental right to data protection: Little shop of horrors? *Maastricht Journal of European and Comparative Law, 23*(5), 812–841. https://doi.org/10.1177/1023263X1602300505

Dewar, K. (2017). The value exchange: Generating trust in the digital world. *Business Information Review, 34*(2), 96–100. https://doi.org/10.1177/0266382117711330

Goddard, M. (2017). The EU General Data Protection Regulation (GDPR): European regulation that has a global impact. *International Journal of Market Research, 59*(6), 703–705. https://doi.org/10.2501/IJMR-2017-050

Laybats, C., & Davies, J. (2018). GDPR: Implementing the regulations. *Business Information Review, 35*(2), 81–83. https://doi.org/10.1177/0266382118777808

Nova Poshta LLC. (2018). Official press release Nova Poshta LLC. Retrieved March 13, 2021, from https://novaposhta.ua/news/rubric/2/id/4813

O'Brien, R. (2016). Privacy and security: The new European data protection regulation and it's data breach notification requirements. *Business Information Review, 33*(2), 81–84. https://doi.org/10.1177/0266382116650297

Order of President of Ukraine. (2015). On approval National Human Rights Strategy, No. 501/2015, 25.08.2015. Retrieved March 13, 2021, from https://zakon.rada.gov.ua/laws/show/501/2015#n15

Tarasyk, A. (2018). The influence of General Data Protection Regulation on controllers and processors of personal data - Residents of Ukraine. *Information and Law Journal, 1*(24), 28–35. http://ippi.org.ua/sites/default/files/5_8.pdf

Part III
IT Security

IT Security Measures and Their Relation to Data Protection

Christoph Sorge

Contents

Abstract Data protection, which aims at protecting the right of individuals to informational self-determination, requires appropriate IT security measures. However, the relation between data protection and IT security is more complex: Protection goals from both fields are identical in some cases, but contradictory in others. Reactive security measures require the processing of vast amounts of personal information, but are often necessary to protect other (personal and non-personal) information. The article attempts to shed some light on this complex relationship.

C. Sorge (✉)
Saarland University, Saarbrücken, Germany
e-mail: christoph.sorge@uni-saarland.de

G. Borges, C. Sorge (eds.), *Law and Technology in a Global Digital Society*,
https://doi.org/10.1007/978-3-030-90513-2_10

1 Introduction

The importance of IT security was underestimated for a long time, though security incidents have been known to happen for decades. This has changed in recent years: Universities all over the world offer study programs in cybersecurity, and legislation such as the European Directive on security of network and information systems[1] indicates that threats to the security of critical infrastructures are now taken seriously on a political level as well. IT security is a very diverse field: Security is often, and correctly, associated with cryptography. However, only a small subset of security researchers are actually cryptographers. Security flaws can originate in hardware,[2] or be related to software design.[3] Social interaction[4] can be exploited by attackers, as can flaws in network protocols[5] and their implementations. Application requirements have an impact on the design of security measures, so a close collaboration between domain experts and security consultants (or security researchers) is often necessary to design secure systems. Within this chapter, we can therefore investigate only a small subset of the field of IT security.

When looking at the relation between IT security on the one hand, and data protection[6] (or data privacy) on the other, the two often share similar goals. Most obviously, ensuring the confidentiality of data is a common security goal, and can contribute to data protection as well.

However, securing IT systems may encompass the processing of personal data. In this case, there is a tension between data protection and IT security. In case the security measures serve the protection of other personal data, this tension is even inherent to data protection itself. To strike a balance between the data to be protected, and the data processing to that end, one needs to understand the necessity of these measures in more detail.

In most of the sub-fields of IT security mentioned above, there is not typically a need to process personal data to improve security. For example, hardware and software bugs can be fixed by only using design documents and source code—though discovering the bugs could be simplified by observing the systems' behaviour (thus, potentially, some of the user input) at runtime. In the article at hand, we focus on areas in which the impact of IT security measures and data protection is obvious and of significant practical relevance.

---

[1] Directive (EU) 2016/1148 of the European Parliament and of the Council of 6 July 2016 concerning measures for a high common level of security of network and information systems across the Union.

[2] For an example, see Lipp et al. (2020).

[3] Acar et al. (2017).

[4] Krombholz et al. (2015).

[5] For one out of many examples, see MacFarland (2017).

[6] In this article, the term data protection refers to all legal and technical measures aimed at protecting the right to informational self-determination.

The remainder of this article is structured as follows: At first, we provide an overview of some core concepts in IT security (Sect. 2). We then discuss some general aspects of the relation between data protection and IT security, in particular Article 32 of the EU's General Data Protection Regulation (GDPR) and the relevance of protection goals in data protection (Sect. 3). In Sect. 4, we investigate the relation between reactive security and data protection. The article is concluded in Sect. 5.

2 Basics of IT Security

IT security deals with the protection of IT systems against attacks. Related terms, such as computer security or information security, focus on particular aspects of security. In information security, as the term suggests, the focus is on security of information (which is usually, but not always, processed in IT systems—which implies a huge overlap between both fields). There is, however, no generally accepted and precise delineation between the terms.

Security is not equivalent to safety or reliability: A random failure of an IT system (e.g., due to a power outage) may have similar consequences to a security incident. Still, since it is not due to an attack, it is not considered as such. The differentiation makes sense due to the different requirements on protection measures. After decades of experience with IT infrastructures, the risk of random failures can usually be assessed with sufficient precision to decide about required mitigation measures. For example, the probability of a hardware failure, e.g. in a hard disk being used under certain conditions, is well-known; it can be mitigated by using appropriate strategies for redundant data storage. The risk of flooding caused by a nearby river is independent, and does not increase due to redundant data storage. Management can assess the potential damage caused by an incident, and define thresholds for acceptable occurrence probabilities for different types of incidents based on that assessment.

When considering security, the situation is different: If there are ten potential entry points for an attacker, of which nine are appropriately mitigated, the overall risk of a successful attack might not have decreased at all. In the field of IT security, we have to assume an intelligent attacker looking for the weakest link in a chain of measures.

This does not imply that there is no overlap between measures that ensure security or reliability, respectively. For example, a state-of-the-art backup strategy can prevent data loss, whether it is caused by flooding or by a malicious attack. However, securing IT systems against malicious attacks is much more difficult than protecting against random failure. A multitude of measures is typically necessary to prevent or mitigate attacks. Some of these measures involve the processing of personal data and are thus relevant to the article at hand.

2.1 Protection Goals

Similar to other engineering disciplines, a detailed analysis of requirements is necessary in order to design and maintain secure IT systems. The differentiation of protection goals is a core building block for the description of these requirements. Depending on the focus of an analysis, the protection goals may vary. There is, however, a broad consensus about the relevance of the following protection goals (with slight variations in the definitions found in different textbooks[7] and standards[8]):

- Confidentiality of information means that no unauthorised party is able to learn the information (or any part of the information).
- Authenticity of information means that the information actually originates from the claimed sender or author.
- Integrity of information means that the information has not been changed without authorisation. Integrity is closely linked to authenticity—but there may be cases in which the origin of information is irrelevant, while alteration of the information would be problematic. In the context of security, the definition refers to changes by an attacker, but when taking the perspective of reliability, integrity is just as relevant: Information may also be changed due to random outside influences, or technical failure.
- Non-repudiation means that the sender or author of a message cannot plausibly deny to have created/sent that message. This goal differs from integrity: The latter is achieved if the receiver of a message can be certain that it has not been altered. Non-repudiation, on the other hand, is only achieved if integrity can be proven to a third party. The distinction is not purely academic. Many security protocols achieve integrity using symmetric cryptography, in which both parties share the same cryptographic key. The receiver therefore knows whether a message has been sent by the expected sender; any third party can only verify that the message has been sent by either of the two parties (and only if the key is revealed to that party, which may be problematic in itself).
- Availability of an IT system means that provision of its service to authorised users cannot be prevented by an attacker. This definition can be generalized to include randomly occurring events instead of just malicious attackers.

Note that the definitions given here are suitable for a rough understanding of security requirements. When developing secure systems, a lot more detail may be required. For example, does confidentiality of information include confidentiality of metadata (e.g., the times at which different authorised users accessed the information, which may be as relevant to an attacker as the content itself)? Which attacker model is used when analysing a system's security—in other words, what capabilities

[7] See Andress (2011), pp. 4–9; Anderson (2020), p. 32; Song et al. (2017), pp. 2–3.

[8] In particular, see ISO (2018), pp. 1–6, and Rescorla and Korver (2003) (RFC 3552), sections 2.1 and 2.2.

is the attacker assumed to have? In cryptography, formal (mathematical) definitions are used to describe the security of an algorithm. However, the complexity of computer hardware, software, and the environment they are used in does not allow rigorous security proofs for IT systems as a whole. The use of relatively vague definitions, such as the ones given above, is therefore common.

Moreover, IT security measures usually come with a cost—if not monetary, then in terms of added complexity, reduced usability, or additional effort to be invested by the users. For this reason, it is usually not possible to achieve the highest level of security that would theoretically be conceivable. Instead, the implementation of security measures is usually based on a risk assessment.

Some of the definitions above refer to authorised parties. Cryptographic definitions usually assume that anyone with the respective cryptographic key is authorised. In practice, managing these authorisations can be a challenging task in itself. In large organisations in particular, responsibilities change frequently; new employees are hired, while others quit their jobs or retire. Managing authorisations is an interdisciplinary task that needs expertise both from a data protection and an IT security perspective, and is closely linked to the organizational structure of a company.

3 On the Relation Between IT Security and Data Protection

Data protection legislation aims at protecting individuals against violations of their privacy by processing of their personal data. This implies a close connection to the IT security protection goals listed above, e.g. concerning the confidentiality of personal data. However, data protection is not the only motivation for IT security measures, and IT security measures are neither always necessary nor always sufficient to fulfil the requirements of data protection.

The close relation between IT security and data protection is reflected in the GDPR. Article 5 (f) requires data processing procedures to ensure "appropriate security of the personal data, including protection against unauthorised or unlawful processing", along with measures to ensure reliability.

3.1 Article 32 GDPR

This principle is concretised in Article 32, which requires controllers and processors of personal data to implement "appropriate technical and organisational measures to ensure a level of security appropriate to the risk" (Art. 32 (1) GDPR). As noted above, it is not generally feasible—or even desirable—to achieve perfect security. In this respect, Art. 32 (1) reflects the technical and economic reality when calling for a risk assessment, and it states several criteria to be taken into account for the selection of appropriate measures. At the same time, it leaves considerable leeway for the actual implementation of security measures.

Standards for information security management, such as the ISO 27000 family of standards, can help filling this gap to a certain extent. More specific standards and guidelines are available on a national level, e.g. from the German Federal Office for Information Security (Bundesamt für Sicherheit in der Informationstechnik, BSI).[9] By referring to certification mechanisms—which are usually based on these or similar standards—, Art. 32 (3) makes the link to standardisation explicit. However, these standards do not replace the data controller's or processor's own expertise and judgment, as illustrated by security incidents that occur despite adherence to standards.

Article 32 (1) further lists some individual protection measures. In Art. 32 (1) lit. a, the measures of pseudonymisation and encryption are mentioned. From a technical perspective, these measures are only loosely related. Pseudonymisation is defined in Art. 4 (5) GDPR as "the processing of personal data in such a manner that the personal data can no longer be attributed to a specific data subject without the use of additional information, provided that such additional information is kept separately and is subject to technical and organisational measures to ensure that the personal data are not attributed to an identified or identifiable natural person". In practice, this means replacement of identifying information by a pseudonym, e.g. a random number, and storing the relation between pseudonym and identifying information separately from the data of interest being processed.

Encryption, on the other hand, means applying a transformation, dependent on an encryption key, to the data. That transformation can be reversed using a decryption function and a decryption key, which may[10] or may not[11] be identical to the encryption key. Technically, this fulfils the definition of pseudonymisation if the decryption key is separated from the encrypted data: any (correctly applied) state of the art encryption algorithm ensures the confidentiality of encrypted data, which should therefore not be considered as personal data for anyone without access to the decryption key. However, the point of pseudonymisation is usually to enable meaningful processing activities (beyond storage and transmission of the data). Most encryption algorithms which are used in practical applications make this impossible.

It should be noted that encryption, per se, is only a suitable measure to protect confidentiality of data. In some instances,[12] authenticity and integrity might be achieved as well, but this cannot be taken for granted. There are secure encryption algorithms which ensure confidentiality, but allow trivial and undetectable changes to the encrypted data.

[9]For example, Bundesamt für Sicherheit in der Informationstechnik (2019).

[10]In the case of symmetric cryptography.

[11]In the case of asymmetric cryptography.

[12]Authenticated Encryption, which combines authentication and encryption, is the subject of numerous research papers; there are now modes of operation for block ciphers which achieve authenticated encryption in a secure manner. See Anderson (2020), pp. 181f., for a more detailed discussion of the topic.

By itemizing it in lit. b, Art. 32 (1) refers to "the ability to ensure the ongoing confidentiality, integrity, availability and resilience of processing systems and service" as another measure. There is no specification how this "ability" needs to be achieved. While the first three protection goals in the list have been listed above in this article, "resilience" is not an elementary protection goal in information security.[13] It could be considered as a specification to the goal of availability: Functionality of systems should be maintained even in case of failures or incidents.[14] A common means of achieving resilience, and availability, is to have redundant systems available. The use of intrusion detection systems, which will be discussed in more detail below, is another one.[15]

By referring to the protection goals, instead of specific measures, the entire toolbox of IT security is made available to processors and controllers of personal data. For example, the confidentiality of personal data can be ensured by encryption in some cases, while traditional access control methods might be more suitable in others. Given the diverse nature of IT systems and tools for their protection, this technology-neutral approach appears adequate.

Art. 32 (1) lit. c is about the ability to restore the availability and access to personal data in case of an incident. The norm is closely related to the concept of resilience, which has been discussed above. It acknowledges the fact that incidents can never be excluded with absolute certainty. The most common means for fulfilling the requirement of Art. 32 (1) lit. c is the regular creation and testing of backups.

Art. 32 (1) lit. d requires the implementation of a "process for regularly testing, assessing and evaluating the effectiveness" of security measures. This process is a core task of IT security management. There is a broad consensus among security researchers and practitioners to consider IT security, or information security, as a process rather than a static property—at least outside of a few clearly delineated, more static areas.

Overall, Article 32 of the GDPR captures the most important aspects of IT security quite well, acknowledging the shortcomings, and leaving enough leeway for sensible solutions instead of trying to regulate specific measures in too much detail. This assessment stands despite the issue of legal uncertainty, as too vague regulations can pose a risk to the addressees of the norm.[16]

---

[13] Hansen (2019), mn. 42.

[14] Hansen (2019), mn. 42.

[15] Hansen (2019), mn. 45.

[16] A more comprehensive discussion is provided by Hansen (2019), mn. 22–25.

3.2 Protection Goals and Data Protection

The protection goals for an application are not always clear-cut. This becomes obvious when looking at the protection goals of authenticity and non-repudiation. Both OpenPGP[17] and S/MIME,[18] which are commonly used for e-mail security, support digital signatures, fulfilling both protection goals. This is desirable in many cases, as both involved parties can rely on the message content to a certain extent.

3.2.1 Plausible Deniability

However, for instant messaging, researchers have argued for the opposite of non-repudiation—plausibly deniability, or message repudiation. Similar to an in-person interaction, each communication partner should be authenticated—i.e., the other party should be certain who they are communicating with. On the other hand, neither should be able to prove anything about the communication content to a third party. The OTR (off-the record) messaging protocol[19] was designed to ensure that tampering with messages is simple with respect to a third party attempting to verify the sender of a message,[20] while message authenticity and integrity are still ensured between the communication partners. The principle is now applied in the Signal protocol,[21] which is used by a number of instant messaging applications (such as Signal and WhatsApp). From a data protection perspective, this attempt at achieving plausible deniability in communication protocols may be seen as a technical measure (in terms of Art. 32 of the GDPR). However, it does not lead to an anonymisation of data: Even if participation in a conversation cannot be cryptographically proven, a third party may be sufficiently convinced that a specific natural person can be linked to the exchanged message.

3.2.2 Specific Protection Goals of Data Protection

While the protection goal of plausible deniability may seem relevant only for specific application scenarios, it illustrates that data protection is not limited to the same protection goals as IT security. This is reflected in Article 25 (1) of the GDPR, which calls for technical and organisational measures "designed to implement data-protection principles". A number of protection goals related to data protection

[17] Callas et al. (2007) (RFC 4880). In case of e-mails, the OpenPGP standard is usually combined with PGP/MIME, see Elkins (1996) (RFC 2015) and Elkins et al. (2001) (RFC 3156).

[18] Schaad et al. (2019) (RFC 8551).

[19] Borisov et al. (2004).

[20] This, in turn, makes it plausible that a message has been tampered with, reducing its value as evidence and thus achieving plausible deniability.

[21] Cohn-Gordon et al. (2017).

have been suggested in literature. Hansen et al.[22] propose six protection goals: In addition to confidentiality, integrity, and availability, as introduced above, these goals include unlinkability, transparency, and intervenability.

Unlinkability, as defined by Hansen et al., means that "privacy-relevant data cannot be linked across domains that are constituted by a common purpose and context". That goal is closely related to anonymity, which can be broken by linking data relating to a person across different domains. Unlinkability is, in many contexts, easier to formalise than anonymity. Methods from the field of IT security can be applied to achieve and verify unlinkability.

Transparency means "that all privacy-relevant data processing [...] can be understood and reconstructed at any time".[23] Transparency is a key principle of data protection, as reflected in Art. 5 (1) lit. a GDPR. There are a number of technical means to improve transparency; so-called privacy dashboards have, in recent years, become a popular tool for this purpose.

The protection goal of *intervenability*, which means that "intervention is possible concerning all ongoing or planned privacy-relevant data processing",[24] is not as commonly supported by technology. Privacy dashboards often allow influencing the processing of personal data rather than just making it transparent; in that respect, they can be seen as a tool for intervenability.

3.2.3 Reconciling Protection Goals

The OTR messaging and Signal protocols illustrate that, while protection goals of IT security and data protection might be in conflict in some cases (such as non-repudiation and plausible deniability), other seemingly contradictory protection goals can sometimes be achieved simultaneously (such as authenticity and integrity on the one hand, and plausible deniability on the other hand).

Technical data protection literature is full of similar examples. Anonymous credentials[25] allow to prove an authorization (or group membership) without revealing one's identity. In other words, the goal of unlinkability is achieved while the verification ensures authenticity in the sense that the communication partner belongs to the group of authorized parties. Group signatures[26] allow any group member to sign a message on behalf of the group, without revealing the individual signer's identity; this means that authenticity, integrity, and non-repudiation can be achieved while also ensuring unlinkability (or, in other words, anonymity).

[22] Hansen et al. (2015).

[23] Hansen et al. (2015), p. 160.

[24] Hansen et al. (2015), p. 160.

[25] For one out of many examples, see Blömer and Bobolz (2018).

[26] For example, Ezerman et al. (2020).

4 Reactive Security

In the previous section, we have seen the relation between the abstract goals of IT security and data protection; we now focus on a specific field of high practical relevance, which further illustrates the complex relation between both fields.

Despite significant process since the early days of digitisation, proactive IT security measures are still unable to guarantee that the above-mentioned protection goals are fulfilled under all circumstances. Given the complexity of IT systems and their embedding in organisations, and the inherent limitations of technology, this is unlikely to change in the foreseeable future. The situation might be compared to road safety: Proactive measures, such as the anti-lock brake system, electronic stability program, or lane change assistants, have helped to significantly reduce the number and severity of accidents per driven kilometre in the last few decades. There is still a need for measures that reduce the impact of an accident, such as seatbelts and airbags.

4.1 Logging and SIEM

In the case of IT systems, the simplest and most common reactive security measure is logging. A typical web server, for example, will log all HTTP queries it receives. This means that every time someone looks at a web page, their IP address, the URL (address) of the page, as well as some technical information, is stored along with a precise time stamp. Similar logs are kept by other servers, such as e-mail servers, DNS servers, and many others. These logs can be useful in a forensic investigation once an attack has been noticed and considered severe enough to warrant such an investigation. One should note, though, that servers which can be reached from the public internet are typically attacked very frequently, and that most of these attacks are without any consequence. Besides the above-mentioned servers, firewalls and intrusion detection systems (which will be discussed below) log events as well.

The large number of logged events implies that log files are not typically checked in a manual process, but that specific log management tools are employed. Among others, these tools can be used to identify events that may relate to an attack. They can be part of comprehensive security information and event management (SIEM) solutions. A detailed discussion of SIEM is out of this article's scope; note, however, that the log files processed by log management and SIEM tools will, in many cases, contain personal data, such as IP addresses, user identifiers, and the like.

4.2 *Intrusion Detection Systems*

Intrusion Detection Systems (IDS) are a common implementation of reactive security measures which goes beyond simple logging. IDS are supposed to detect attacks on computer systems or networks. Intrusion Prevention Systems (IPS) are similar to IDS in that they try to identify attack patterns, but they filter out data traffic corresponding to those patterns when possible. In the paper at hand, we focus on IDS.[27]

They can be roughly categorised into host-based IDS (monitoring activities on an individual computer, or host), and network-based IDS (focusing on the network level). Obviously, both approaches can be combined, i.e. an IDS is not limited to considering only information from individual hosts, or only information derived from network traffic. The common goal of both IDS types is to detect attacks (or "intrusions" into the host or network). Different methods can be used to that end:

- Signature-based detection. Once the designer of an IDS learns of a type of attack, they derive typical patterns that occur when the attack is executed. For example, if a specific malware is installed on a computer, it might be known to access certain files in a specific order (which can be detected by a host-based IDS). It might also communicate with certain servers (e.g., with command and control servers, which send instructions to the individual malware installations). This communication might be noted by a network-based IDS. Similarly, denial-of-service (DoS) attacks, which try to disrupt the service offered by a certain computer or network, might be detected if the ratio of certain packets within the network traffic is higher than a defined threshold.[28] A large number of parameters can be part of a signature that is used by an IDS.
- Anomaly detection. The IDS "learns" the normal behavior of an IT system, enabling it to detect unusual or anomalous behavior. Once again, this is possible both on an individual host and in a network. A multitude of parameters can be analyzed. In a network, the overall outbound and inbound traffic, and the relation between the two, might be such parameters. An unusual increase of inbound traffic, without corresponding outbound traffic, could indicate a DoS attack. On the other hand, unusually high outbound traffic, without corresponding inbound traffic, might be a sign of a compromise of internal systems. While an anomaly can indicate an attack, it might also be due to the introduction of new applications

[27] For a survey of the state of the art in intrusion detection, see Khraisat et al. (2019).

[28] A common type of DoS attack is the SYN flood attack. A high number of TCP segments with the "SYN" flag set is sent to the victim. This indicates attempts to open TCP connections, but the attacker does not continue after that initial communication. As a consequence, the number of open TCP connections on the victim's side gety very high, potentially reaching the limit of allowed connections, and preventing communication with other systems. Hence, a high percentage of TCP segments with the SYN flag set is an indication that can be used by an IDS to detect this type of DoS attacks.

or services, or changed user behaviour. False positives are to be expected when using an anomaly detection system.

- Honeypots. A honeypot is a system that appears both vulnerable to an attack, and relevant enough to warrant an attack. However, the system is secured to ensure that the consequences of a successful attack are limited, and that other systems are not affected (or, in some cases, affected only to a limited degree). So-called honeynets are honeypots which focus on attacks that affect networks instead of individual hosts only.

  A honeypot is closely monitored. It can be used as a simple warning system to indicate that an attack is taking place, or it can be used to closely observe the exact the attack methodology. By learning about the attacker's strategy, system operators can adapt their defense mechanisms; moreover, honeypots can also be a tool for IT security researchers.29 A careful design and placement of honeypots is required: It is common for IT systems connected to the internet to be attacked several times per day. A honeypot should be vulnerable enough to be attacked, but should enable a differentiation between well-known attack patterns and actually dangerous ones.

While the data collection by honeypots usually becomes relevant only in the case of an actual attack, both anomaly detection and signature-based detection constantly evaluate patterns that are, at least in part, based on the behaviour of legitimate users. If anomalous behaviour, or behaviour matching an attack signature, is detected, some of that data may be analysed by a system administrator as well.

From a data protection perspective, data processing by IDS is worth a closer look. A network IDS will usually be able to analyse all incoming and outgoing traffic in a network, though internal traffic might not be monitored. Detailed statistics can be derived, detailing how much traffic there is on which days and during which times. This traffic can be attributed to specific communication partners, both on the level of subnets or specific IP addresses. Furthermore, individual services (e.g., web traffic, ssh connections, or VoIP communication) can be differentiated using either the respective transport layer port number or application layer information from within the packets.

In recent years, detection based on the domain name system (DNS) has become more and more common.[29] The domain name system is responsible, among others, for the retrieval of IP addresses belonging to domain names (such as *mail.example.com* or *www.legalinf.de*).[30] Knowledge of the communication partner's IP address is necessary for any communication in the internet, but users very rarely memorise IP addresses; as a consequence, the DNS is a vital infrastructure for most internet applications.

All DNS queries within a company will usually be directed to an internal DNS server, which either answers them based on its cache, or queries the authoritative

---

[29] For an overview of the use of DNS data in IDS, with a focus on botnet detection, see Alieyan et al. (2017).

[30] Mockapetris (1987) (RFC 1035).

servers in charge of the respective domain. Unless the respective IP address for a domain is already cached locally (or the IP address has been manually entered), each communication in the internet is usually preceded by a DNS query.

Evaluating DNS queries within a network can be very helpful to recognise certain types of attacks. For example, botnets often use the domain name system to locate command & control servers. Those servers are used to relay commands, upload additional malware components, or more generally, to coordinate the activities of the botnet. If DNS queries are found that are typical for a certain botnet, a malware infection can be identified even if it is not detected locally on the infected systems. Even if the specific domain names used by a botnet are not known, anomalous DNS query patterns may be detected, which may have been caused by a botnet as well.

Host-based IDS can gather even more fine-grained information, e.g. by tracking which files are accessed at which time, and by looking for other indicators of a potential malware infection.

Unfortunately, attackers—including the developers of malware—are aware of signatures used by signature-based detection systems, and also of the common approaches used to detect anomalies. As far as technically feasible (and within the technical competence of the attacker), they will adapt their strategies to avoid detection. To keep up with this arms race, IDS are becoming more sophisticated as well, and may attempt to use as many data sources as possible to keep their ability to detect attacks.

4.3 *The Role of IDS in Data Protection*

Overall, the role of IDS in data protection is two-faced. On the one hand, IDS are an essential tool to improve IT security in organisations. As pointed out above, preventing attacks may be preferable, but is not always realistic. Consequently, risk assessments (within the scope of Art. 32 GDPR) will often lead to the conclusion that a systematic approach to attack detection is necessary. IDS have become a standard instrument of IT security, and will still be required in the foreseeable future.

In particular when significant processing of special categories of personal data (Art. 9 GDPR) takes place on computer systems exposed to the internet, not using an IDS should be an exception that needs to be justified. This assessment is in line with expert recommendations. For example, the German Federal Office for Information Security (BSI) recommends security measures for "general servers" ("all IT systems which make services available to other IT systems, such as clients or other servers"[31]). In case of increased protection needs, recommendation SYS.1.1.A27[32]

---

[31] Bundesamt für Sicherheit in der Informationstechnik (2019), chapter SYS.1.1: General Server, page 1.

[32] Bundesamt für Sicherheit in der Informationstechnik (2019), chapter SYS.1.1: General Server, page 9. While the English version is only available as a draft, the final 2021 German version

states that host-based IDS should be used to monitor the system for abnormalities and misuse. The situation is less clear for network-based IDS, which are typically more relevant for larger organisations with significant security requirements, but might not be needed for individual servers protected by proactive security measures.[33] Detailed logging of security incidents is required in any case.[34]

On the other hand, IDS usually process personal data. For host-based IDS, this is obvious if the respective computer is assigned to a known person. There is not much of a problem if a host-based IDS only collects data from servers—where they are typically most useful, as described above.

Network-based IDS typically learn at least the IP addresses of communication partners. This includes the above-mentioned analysis of DNS queries, as their respective source IP address is required in case of a suspected attack. In a general setting, IP addresses are usually considered as personal data. While legal literature does not typically differentiate, there are many cases in which IP addresses cannot actually be traced back to an individual person.[35] However, if a company runs an IDS in its own network, the respective user of each internal computer will usually be known.

In summary, the data collected and analysed by an IDS could also be used to establish profiles that describe all of a user's activities, such as the times they log on and off, the software they use, the files they work with, and the websites they interact with (including precise timing information). From a network perspective, the analysis may include application layer data, such as Javascript code on websites a user is browsing. While the actual content of websites is not typically analysed, collection of all the above-mentioned information will be considered as problematic by many users, and requires sufficient justification. Moreover, technical and organisational security measures should be taken to prevent their misuse.

4.4 Pseudonymisation

Attack detection does not generally require storing direct identifiers. While IP addresses have to be *processed* in order to enable network-based IDS, there is no need to actually store them. Pseudonymisation is generally possible, both for IP addresses from external networks and for internal IP addresses. Pseudonymisation

---

includes the same requirement, see Bundesamt für Sicherheit in der Informationstechnik (2021), chapter SYS1.1: Allgemeiner Server, page 6.

[33] The BSI provides a guideline on the introduction of IDS, see Bundesamt für Sicherheit in der Informationstechnik (2002).

[34] IT-Grundschutz Compendium, measure NET.1.2.A26 (Bundesamt für Sicherheit in der Informationstechnik, 2019, chapter NET.1.2: Network Management, page 7; Bundesamt für Sicherheit in der Informationstechnik, 2021, chapter NET.1.2: Netzmanagement, page 6).

[35] The most notable exception is the use of carrier-grade NAT (network address translation) for IPv4 addresses, which is currently widespread.

can be the first processing step, so only pseudonymised IP addresses have to be processed afterwards.

There are various methods for IP address pseudonymisation. Usually, each IP address is assigned to exactly one pseudonym which does not change over time. To meet the definition of Article 4(5) of the GDPR, deriving the IP address (which we have to assume to be personal data) from the respective pseudonym should not be possible without the use of additional information. A straightforward way of achieving this pseudonymisation is the use of a keyed hash function. Cryptographic hash functions,[36] as such, are resistant to preimage attacks: There is no efficient method of calculating a preimage to a given image, i.e. an input that produces a given output. In the case of IP addresses, in particular when using IPv4, even inefficient brute-force attacks can be easily carried out. As a consequence, some secret data (i.e., a key) needs to be used as additional input into the hash functions.

Pseudonymisation, however, comes with two main problems. The first one concerns the functionality of reactive security measures. Pseudonymisation still allows detecting patterns in the communication from individual IP addresses, but prevents the detection of patterns based on IP address ranges (subnets, autonomous systems, or countries). While the issue can be mitigated by pseudonymising only parts of IP addresses (e.g., the last octet of IPv4 addresses), this mitigation may simplify the re-identification of individuals. For example, a German company might have only one employee connecting to the company VPN from France (and, thus, from an IP address belonging to a French internet service provider).

This issue leads to the second main problem of pseudonymisation. The use of persistent pseudonyms means that profiles of the communication patterns can be built over time. With every additional interaction, an additional data point is created that can be used to reveal the identity of the respective user—even if IP addresses are pseudonymised as a whole. Whether or not the identification is successful depends on the background knowledge that is available to the person trying to reveal the identity "behind" the pseudonym. In general, modelling this background knowledge, and thus enabling a formal analysis of identification risks, is very difficult. For example, knowing the usual working times of some colleagues might already be sufficient to identify them in a data set that contains time stamps. A number of research papers have shown how to successfully identify users in pseudonymised databases.[37]

As a consequence, pseudonymisation is not generally considered as a sufficient protection measure in the technical privacy research community. From a practical perspective, pseudonymisation may be an imperfect measure—but given the lack of better alternatives, working with pseudonyms still makes sense. This is reflected in Article 32 (1) lit. a GDPR, which lists pseudonymisation as a technical measure. As, in many cases, pseudonyms can be uncovered with a relatively low effort,

[36] Also referred to as "fingerprints" or "digital fingerprints". For an overview of the properties of cryptographic hash functions, see Stallings (2017), p. 349.

[37] Among the most prominent examples are Sweeney (2002) and Narayanan and Shmatikov (2008).

pseudonymised data should still be treated as personal data—unless there is evidence of pseudonymisation being sufficient in a particular application scenario.

Overall, this implies that data processing by an IDS, or in the context of other reactive security measures, requires a legal basis even if pseudonymisation is applied.

4.5 Decryption of TLS Communication

A major challenge, both for proactive and reactive security measures, lies in the encryption of communication connections. This may seem paradoxical, since encryption is among the most important security measures in communication. However, encrypted communication cannot be scanned for malware, or for any attempts at exploiting security vulnerabilities. Any attack patterns within the encrypted communication have to remain undetected unless that communication is decrypted.

TLS (transport layer security)[38] is probably the most well-known protocol for encrypted communication, and can ensure confidentiality, integrity, and authenticity. The protocol is often referred to as SSL (Secure Sockets Layer), which was the name of previous versions of the standard in the 1990s. TLS provides end-to-end encryption on top of the transport layer in the TCP/IP layer model.[39] This means that a secure channel is provided between two applications running on any computers connected to the internet. An attacker intercepting the communication can see any information from the transport layer and below, such as IP addresses and port numbers (i.e., application addresses on the communicating systems which usually reveal the type of application). Application-specific information, such as application headers, are not revealed.

Any type of attack detection, including IDS, which are unable to decrypt TLS-protected communication, would therefore miss a majority of potential attacks. As a consequence, the capability to decrypt TLS traffic is now common for such systems in corporate environments. This is only possible if system administrators can change the configuration of one of the communication endpoints (i.e., the computer systems within the company), allowing the IDS to perform a "man-in-the-middle" attack: Due to the configuration change, the IDS can successfully authenticate as the intended communication partner, decrypt and inspect the communication, and forward all data afterwards.

From a data protection perspective, this procedure can be considered as problematic. If properly configured, the user within the company does not usually see a difference between a TLS connection that is actually end-to-end encrypted, and one which has been intercepted by some sort of attack detection system. Transparency of

[38] The current version 1.3 is standardised in Rescorla (2018) (RFC 8446).

[39] See Braden (1989) (RFC 1122).

processing, which is one of the key principles of data protection (as laid down in Article 5 (1) lit. a GDPR), has to be established by other (non-technical) means. Additional considerations might arise from telecommunications law, but are out of scope of this article.

5 Conclusion

The relation between IT security measures and data protection seems obvious at first glance: The measures, which serve (among others) the fulfilment of legal obligations from Article 32 of the GDPR, protect individuals against potential misuse of their personal data. In reality, this relation is more complex: As shown in the article at hand, there is a large overlap of the protection goals of IT security and data protection—but there are also scenarios in which privacy engineers deviate from the traditional IT security protection goals. A reconciliation of the different protection goals is often possible, but complex—and practical implementations are rare despite a plethora of research papers.

The complex relation of IT security measures and data protection goals is most clearly visible in the field of reactive security, as demonstrated in this article. Further interdisciplinary research between data protection law and privacy engineering is needed to work out both the legal limitations of security technology, and the legally required standard of security.

References

Acar, Y., Stransky, C., Wermke, D., Weir, C., Mazurek, M., & Fahl, S. (2017). Developers need support, too: a survey of security advice for software developers. In *Proc. 2017 IEEE Cybersecurity Development (SecDev)* (pp. 22–26). IEEE.

Alieyan, K., Almomani, A., Manasrah, A., & Kadhum, M. (2017). A survey of botnet detection based on DNS. *Neural Computing and Applications, 28*(7), 1541–1558.

Anderson, R. (2020). *Security engineering: A guide to building dependable distributed systems* (3rd ed.). Wiley.

Andress, J. (2011). *The basics of information security*. Elsevier.

Blömer, J., & Bobolz, J. (2018). Delegatable attribute-based anonymous credentials from dynamically malleable signatures. In B. Preneel & F. Vercauteren (Eds.), *Applied cryptography and network security. ACNS 2018* (pp. 221–239), Lecture Notes in Computer Science, vol 10892). Springer.

Borisov, N., Goldberg, I., & Brewer, E. (2004). Off-the-record communication, or, why not to use PGP. In *Proceedings of the 2004 ACM workshop on privacy in the electronic society* (pp. 77–84). ACM.

Braden, R. (1989). *Requirements for Internet Hosts – Communication Layers*. RFC 1122 of the Internet Engineering Task Force.

Bundesamt für Sicherheit in der Informationstechnik. (2002). *BSI-Leitfaden zur Einführung von Intrusion-Detection-Systemen*. https://www.bsi.bund.de/DE/Service-Navi/Publikationen/Studien/IDS02/index_htm.html

Bundesamt für Sicherheit in der Informationstechnik. (2019). *IT-Grundschutz Compendium, English Version, Final Draft, 1 February 2019.* https://www.bsi.bund.de/SharedDocs/Downloads/EN/BSI/Grundschutz/International/bsi-it-gs-comp-2019.pdf?__blob=publicationFile&v=1

Bundesamt für Sicherheit in der Informationstechnik. (2021). *IT-Grundschutz-Kompendium.* https://www.bsi.bund.de/SharedDocs/Downloads/DE/BSI/Grundschutz/Kompendium/IT_Grundschutz_Kompendium_Edition2021.pdf?__blob=publicationFile&v=6 (in German).

Callas, J., Donnerhacke, L., Finney, H., Shaw, D., & Thayer, R. (2007). *OpenPGP Message Format.* RFC 4880 of the Internet Engineering Task Force.

Cohn-Gordon, K., Cremers, C., Dowling, B., Garratt, L., & Stebila, D. (2017). A formal security analysis of the signal messaging protocol. In *2017 IEEE European Symposium on Security and Privacy (EuroS&P)* (pp. 451–466). IEEE.

Elkins, M. (1996), *MIME Security with Pretty Good Privacy (PGP).* RFC 2015 of the Internet Engineering Task Force.

Elkins, M., Del Torto, D., Levien, R., & Roessler, T. (2001). *MIME Security with OpenPGP.* RFC 3156 of the Internet Engineering Task Force.

Ezerman, M. F., Lee, H. T., Ling, S., Nguyen, K., & Wang, H. (2020). Provably secure group signature schemes from code-based assumptions. *IEEE Transactions on Information Theory, 66*(9), 5754–5773.

Hansen, M. (2019). Art. 32 DSGVO. In S. Simitis, G. Hornung, & I. Spiecker gen. Döhmann (Eds.), *Datenschutzrecht*. Nomos. (in German).

Hansen, M., Jensen, M., & Rost, M. (2015). Protection goals for privacy engineering. In *2015 IEEE Symposium on Security and Privacy Workshops* (pp. 159–166). IEEE Computer Society.

ISO. (2018). International Standard ISO/IEC 27000, Fifth ed. *Information technology — Security techniques — Information security management systems — Overview and vocabulary.*

Khraisat, A., Gondal, I., Vamplew, P., & Kamruzzaman, J. (2019). Survey of intrusion detection systems: Techniques, datasets and challenges. *Cybersecurity, 2*(20).

Krombholz, K., Hobel, H., Huber, M., & Weippl, E. (2015). Advanced social engineering attacks. *Journal of Information Security and Applications, 22*, 113–122.

Krupp, J., Karami, M., Rossow, C., McCoy, D., & Backes, M. (2017). Linking amplification DDoS attacks to booter services. In M. Dacier, M. Bailey, M. Polychronakis, & M. Antonakakis (Eds.), *Research in attacks, intrusions, and defenses. RAID 2017* (Lecture Notes in Computer Science) (Vol. 10453, pp. 427–449). Springer.

Lipp, M., Schwarz, M., Gruss, D., Prescher, T., Haas, W., Horn, J., Mangard, S., Kocher, P., Genkin, D., Yarom, Y., Hamburg, M., & Strackx, R. (2020). Meltdown: Reading kernel memory from user space. *Communications of the ACM, 63*(6), 46–56.

MacFarland, D., Shue, C., & Kalafut, A. (2017). The best bang for the byte: Characterizing the potential of DNS amplification attacks. *Computer Networks, 116*, 12–21.

Mockapetris, P. (1987). *Domain Names – Implementation and Specification.* RFC 1035 of the Internet Engineering Task Force.

Narayanan, A., & Shmatikov, V. (2008). Robust de-anonymization of large sparse datasets. In *2008 IEEE Symposium on Security and Privacy (S&P 2008)* (pp. 111–125). IEEE Computer Society.

Rescorla, E. (2018). *The Transport Layer Security (TLS) Protocol Version 1.3.* RFC 8446 of the Internet Engineering Task Force.

Rescorla, E., & Korver, B. (2003). *Guidelines for Writing RFC Text on Security Considerations.* RFC 3552 of the Internet Engineering Task Force.

Schaad, J., Ramsdell, B., & Turner, S. (2019). *Secure/Multipurpose Internet Mail Extensions (S/MIME) Version 4.0 Message Specification*. RFC 8551 of the Internet Engineering Task Force.

Song, H., Fink, G., & Jeschke, S. (2017). *Security and privacy in cyber-physical systems: Foundations, principles, and applications*. Wiley.

Stallings, W. (2017). *Cryptography and network security: Principles and practice* (7th ed.). Pearson.

Sweeney, L. (2002). k-Anonymity: A model for protecting privacy. *International Journal of Uncertainty, Fuzziness and Knowledge-Based Systems, 10*(5), 557–570.

The Legal Framework for IT Security in the "Industry 4.0"

Georg Borges

Contents

Abstract IT security is a central basis for industrial networks, so-called "Industrie 4.0", as indicated by current security incidents and criminal attacks on IT systems, some of which have caused dramatic damage. The significance of the law for IT security and the design of a legal framework for IT security have not been clarified yet. This is probably also related to the fact that IT security has been understood traditionally as a technical task.

G. Borges (✉)
Institute of Legal Informatics, Saarland University, Saarbrücken, Germany
e-mail: georg.borges@uni-saarland.de

G. Borges, C. Sorge (eds.), *Law and Technology in a Global Digital Society*,
https://doi.org/10.1007/978-3-030-90513-2_11

This paper analyzes, on the basis of German and European law, the potential of the various legal instruments for ensuring IT security. This analysis relates to the two central tasks of law in in the area of IT security, the formulation of normative requirements and the enforcement of such requirements. Finally, conclusions are drawn for the further development of the legal framework for IT security.

1 Introduction

There is no doubt that security is one of the foundations of the industrial internet. The English terms "Internet of things" or "industrial internet" or their German equivalent, "Industrie 4.0", describe one of the main trends in current industrial activity, which centres on the connection of industrial facilities through communication networks as well as the autonomous control of industrial facilities and machines through software. This paper will focus on issues connected with networks.

1.1 IT Security for the Industrial Internet: Recent Challenges

The global networking of industry can only succeed if a sufficient level of security for the connected machines and facilities can be guaranteed. This includes the issue of securing facilities against third-party attacks, a possibility which is very much in the focus of current public attention. One of the crucial problems identified has been termed 'advanced persistent threats' and describes the situation where attackers target the network of an industrial company, infiltrate it, and are active inside the network for a significant period until they finally take control of all, or part, of the company's IT system.

Two examples can be used to illustrate this current threat. The German Federal Office for Information Security (BSI) reported an attack on a steel plant in its annual report of 2014 in which the perpetrator had initially gained access to the office network of the company, most likely by way of a targeted phishing attack known as spear phishing, and later managed to enter the production network.[1] This eventually enabled the attacker to disrupt the control system of the steel plant to such a degree that one furnace could no longer be shut down in a controlled manner, thus causing extensive damage to the plant.

In recent years, businesses and organisations in Europe and many other countries have been confronted with so-called ransomware attacks. The term "ransomware" describes a type of malicious software (malware) with the ability to block the access to a computer system itself or to specific data until a demanded sum of money (the

[1] BSI (2019).

"ransom") is paid.[2] Ransomware is typically installed—by chance or well-directed—into a user's computer system by either an attachment of an e-mail or by loading a compromised website (drive-by-attacks).[3] Most of the attacks are carried out by means of ransomware with encryption function (Crypto-Trojans).[4] In most situations, ransom payments are the only solution to regain access to data which has not been saved in a separate system.[5]

In 2016, a first big wave of ransomware attacks hit health care institutions affecting hospitals in the United Kingdom and Germany. In the German state of North Rhine-Westphalia, 28 hospitals were attacked.[6] One of the best-reported ransomware attacks aimed at a hospital in Neuss, Germany, the Lukas hospital (Lukas Krankenhaus).[7] The whole IT-system of the Lukas hospital was infected by malware. By way of an attachment of an e-mail sent to a hospital staff, the malware infiltrated into the hospital's network. On 10th February 2016, the attack was noticed as systems were working exceptionally slow, unknown errors and messages suddenly appeared on several computer screens announcing that the hospital's important data had been encrypted and that a specific ransom was required to regain the data. The Lukas hospital decided to not pay the ransom. Instead, all systems were shut down in order to secure the patients' sensitive data and the hospital continued its work without internet connection, using paper and pencil, stand-alone devices etc.[8]

It took almost one week to reboot some of the computers, and the hospital's operation was severely affected for several weeks. Fortunately, in spite of those enormous restrictions, only some of the major surgeries had to be postponed and most of the patient care could be carried out unimpaired.[9]

The crypto ransomware "WannaCry", also known as WannaCrypt, WannaCrypt0r, WRrypt and WCRY, is a variant of ransomware that encrypts user

[2] Beukelmann (2017), p. 376; BSI (2018), pp. 35, 97; Eckert (2018), p. 23; Salomon (2016), p. 575; Stocklas (2016); Weidemann (2019), mn. 3.1.

[3] Salomon (2016), p. 575; Trinks (2016), p. 264; Ceffinato (2016), p. 467; BSI (2016), p. 14.

[4] BSI (2016), p. 14; BSI (2018), p. 35.

[5] Salomon (2016), p. 575; Europol, Wannacry Ransomware Overview, available at https://www.europol.europa.eu/wannacry-ransomware. Accessed 1 September 2021.

[6] Smith, *Ärzte Zeitung online*, available at https://www.aerztezeitung.de/praxis_wirtschaft/digitalisierung_it/datenschutz/article/971128/cybersicherheit-wenn-medizin-verwundbar.html. Accessed 1 September 2021; Krämer and Dahmen (2017).

[7] Becker's Health IT & CIO Report, available at https://www.beckershospitalreview.com/healthcare-information-technology/12-healthcare-ransomware-attacks-of-2016.html. Accessed 1 September 2021.; Leyden, The Register, available at https://www.theregister.co.uk/2016/02/26/german_hospitals_ransomware/. Accessed 1 September 2021.

[8] Krämer and Dahmen (2017); Smith, Ärzte Zeitung online (*supra* note 6); Becker's Health IT & CIO Report, (*supra* note 7).

[9] Smith, Ärzte Zeitung online (*supra* note 6); Becker's Health IT & CIO Report, (*supra* note 7); *Leyden* in The Register, (*supra* note 7).

data and demands a ransom in order to regain the control of the data.[10] Since May 2017 this specific crypto ransomware has massively spread around the world causing an enormous abundance of damages both in the private and in the public sector and infecting up to 300,000 victims in over 150 countries.[11] Also in 2017 and only 1 month after the start of the WannaCry ransomware attack, an updated version of Petya ransomware ("NotPetya") damaged more than 20,000 victims in more than 60 countries, mainly in Europe.[12] According to reports, "more than 50% of the business targeted" were industrial companies.[13]

These attacks clearly demonstrate that, today, IT Security is a major concern for any business connected to the internet. This is particularly true for the industrial internet (Industrie 4.0) which is based on the ongoing connection of industrial businesses via the internet. In 2016, the German Federal ministry of Economics an Energy (BMWi) published a scientific study on IT security in Industry 4.0. Among other measures, the authors of the study called for the further development of the legal framework for IT security.[14]

1.2 Law and IT Security: Initiatives by the European Legislator

Safeguarding IT security requires the interaction of a multitude of technical and organisational measures.[15] In order to orchestrate this interaction and ensure the implementation of the necessary measures, the law is of great importance.[16] This is particularly evident in the increasing amount of legislation dealing directly or indirectly with IT security.[17] In Europe, the EU has taken numerous initiatives to develop a framework of IT security. The European Commission introduced its EU Cybersecurity Strategy in 2013[18] and, in 2015, included the aim to fight cybercrime

[10] Europol, WannaCry Ransomware Overview (*supra* note 5).

[11] IOCTA (2017), p. 10.

[12] IOCTA (2017), p. 20; BSI (2018), p. 91; Jansen & Armbruster, Frankfurter Allgemeine Zeitung, Ransomware-Attacke, available at https://www.faz.net/aktuell/wirtschaft/ransomware-attacke-legt-viele-unternehmen-lahm-15079944.html. Accessed 1 September 2021.

[13] IOCTA (2017), p. 20.

[14] BMWi (2016a).

[15] BMWi (2016a), p. 27 ff.

[16] Borges (2016), pp. 159, 161; Riehm and Meier (2020), p. 571.

[17] Conrad (2019), § 33, mn. 179–320; Spindler (2008), pp. 7, 9 ff.

[18] European Commission, Cybersecurity Strategy of the European Union, Brussels 2013-02-07, JOIN(2013) 1 final, available at http://www.eeas.europa.eu/archives/docs/policies/eu-cyber-security/cybsec_comm_en.pdf. Accessed 1 September 2021.

in a more effective manner as one of the three pillars into the European Agenda on Security.[19]

Regarding EU Legislation, a milestone was reached in 2016: The Directive on security of network and information systems (NIS Directive)[20] was adopted by the European Parliament on 6th July 2016 and came into force in August 2016. The directive aims at ensuring a high common level of security of network and information systems within the EU. It requires appropriate equipment and preparedness from the EU Member States as well as operating cooperation systems between all Member States, so called "Cooperation Groups" and "Computer Security Incident Response Team (CSIRT) -Networks", in order to ensure strategic and effective operational cooperation concerning cybersecurity incidents and potential threats.[21] In addition to these measures addressing an organisational framework for IT security within the Member States, chapter IV of the Directive contains IT security requirements for so-called digital service providers.

The implementation of the NIS Directive has been analysed by the Commission in its Communication[22] to the European Parliament and the Council of October 2017. With this Communication, the Commission analyses and demonstrates the implementations of the NIS Directive within the different Member States so far as well as the work of the European Agency for Network and Information Security (ENISA).[23]

In September 2017, the EU Commission presented a Recommendation[24] on Coordinated Response to Large Scale Cybersecurity Incidents and Crises as well as a Joint Communication[25] concerning Resilience, Deterrence and Defence aiming

---

[19] European Commission, The European Agenda on Security, Strasbourg 2015-04-28, COM(2015) 185 final, available at https://ec.europa.eu/home-affairs/sites/homeaffairs/files/e-library/documents/basic-documents/docs/eu_agenda_on_security_en.pdf. Accessed 1 September 2021.

[20] Directive (EU) 2016/1148 of the European Parliament and of the Council of 6 July 2016, OJ L 194, 2016-07-19, pp. 1–30, available at https://eur-lex.europa.eu/eli/dir/2016/1148/oj. Accessed 1 September 2021.

[21] European Commission, EU cybersecurity initiatives, January 2017, p. 4, available at http://ec.europa.eu/information_society/newsroom/image/document/2017-3/factsheet_cybersecurity_update_january_2017_41543.pdf. Accessed 1 September 2021.

[22] European Commission, Communication from the Commission to the European Parliament and the Council, Making the most of NIS – towards the effective implementation of Directive (EU) 2016/1148 concerning measures for a high common level of security of network and information systems across the Union, Brussels 2017-10-04, COM(2017) 476 final/2, available at https://eur-lex.europa.eu/resource.html?uri=cellar:d829f91d-9859-11e7-b92d-01aa75ed71a1.0001.02/DOC_3&format=PDF. Accessed 1 September 2021.

[23] European Commission, COM(2017) 476 final/2, Brussels 2017-10-04, pp. 2 f.

[24] European Commission, Commission Recommendation on Coordinated Response to Large Scale Cybersecurity Incidents and Crises, Brussels 2017-09-13, C(2017) 6100 final, available at https://ec.europa.eu/transparency/regdoc/rep/3/2017/EN/C-2017-6100-F1-EN-MAIN-PART-1.PDF. Accessed 1 September 2021.

[25] European Commission, Joint Communication to the European Parliament and the Council, Resilience, Deterrence and Defence: Building strong cybersecurity for the EU, Brussels 2017-09-13, JOIN(2017) 450 final, available at https://eur-lex.europa.eu/legal-content/EN/TXT/PDF/?uri=JOIN:2017:450:FIN&rid=3. Accessed 1 September 2021.

at building strong cybersecurity for the EU. With the same date, the Commission introduced a proposal for a Regulation on ENISA, the European Union Agency for Cybersecurity, and on cybersecurity, the so-called "Cybersecurity Act".[26] After an intensive discussion within the European institutions and the member states, the EU Cybersecurity Act was promulgated in June 2019.[27] By the act, ENISA is designated to be a major player in the regulation of IT security. ENISA now has the mandate to give not only advice to different institutions but also to carry out operational and coordination tasks.[28] Moreover, the act is introducing a EU cybersecurity certification framework for ICT products and services in order to enhance trust and security for consumers and businesses, as one aim mentioned also in the Digital Single Market Strategy.

One of the latest developments is the further improvement of the NIS Directive. A legislative proposal to update and repeal the current Directive was presented on 16th December 2020.[29] The proposal aims at eliminating the existing weaknesses of the current regulations and achieving better coordination with other policies in the area of IT security.[30]

Within the Member states, a similar development is going on. Germany anticipated the main concept of the NIS Directive by its IT Security Act[31] of July 2015.[32]

[26] European Commission, Proposal for a Regulation of the European Parliament and of the Council on ENISA, the "EU Cybersecurity Agency", and repealing Regulation (EU) 526/2013, and on Information and Communication Technology cybersecurity certification ("Cybersecurity Act"), Brussels 2017-09-13, COM(2017) 477 final 2017/0225(COD), available at https://eur-lex.europa.eu/legal-content/EN/TXT/?qid=1505737096808&uri=CELEX:52017PC0477. Accessed 1 September 2021.

[27] Regulation (EU) 2019/881 of the European Parliament and of the Council of 17 April 2019, on ENISA (the European Union Agency for Cybersecurity) and on information and communications technology cybersecurity certification and repealing Regulation (EU) No 526/2013 (Cybersecurity Act), available at https://eur-lex.europa.eu/legal-content/EN/TXT/PDF/?uri=CELEX:32019R0881. Accessed 1 September 2021.

[28] EPRS, ENISA and new EU Cybersecurity Act, PE 625.160, March 2019, available at www.europarl.europa.eu/RegData/etudes/ATAG/2019/625160/EPRS_ATA(2019)625160_EN.pdf (accessed 1 September 2021); European Commission, Cybersecurity – EU Agency and Certification Framework, State of the Union 2017, available at https://ec.europa.eu/commission/presscorner/api/files/attachment/854580/Cybersecurity-EU%20agency%20and%20certification%20framework.pdf. Accessed 1 September 2021.

[29] Proposal for a Directive of the European Parliament and of the Council on measures for a high common level of cybersecurity across the Union, repealing Directive (EU) 2016/1148, COM(2020) 823 final, available at https://eur-lex.europa.eu/resource.html?uri=cellar:be0b5038-3fa8-11eb-b27b-01aa75ed71a1.0001.02/DOC_1&format=PDF. Accessed 1 September 2021.

[30] Proposal for a Directive of the European Parliament and of the Council on measures for a high common level of cybersecurity across the Union, repealing Directive (EU) 2016/1148, COM(2020) 823 final, p. 2 *et seq.*, available at https://eur-lex.europa.eu/resource.html?uri=cellar:be0b5038-3fa8-11eb-b27b-01aa75ed71a1.0001.02/DOC_1&format=PDF. Accessed 1 September 2021.

[31] Act to increase the security of information technology systems (Gesetz zur Erhöhung der Sicherheit informationstechnischer Systeme (IT-Sicherheitsgesetz), BGBl. I 2015, p. 1324 *et seq.*

[32] Hornung (2015), pp. 3334 f.

The act was later amended by the law implementing the Directive (EU) 2016/1148 in June 2017.[33]

Currently, further amendments of the legal framework by a so-called IT Security Act 2.0 (IT-Sicherheitsgesetz 2.0) are planned. In December 2020, the German Federal Government introduced a bill[34] which is intended to amend several existing laws, in particular the Act on the German Federal Office for Information Security (BSI-Gesetz). By the IT Security Act 2.0, the Federal Office shall be equipped with more extensive competences, in particular, authorising inspection and control of IT products with regard to their IT security-relevant properties.

2 Tasks and Tools in Regulating IT Security

2.1 Variety of Approaches in the Current Development

With its 2019 Cybersecurity Act, the European legislator is aiming at strengthening the EU Agency for Network Information Security (ENISA) by providing more resources, a permanent mandate and an enhanced role with the ability not only to advise but also to perform operational and coordination tasks.[35] At the same time, the act aims at establishing a European System of technical standards and certification. By providing a common certification scheme, the framework aims at guaranteeing a common high level of security and at preventing a multiplication of different overlapping national cybersecurity certification schemes, which also lead to the risk of certification shopping.[36]

On the other hand, liability is considered to be an important player. In its joint communication on cybersecurity of September 2017, the EU Commission and the High Representative of the Union for Foreign Affairs and Security, while examining

[33] "Gesetz zur Umsetzung der EU 2016/1148 des Europäischen Parlaments und des Rates vom 6. Juli 2016 über Maßnahmen zur Gewährleistung eines hohen gemeinsamen Sicherheitsniveaus von Netz- und Informationssystemen in der Union" dated 2017-06-23, BGBl 2017 Part I No. 40, pp. 1885 ff., available at https://www.bgbl.de/xaver/bgbl/start.xav?startbk=Bundesanzeiger_BGBl&jumpTo=bgbl117s1885.pdf#__bgbl__%2F%2F*%5B%40attr_id%3D%27bgbl117s1885.pdf%27%5D__1556537298570. Accessed 1 September 2021.

[34] Government draft for a Second Law for Enhancement of the Security of IT-Systems ("(IT-Sicherheitsgesetz 2.0") dated 2021-01-25, BT-Drs. 19/26106, available at https://dip21.bundestag.de/dip21/btd/19/261/1926106.pdf. Accessed 1 September 2021.

[35] European Parliament, EU Cybersecurity Act, Consolidated text, 2019-03-12, recitals 65 ff., available at http://www.europarl.europa.eu/doceo/document/TA-8-2019-0151_EN.html?redirect#title2 (accessed 1 September 2021); EPRS, ENISA and new EU Cybersecurity Act, PE 625.160, March 2019, available at www.europarl.europa.eu/RegData/etudes/ATAG/2019/625160/EPRS_ATA(2019)625160_EN.pdf. Accessed 1 September 2021.

[36] European Parliament, EU Cybersecurity Act, Consolidated text, 2019-03-12, recitals 69 f., available at http://www.europarl.europa.eu/doceo/document/TA-8-2019-0151_EN.html?redirect#title2. Accessed 1 September 2021.

the potential as well as risks of new technologies, explicitly refer to liability and its role within IT security.[37]

The role of liability as an instrument for ensuring IT security is closely linked to the applicability of product liability to software. In spring 2018, the European Commission established an Expert Group on liability and new technologies (E03592) with the task to provide the commission with expertise on the applicability of the Product Liability Directive to new technologies and with assistance in developing principles for potential changes in the law relating to new technologies. The Expert Group is working in two formations: one formation concerning the Product Liability Directive ("Product Liability Directive formation") and another formation regarding new technologies ("New Technologies formation").[38]

One of the topics discussed in both formations of the Expert Group is IT security. In the PLD formation, a controversially debated aspect was the applicability of the PLD to non-embedded software. Determining the status of non-embedded software, the experts especially emphasize the difficulty to distinct such type of software from a service, which does not fall within the scope of the directive.[39]

Regarding embedded software, the discussion leads to a different result. Here, while a minority stated that a tangible good with embedded software could not be regarded as a "finished product" in the sense of the PLD concerning the constantly provided updates and upgrades,[40] most of the experts agreed that embedded software should be regarded as a product within the PLD. The need of updating software should not be considered as an obstacle to this classification, as similarly in the analogous world, the fact that products can be upgraded, improved or changed afterwards, may not lead to the conclusion that those products are not subjects to the PLD.[41] A final decision on this discussion by the Commission's Expert Group on Liability and New Technologies has not been published yet.

This discussion corresponds to a similarly controversial debate in literature. While many statements support the view that non-embedded software does not fall

[37]European Commission, JOIN (2017) 450 final, Brussels 2017-09-13, pp. 1, 6, available at https://eur-lex.europa.eu/legal-content/DE/ALL/?uri=CELEX%3A52017JC0450. Accessed 1 September 2021.

[38]Expert Group on liability and new technologies (E03592), available at http://ec.europa.eu/transparency/regexpert/index.cfm?do=groupDetail.groupDetail&groupID=3592&NewSearch=1&NewSearch=1). Accessed 1 September 2021.

[39]European Commission, Meeting of the Expert Group on Liability and New Technologies – Product Liability Formation, Brussels, 2018-11-26, p. 2, available at https://ec.europa.eu/transparency/expert-groups-register/core/api/front/document/29988/download. Accessed 1 September.

[40]European Commission, Meeting of the Expert Group on Liability and New Technologies – Product Liability Formation, Brussels, 2018-11-26, p. 2, available at https://ec.europa.eu/transparency/expert-groups-register/core/api/front/document/29988/download. Accessed 1 September.

[41]European Commission, Meeting of the Expert Group on Liability and New Technologies – Product Liability Formation, Brussels, 2018-11-26, p. 2, available at https://ec.europa.eu/transparency/expert-groups-register/core/api/front/document/29988/download. Accessed 1 September.

within the scope of the PLD,[42] in recent years a majority of authors have favoured the applicability of the PLD to software.[43]

This raises the question as to which role civil liability of software producers can play in the context of IT security. The consequence of being able to establish liability is that the injured party is able to make a claim for damages against the injuring party. Thus, under most legal systems, liability is primarily an instrument for granting compensation for damage which has already occurred.[44] However, the liability rule is also an effective tool for steering behaviour[45] as the threat of potential liability provides an incentive for the addressees to implement suitable measures to avoid the occurrence of damage.[46] Thus, civil liability can also be implemented to improve IT security (infra Sect. 4.1).[47]

2.2 *Tasks and Measures to Regulate IT Security*

In the regulation of IT security, all means of regulating technology are available. Besides national legislators, who can create new requirements by enacting laws such as the NIS Directive, jurisprudence is also able to specify legal requirements or deduce them from general principles. Supervisory authorities are of similar importance as, in applying the laws, they are able to clarify their application to individual circumstances and ensure that the legal requirements are implemented. As a contrast to regulation, relying on market mechanisms can be considered as an option.

---

[42] European Commission, Meeting of the Expert Group on Liability and New Technologies – Product Liability Formation, Brussels, 2018-11-26, p. 2, available at https://ec.europa.eu/transparency/expert-groups-register/core/api/front/document/29988/download. Accessed 1 September.; Schmon (2018), pp. 255 ff.; Streatman and Verhoeven (2016), p. 47; Schneider and Witzel (2015), p. 168; Wagner (2020a), mn. 21 ff.; Cahn (1996), pp. 2899 ff.; Müller-Hengstenberg (1994), p. 3131; Spindler (1998), p. 121; Beckmann and Müller (1999), pp. 14 ff.; Förster (2021b), mn. 22 f.; Oechsler (2018), mn. 64 f.; Schlutz (1994), p. 791; Wilhelmi (2020a) mn. 2; Sommerlad (1997), No. 18, pp. 5 ff. (6).

[43] Magnus (2016a), p. 245; Wagner (2020a), mn. 22 ff., 27; Zech (2012), pp. 342 f.; Lehmann (1992), p. 1724; Meier and Wehlau (1998), p. 1589; Littbarski (2020), mn. 120; Redeker (2020), mn. 878; Rebin (2021) mn. 59; Meyer (1997), p. 33; Fairgrieve and Rajneri (2019), pp. 24 ff.; Cahn (1996), p. 2904; Taeger (1996), p. 262; Morscher (1999), p. 273; Hamdan and Günes (2020), mn. 9; Rothe (1993), p. 312; Henssler (1993), p. 494; Kort (1990), p. 174; Riehm (2014), p. 114; Plath (2001), p. 275; Reusch (2019), pp. 905 f.

[44] Fleming (1992); Howarth et al. (2015), p. 173; Schwarze, Vorbemerkungen zu §§ 280–285 (2019), mn. A2; Hager (2017), mn. 9; Wagner (2020b), mn. 43; Schäfer & Ott, pp. 166, 565.

[45] Hager (2017), mn. 13; Wagner (2020c), mn. 53 f.; Oetker (2019), mn. 9; Förster (2021a), mn. 9–11; Magnus (2016b), mn. 11 f.; Schäfer and Ott (2020), pp. 166, 169, 565.

[46] Hager (2017), mn. 10; Wagner (2020b), mn. 45 f., 47; Oetker (2019), mn. 9; Förster (2021a), mn. 9–11; Magnus (2016b), mn. 11 f.; Schäfer and Ott (2020), p. 166, 169.

[47] See BMWi (2016b), p. 171.

Self-regulation can also be of great significance and can be achieved, for example, by contractual stipulation of IT security requirements. Technical standards are an essential instrument in formulating specific technical and organisational requirements, especially those relating to security. In addition, certification plays a central role for security requirements.

The tasks of legal regulation in the area of IT security must be clarified and the factors guaranteeing IT security, as well as its threats, must be analysed to enable these factors to be addressed by legal norms. The factors guaranteeing IT security against third-party attacks are manifold and cannot be listed exhaustively. However, one factor for security is warding off attacks by adopting technical measures such as cryptography and firewalls, and organizational measures such as limiting the number of people having access to devices. Another is the ability of persons to recognize attacks and to react to them appropriately. Prohibiting attacks and preparatory acts can have a deterrent effect as well as many other legal consequences such as an obligation to pay damages pursuant to tort law.

One challenge of regulating IT security would seem to be a lack of understanding regarding the form of suitable regulation. The way in which the various factors such as awareness, technical and organizational measures can be influenced by legal norms is not yet clear. By way of example, the possibility of legal norms being able to prevent an attack using social engineering should be considered. Outlawing attacks does not necessarily mean that they will stop and attacks such as the one on the steel plant are already illegal. An additional difficulty would seem to be that there is still insufficient knowledge of how the various regulatory measures interact with each other.

2.3 *The Two Task Areas in Regulating IT Security*

Concerning the tasks of IT security regulation, two main areas should be distinguished.[48] Firstly, normative requirements for IT security have to be established, meaning stipulation of measures to be implemented by the addressee of the norm (Sect. 3). Secondly, there is the necessity to establish mechanisms to enforce these normative requirements in order to ensure that the legally prescribed measures are actually implemented (Sect. 4).[49] The successful regulation of IT security requires adequate solutions for both tasks, which will be analysed in greater detail below.

[48] Borges (2016), p. 161.

[49] Borges (2016), p. 161.

3 Establishing Normative Requirements

Already, there is an existing number and variety of normative requirements concerning IT security encompassing technical norms on the security of data centres as well as the duty of heads of companies to ensure adequate IT security as part of risk management.

The establishment of normative requirements takes place in very different ways using a broad range of legal instruments. Therefore, from a systematic perspective, it is of great interest to know how effective the various measures, such as legislation, standards and others, would be with regard to the establishment of normative requirements for IT security.

3.1 Legislation

3.1.1 Lack of Broadly Applicable Horizontal Legislation

Regulation of IT security through legislation is highly complex. It should be noted that in most jurisdictions, IT security is not subject to a broadly applicable horizontal legislation. This is true for EU law as well as for national legal frameworks such as Germany, which do not have a specific security law that applies to all sectors, nor is there specific legislation for the industrial internet. This can be demonstrated at the example of the European NIS Directive of 2016. As stated before, the Directive was regarded as a milestone in IT security regulation.[50] Also, in its Art. 14, it obliges operators of so-called essential services to "take appropriate and proportionate technical and organisational measures to manage the risks posed to the security of network and information systems which they use in their operations". Similarly, in its Art. 16, the NIS Directive obliges so-called digital services providers to "take appropriate and proportionate technical and organizational measures to manage the risks posed to the security of network and information systems which they use in the context of offering services". By establishing a duty of service providers related to IT security, the NIS Directive takes an important step.

However, this step is smaller than it might appear at first reading as the scope of both articles is rather limited. Art. 14 addresses operators of essential services. This term is defined in Art. 4 para. 4 and Annex II of the Directive and refers to large suppliers in the fields of energy, transport, banking, financial markets infrastructures, health sector, drinking water supply and distribution as well as digital infrastructure such as internet exchange points (IXP), DNS providers and TLD registries. The European legislator intentionally limits the scope of this duty to a small number of businesses. In Germany, a number of 500–2000 entities was estimated to fall within this notion, with a majority of water suppliers. Similarly, Art. 16 which addresses

[50] See supra, Sect. 1.2.

"digital services providers" only applies to services "referred to in Annex III". Annex III, however, contains a surprisingly short list expressly and exclusively referring to online marketplaces, online search engines and cloud computing services. Most interestingly, notwithstanding a few exceptions, industry is not seen as part of 'critical infrastructure'. This highlights the fact that the NIS Directive is of limited importance for the industrial internet.

The 2019 Cybersecurity Act[51] does not contain, in a certain contrast to its title, any substantive requirements on IT security. Instead, it focusses on strengthening the role of ENISA and on establishing a common certification framework.

As an interim conclusion the European NIS Directive of 2016 as well as the 2019 Cybersecurity Act can be considered to be elements of an evolving IT Security law. However, a broadly applicable legal framework of IT security covering industry is not in place yet.

3.1.2 IT Security Requirements in General Norms

However, several more general norms do refer to IT security. The head or the board of directors of a company are legally required to implement adequate risk management processes to protect the company and this requirement encompasses IT security. Under German law, for instance, this duty means to establish an adequate risk management system regulated in § 91 II of the AktG.[52] While the statutory duty refers to risks in a general way and IT security is not explicitly mentioned, there is no doubt that IT risks are essential for any company and that a proper management of IT security is an essential task of the board.[53]

Similar duties can be found in banking supervisory law for financial institutions.[54] Even though the legal requirements are specified by the German Federal Financial Supervisory Authority (Bundesanstalt für Finanzdienstleistungsaufsicht—BaFin) in the Minimum Requirements for Risk Management (MaRisk) of 27.10.2017,[55] it is difficult to deduce duties for the implementation of a specific protective measure from these provisions and such duties mainly serve to protect the company itself rather than protecting third parties.

Normative requirements for IT security in the industrial internet protecting third parties can partly be deduced from the general norms set out in tort law.[56] § 823 para.

[51] See supra, Sect. 1.2.

[52] LG München I, 05.04.2007 - 5 HK O 15964/06; Fleischer, 2021, AktG § 91 at mn. 36.

[53] Beucher and Utzerath (2013), pp. 362, 366; Conrad and Streitz (2019), § 33 at mn. 41 ff.; Kipker et al. (2017), pp. 651, 657, 658; Nolte and Becker (2008), Annex No 005, 23; Sattler (2018), pp. 2243, 2247; Schmidt (2009), pp. 1295, 1295 ff.; Schmidl (2016), mn. 469.

[54] Frisse et al. (2018), pp. 178 ff.; Riehm and Meier (2020), p. 572.

[55] BaFin (2017).

[56] Borges (2016), p. 171; De la Durantaye (2020), mn. 44 ff.

1 of the German Civil Code (BGB) for instance, which is the general liability clause in German tort law, could be of significance regarding security requirements for all objects capable of posing threat for third parties.[57] This also applies to industrial facilities. Therefore, requirements for IT security of industrial facilities can be discerned from § 823 para. 1 BGB.[58]

Such general clauses require specification, however specification mainly occurs through jurisprudence in liability lawsuits.[59] Thus, if general norms such as § 823 para. 1 BGB formed the legal basis for IT security, this would, in effect, transfer the task of specifying the regulatory duties on IT security completely from the legislator to the courts.[60]

Legal duties to install IT security measures can be derived from safety requirements. Safety is often regulated in laws for governing specific sectors or products contain security requirements. All legal security requirements for motor vehicles, trains and aircraft are stipulated in specific laws and the same applies to production facilities such as energy production and supply facilities, especially for nuclear energy. Numerous other examples can be found.

In most cases, however, there is considerable uncertainty regarding the existence and content of such duties. The European Machinery Directive[61] which requires producers to guarantee a sufficient level of safety, can be taken as an example for this difficulty. Whilst IT security is not explicitly mentioned, it can be stated that connected devices must be equipped with security features to avoid misuse. Some authors, however, support the view that protection against cyberattacks is not addressed by the Directive.[62]

Among the laws protecting certain goods, the employee protection laws contain comprehensive requirements for the security of industrial facilities.[63] IT security is included, but not explicitly addressed.

Thus, the most important law on IT security in Europe might be the European Data Protection Regulation (GDPR). The GDPR is to be considered as a law of almost general application as it applies to public as well as to private entities including individuals in some cases. As the processing of personal data is executed by almost any entity on a daily basis, the GDPR is a legislation covering the broadest area of applicability.

Therefore, it is of particular importance that the regulation of IT security is one of the core elements of the GDPR. Art. 32 para. 1 of the GDPR obliges any controller or processor of personal data to implement appropriate technical and organizational

---

[57] De la Durantaye (2020), mn. 44; Spindler (2021), mn. 395.

[58] Borges (2016), p. 171; De la Durantaye (2020) mn. 44 ff.

[59] Staudinger (2019), mn. 8, 10.

[60] For a critique see also BMWi (2016b), pp. 172 ff.

[61] Directive 2006/42/EC of the European Parliament and of the Council of 17 May 2006 on machinery, and amending Directive 95/16/EC, OJEU No. L 157, pp. 24 *et seq.*

[62] Neudörfer (2014), p. 178.

[63] Wilrich (2015), p. 175.

measures to ensure an appropriate level of security. Thus, the GDPR includes a general duty to guarantee IT security by means of a general clause.

3.1.3 Conclusion

As an interim conclusion, it can be said that that many jurisdictions such as the Germany legal system do not lack laws dealing with IT security. In some cases, such as in the European recent legislation on the security of financial institutions, IT security is explicitly addressed and, in data protection law, this is not a new phenomenon. In many cases, however, specific requirements for IT security must be inferred from general security norms.

In all cases, specification of the requirements for IT security is necessary and this task is, in the main, undertaken by the supervisory authorities although they are assisted in this by the courts.

3.2 Government Supervision

The regulation of trade and industry by supervisory authorities has a long tradition in many jurisdictions and is still of great importance today. Supervision can, in a similar way to legislation, be differentiated with respect to the application of certain safety hazards or to the goods protected.

Government supervision in Germany is mostly sector-specific and most sectors, for example the energy sector,[64] the financial sector,[65] the aviation sector,[66] and the rail sector,[67] have their own supervisory authorities. Furthermore, there is the General Trade Supervisory Authority for the trade and industrial sectors whereas the safety of buildings is supervised by Construction Supervision and the safety of motor vehicles by the Federal Motor Vehicle Office, and so on.

There are also special supervisory authorities for the protection of certain goods and institutions, such as, by way of example, the Antitrust Supervision to ensure fair competition and the data protection authorities to protect privacy rights against automated data processing.

With the establishment of the BSI, the German Federal Office for Information Security, Germany has implemented a special agency for IT security. For the issue at hand it would therefore be interesting to know the BSI's role in the legal regulation of IT security.

---

[64] Federal Network Agency (BNetzA).

[65] Federal Financial Supervisory Authority (BaFin).

[66] Federal Air Traffic Control Office (BAF).

[67] Federal Railway Authority (EBA).

The duties and powers of the BSI are stipulated in the BSI Act. § 3 of the Act lists the numerous duties assumed by the BSI. However, it is important to note that the supervision of the IT security of commercial enterprises is not among the duties listed in this statutory provision. Consequently, the BSI had, until recently, no authority to impose measures in this area. This has only recently changed by the introduction of the IT Security Act, which provides the BSI with certain supervisory powers for the area of critical infrastructures, meaning the operators of 500–2000 facilities across Germany. However, as already stated, the BSI has no supervisory authority for the area of the industrial internet as this area remains outside the ambit of the IT Security Act.

This leads to the conclusion that there is no general supervisory authority for IT security in Germany, and, in particular, no general supervisory authority for the industrial internet.

Data protection authorities can have a specific relevance to IT security. This is particularly true in the case of the European GDPR. Art. 32 GDPR obliges controllers and processors to undertake adequate measures to guarantee IT security whilst processing personal data.

Under the GDPR, data protection authorities are charged with the supervision of data protection and supervise the compliance of private businesses and thus also the industrial sector, by virtue of powers given to them by Art. 57 GDPR. In particular, the authorities have strong supervisory powers in the execution of their duties.

Art. 32 GDPR contains a general clause on data security and it is the responsibility of the data protection authorities to precisely clarify how this legislation should be interpreted and applied. Thus, in trying to assess the effectiveness of IT security regulation in the area of data protection, the measures implemented by the data protection authorities are extremely important.

Similar clauses on data security were already part of the predecessor of the GDPR, the European Data Protection Directive of 1995,[68] and were also included in earlier national regulations such as the data protection laws in Germany which exist since the 1970s. Also, data protection authorities with extensive powers of intervention exist since the early 1990s. Thus, German data protection law and practice provides a sufficiently long time frame for analysis.

The results of such analysis were unexpected. German data protection authorities have been very active in issuing recommendations on numerous aspects of data protection, including data security, both on national as well as on European level. At national level, the Conference of Independent German Federal and State Data Protection Supervisory Authorities (DSK), has served as a leading institution in the field of elaborating recommendations. Since 1992, the DSK published hundreds

---

[68] Directive 95/46/EC of the European Parliament and of the Council of 24 October 1995 on the protection of individuals with regard to the processing of personal data and on the free movement of such data, OJEC No. L 281, pp. 31 *et seq*. Official Journal L 281, 23/11/1995 P. 0031 – 0050.

of resolutions and several position papers[69] giving guidance and advice to the practice as well as to policy makers.

However, despite the German Federal Data Protection Act with its strong supervisory authorities having been in place since 1991, there has been, until 2018, no measurable regulation by the German data protection authorities. There have been almost no published court rulings on supervisory measures by any data protection authorities concerning data security and there are only very few formal decisions by data protection authorities in this area and hardly any supervisory procedures.

This situation has changed dramatically under the GDPR. The GDPR was applicable from 25th May 2018, and ever since, a remarkable increase in individual decisions made by Data Protection Authorities can be seen. Those decisions in individual cases cover a broad range of aspects, including data security.

It is interesting to note that insufficient guarantee of technical organisational measures and inadequate IT security design are the second most common reason for a GDPR breach.[70] Also, some of the highest fines were imposed by authorities for violation of data security requirements. A fine of 22.4 million euros, the fourth highest fine by end of 2020, was imposed on British Airways for inadequately securing the personal data of over 400,000 customers.[71] The fifth highest fine, amounting to 20.45 million euros, was imposed on Marriott International Inc. for years of not taking technical and organisational security measures for millions of hotel guests' data in compliance with the GDPR.[72]

3.3 *Jurisprudence*

Jurisprudence is of great significance, not only in enforcing but also in establishing normative requirements, as it specifies legal norms. This is particularly relevant to IT security in the industrial internet, as the legal regulation of IT security, as previously shown, is mostly limited to general clauses.

Historically, jurisprudence has repeatedly established duties of care regarding IT security based on general norms, such as § 823 para. 1 BGB. In the area of product safety, for example, jurisprudence has developed the entire field of manufacturer liability based on § 823 para. 1 BGB.[73] Manufacturer liability obliges the producers to, among other things, monitor the safety of their products even after placing the

[69] Position papers of the DSK, available at https://www.bfdi.bund.de/DE/Home/home_node.html. Accessed 1 September 2021.

[70] Statistics to "Highest fines by type of violation", available at https://www.enforcementtracker.com/?insights. Accessed 1 September 2021.

[71] Information Commissioner's Office, Case Ref.: COM0783542.

[72] Information Commissioner's Office, Case Ref.: COM0804337.

[73] See, as an overview, Wagner (2020b), § 823 at mn. 949 ff.; Wilhelmi (2020b), § 823 at mn. 111 ff.

product on the market.[74] The rules of manufacturer liability also apply to IT security in the industrial internet.[75]

It is more difficult for jurisprudence, however, to specify technical requirements. In Germany, for instance, there is a distinct lack of published rulings regarding, for example, security requirements to ward off malware attacks. There are individual court rulings which state that internet users must keep their virus protection software up to date. It seems that there is indeed only one published court decision, by the Hamburg Regional Court, granting damages in the context of insufficient IT security. In this case, the defendant was to duplicate floppy disks on behalf of the plaintiff and send them to his customers. In the process, malicious software introduced by third parties was unknowingly sent. The court ruled that the defendant had breached an unwritten contractual duty because he used an outdated version of a virus scanner in its systems.[76]

Such cases would, as result of the fundamental structural problem of IT regulation by jurisprudence, not seem to be suited to establishing the technical security requirements for the industrial internet. As previously stated, not only is there a lack of published court rulings but the overall number of lawsuits is small compared to other areas and lawyers report that almost all these cases end in a settlement.

Without published court decisions, however, jurisprudence cannot establish legal requirements for IT security.

A further problem is the duration of court cases. In Germany, for instance, the decisions of the German Federal Supreme Court (Bundesgerichtshof, BGH) are the main reference points. However, several years can pass between the initial event taking place and a ruling by the Federal Supreme Court being made and often several decisions are necessary to settle a legal issue. This leads to the situation that the technical basis, which the Federal Supreme Court references in its decision, is often no longer relevant due to the pace of technical innovation. Thus, jurisprudence is always a step behind technological developments.

The conclusion can thus be drawn that, although jurisprudence has accrued great merit in establishing duties of care, which are also of importance for IT security, it is only of very limited use in establishing technical requirements. Such conclusion is unsatisfactory as the legislation consists mostly of general clauses and specification of such clauses is urgently required.

---

[74] BGH, case Ref.: V ZR 153/79; BGH, case Ref.: VI ZR 191/79; Borges (2020), mn. 82; Förster (2021a), mn. 731 ff.; Wagner (2020b), mn. 988 ff.

[75] Spindler (2021), mn. 743, 754.

[76] LG Hamburg, Case Ref.: 7 O 3154/01.

3.4 Standards

Standards are of great importance for industry. Standardisation organisations, such as the ISO, are commonly implemented and it can be said that recognised standards are the "law" for industrial production.

Standards for IT security already exist. For example, in Germany there is the so-called BSI Baseline Protection[77] and internationally, the standards of the ISO-27000 Group on the management of IT security are very influential. The ISO Standard 27002 already describes requirements for IT security in a detailed manner. If standards are such a successful instrument in standardising technical requirements in industry and if there are standards for IT security, it would seem possible to see the solution for the task of regulation in the development of standards for IT security in the industrial internet.

Regarding regulation however, meaning the establishment of binding requirements, industry standards show structural problems. Firstly, such standards are not automatically legally binding[78] and are only legally enforceable if they are referenced by legal norms.[79] However, direct legal references to standards are very rare.[80] Although standards are often contractually incorporated,[81] and this might be of great relevance for the industrial internet,[82] this does not create a legal obligation to protect third parties.[83]

As stated, standards are used by courts to specify legal requirements. This is not done in a consistent manner, however, and it is also very unclear under which circumstances technical standards will be recognised as specifications of legal requirements by the courts.[84]

3.5 Interim Result

The interim result shows that currently there is tremendous uncertainty concerning legal requirements for IT security. There is a distinct lack of legal specifications of

[77] See an overview of current standards of the German BSI (2021).

[78] Illies et al. (2020), mn. 83, 35.; Marschollek (2020), § 7 at mn. 83; Kanevskaia (2020), p. 559; focused on European Standardization: van Leeuwen (2017), p. 77.

[79] Illies et al. (2020), mn. 83, 35.; Marschollek (2020), § 7 at mn. 227; Kanevskaia (2020), p. 559; van Leeuwen (2017), p. 77.

[80] For more detail on the problems see Thiele (2020), pp. 309 f.

[81] Kanevskaia (2020), p. 560.

[82] BMWi (2016b), p. 179; Scientific Service of the German Bundestag "DIN-Normen und Rechtssetzung" WD 7 – 3000 – 198/19, p. 5.

[83] On the question of liability towards third parties under contractual agreements, see also Borges (2011), p. 168; Borges (2014), pp. 395 f; Meyer (2014), p. 52.

[84] Borges (2020), mn. 49, 55.

requirements for IT security and there has been no noteworthy specification by the courts. While the supervisory authorities do provide specifications, this is not done in a systematic or comprehensive manner and there are no specifications for the industrial internet specifically.

3.6 Possible Approach: Linking Legal Requirements to Private Norms

In the area of establishing substantive legal requirements for IT security, an interesting approach is currently being pursued in a few cases in Germany. The IT Security Act 2015,[85] which regulates the duties for the maintenance of IT security of critical infrastructures, provides a very interesting example in this context. The duty to maintain IT security is regulated in § 8 a para. 1 of the BSI Act. It reads:

> (1) Operators of critical infrastructures shall take appropriate organisational and technical measures to prevent disruptions to the availability, integrity, authenticity and confidentiality of their IT systems, components or processes that are essential for the functioning of the critical infrastructures they operate no later than two years after the entry into force of the statutory instrument pursuant to Article 10(1). In doing so, the state of the art shall be observed. Organisational and technical precautions are appropriate if the effort required is not disproportionate to the consequences of a failure or impairment of the critical infrastructure concerned.

§ 8a para. 1 BSI Act, by which Art. 14 of the NIS Directive is transposed into German national law,[86] is a classic general clause, comparable to the obligation of data security in Art. 32 GDPR. Therefore, the obligation set forth in § 8a para. 1 requires further clarification and specification.

To offer such clarification, the German legislator added a second paragraph to § 8a BSI. § 8a para. 2 BSI Act reads:

> (2) Operators of critical infrastructures and their sector associations may propose sector-specific security standards to ensure the requirements under paragraph 1. Upon application, the Federal Office shall determine whether these are suitable for ensuring the requirements pursuant to paragraph 1. The determination shall be made
>
> 1. in consultation with the Federal Office of Civil Protection and Disaster Assistance,
> 2. in consultation with the competent supervisory authority of the Federation or in consultation with the otherwise competent supervisory authority.

The approach taken by § 8a para. 2 BSI Act is highly innovative and interesting. Unlike the previous situation, the specification is not left to jurisprudence or the BSI

[85] See supra, Sect. 1.2.

[86] For a deeper analysis, see Müllmann et al. (2020), p. 584 at mn. 7; Schallbruch (2017), pp. 648, 651.

but it is to be determined by security standards developed by industry itself. By complying with these standards, the legal obligation is simultaneously fulfilled.[87]

This regulation creates legal certainty for the companies and the companies that comply with the standards can be certain that they are simultaneously fulfilling their legal obligations.[88] Although this legal effect is, unfortunately, not explicitly stated in the act itself, it would seem to be the clear basis of the legal regulation. This establishes the connection between legal obligations and standards which had previously been lacking. These sector-specific standards then become supervisory minimum requirements, in the way which the government aimed to establish with the IT Security Act.[89]

4 Enforcement of Security Requirements

Companies have a natural interest in IT security in order to protect their own property, and often take adequate measures to do so. However, insufficiency in this area can be observed.

Many companies are not aware of the extent of possible risks. According to a worldwide survey of Bitdefender in October 2019, IT-Security Experts report, that especially upper management in particular often disregards IT security aspects.[90] Similarly, a recent study by the Association of the German Insurance Industry (*Gesamtverband der Deutschen Versicherungswirtschaft*) concluded that companies often underestimate the risks of a lack of IT security, in particular incorrectly evaluate the level of protection provided by existing IT security measures as well as the probability of being attacked.[91]

If companies do not adequately protect their own property by insufficient implementation of IT security measures, they may do so at their own risk. From a legal perspective however, the protection of third parties against harm resulting from such insufficient implementation is the focus of the legislator.

This is of particular importance for the industrial internet, in which the integrated companies rely on their partners to maintain adequate IT security and in which the users of products depend on the product's IT security.[92]

---

[87] Borges (2016), p. 170 f.

[88] Borges (2016), p. 171.

[89] Borges (2016), p. 171.

[90] Bitdefender (2019), pp. 11, 32.

[91] Gesamtverband der Deutschen Versicherungswirtschaft (Ed.), 2020, p. 6.

[92] See, for instance, the description of scenarios related to Industry 4.0, BMWi (2016a), p. 94 f.

4.1 Overview: Enforcement Mechanisms for IT Security

There are various mechanisms which can be used to enforce the material requirements for IT security.[93] One mechanism to enforce the fulfilment of duties, which is highly regarded by economists, is the market itself, because market participants price in compliance with the duties and requirements and react adversely to expected violation.[94]

This mechanism could also be effective for IT security. Impending reputational damage due to insufficient IT security could be an incentive to establish IT security and compliance with high security standards could prove to be a competitive advantage. For instance, in the area of cloud computing, German providers of cloud services advertise the fact that their services are especially safe from access by US government agencies, at least, much safer than those of their US competitors, and this factor seems to be of importance within the market.[95]

Criminal sanctions can also present an incentive to comply with norms although currently no criminal offences specific to providing a certain level of IT security have been defined.

Another essential enforcement mechanism is supervision, where government agencies can enforce measures for IT security with administrative acts in individual cases and have the full range of governmental coercive instruments at their disposal.

As stated previously (supra Sect. 2.1), civil liability can be an important enforcement instrument, as the obligation to compensate for damage when such damage occurs provides an incentive for the potentially liable party to prevent damages by employing appropriate security measures.

From the perspective of the legal regulation of IT security, the question of which measure or combination of enforcement instruments would prove most effective in the area of IT security remains to be answered. In the following, a brief analysis of the importance of civil liability as an enforcement instrument for IT security will be given.

---

[93] Riehm and Meier (2020), pp. 571 ff.

[94] Cf. Shabbir and Padgett (2008), pp. 167 f.; generally on the market mechanism and reputation Klöhn and Schmolke (2015), pp. 689 f.

[95] As an example, A1 Digital advertises German servers, DSGVO compliance and that they are not subject to the US Patriot Act, available at https://www.a1.digital/loesungen/virtuelle-infrastruktur/virtuelle-infrastruktur/exoscale-europaeische-cloud-infrastruktur/ (accessed 1 September); Global Access, promises protection from access by the NSA and all other security agencies in the world through its server location and IT infrastructure, available at https://www.global.de/unternehmen/. Accessed 1 September.

4.2 *Factors of Effective Norm Enforcement Through Liability*

As the initial example of the current discussion concerning product liability of the producers of software for IT security shows, the enforcement of security requirements through civil liability is currently being debated in Germany/Europe.

From the perspective of IT security regulation, this leads to the question of whether the enforcement of IT security requirements can effectively be guaranteed using civil liability instruments.

A possible starting point would seem to be the notion that, in order to effectively enforce legal norms through liability, clear conditions for said liability are required. This would lead to two key requirements. Firstly, clear substantive legal conditions for liability must be established as uncertainties regarding the substantive conditions for liability would lead to a significant litigation risk for the claimant. Secondly, the conditions for liability must be easily ascertainable for the claimant in a lawsuit. Under German law, the claimant bears the burden of proof regarding the factual basis of his claim requiring him to present and prove all the facts from which his claim results.

The current situation concerning these requirements in the case of liability for insufficient IT security must be examined.

Unfortunately, clear material conditions of liability simply do not yet exist and there is a lack of clear legal requirements for IT security under German law. This fact alone would seem to turn a liability lawsuit into a significant risk.

Secondly, the factual basis for liability has to be proven in a civil lawsuit. The norms which have been applied to IT security to date, such as § 823 para. 1 BGB, require a breach of duty.

Proving such a breach of duty can often be very difficult, even if there are clear conditions for liability, as it would have to be shown how the damages occurred and how the defendant's behaviour contributed to the damages.

In the case of IT security flaws, this would require a comprehensive technical investigation. Whether this is a feasible possibility for a claimant in a civil lawsuit depends, amongst others, on the rules of procedural law in the respective country. While in many Common Law jurisdictions,[96] such as US,[97] the United Kingdom[98] or Ireland[99] there are procedures like the pre-trial discovery which allows for

---

[96] For the distinction of Common Law and Statutory Law tradition, see Kischel (2019), p. 242 f.

[97] Schack (2020), rec. 110 ff., 143; Landwehr (2017), p. 78, 107 ff.; Bareiß (2014), pp. 14 ff.; Klein (2019), pp. 7 ff.; Stempfle (2017), § 36 rec. 43 f.; U.S. Supreme Court, Hickman v. Taylor, 329 U.S. 495, pp. 507 ff.; Hazard and Taruffo (1993), pp. 115 ff.; Tolman (1958), pp. 498 ff.; Rosenberg (1958), pp. 480 ff.; Kaplan and Clermont (1984), p. 17.

[98] Art. 31 ff. of CPR (Civil Procedure Rules); Kaplan and Clermont (1984), p. 14 f.; Sydow (2019), p. 70.

[99] Art. 31 of RSC (Superior Court Rules); Moriarty, Evidence in Civil Law – Ireland (2015), pp. 21 ff.

extensive pre-trial fact finding, such instruments are not available in many Civil Law jurisdictions such as Germany[100] or France.[101] Outside pre-trial discovery, collecting the facts to establish the basis of a claim for damages is hardly ever a feasible possibility for a claimant in a civil lawsuit. In addition, proving the amount of damage can also be very difficult.

Another, and often possibly decisive, obstacle in cases of IT security flaws is the need to prove causation. It has to be proven that the damage would not have occurred if the unimplemented measure had in fact been implemented. Considering the multitude of possible attack scenarios, this is often impossible. By way of example, under German law, it is an established principle that the users of online banking have to keep their anti-virus protection up to date.[102] If damage occurs in online banking as a result of malware and it can be shown that the anti-virus protection was not up to date, then a breach of duty has been proven. However, it must be established that the malware program in question would have been recognized and eliminated by up-to-date anti-virus protection software. Such proof is often very difficult, as the malware can either often not be unequivocally identified, or such identification only be established with tremendous effort.[103]

The interim result is thus that the enforcement of IT security through liability is currently not effective.

5 Reflections De Lege Ferenda

5.1 Necessity of Legal Measures

The final part of this paper will contain an analysis of options for future legislation. As stated above, the German Ministry of Economic Affairs and Energy published a study in 2016 on the state of IT security in the industrial internet, which proposes, among others, measures to improve the legal situation. Among the measures set out in the study is the improvement of the legal framework for IT security through legislative measures. It proposes that the legislator promotes the development of standards, for example by expanding the mechanisms introduced in the IT Security Act to other areas of application. Other proposals include the legal regulation of IT Security Certification. The expansion of the powers of governmental authorities in

[100] Adler (2014), p. 138; Bareiß (2014), pp. 1 ff.; Dombrowski (2016), pp. 319 f.; Kern (2010), pp. 78, 84; Landwehr (2017), p. 78; Müller-Stoy (2005), pp. 563 f.; Schönknecht (2011), p. 1000.

[101] Tribunal de grande instance de Paris, Chambre civile 3, 01.02.2008, n° 06/12688; Cour de cassation, Chambre civile 2, 30.04.2009, n° 08-14.883.

[102] LG Berlin, Case Ref. 37 O 4/09; LG Köln, Case Ref. 9 S 195/07; Bender (2008), p. 2054; Borges (2011), p. 145; Borges et al. (2011), pp. 284 f.; Schulte am Hülse & Klabunde (2010), p. 87; Riehm and Meier (2020), p. 573.

[103] Borges (2012), p. 2386; Borges (2011), p. 175 f.; Borges et al. (2011), pp. 291, 313; BMWi (2016b), p. 178; Werner (2010), p. 177 f.

the area of IT security is also suggested. The effect of such proposals is that legislative measures would be necessary not only to establish standards but also to enforce them.

Following the recommendations of the study, improvements of the legal framework's conditions for IT security in the industrial internet are to be achieved, but still the question of by which measures this can be attained remains unanswered. One assessment could be that, to further develop the legal framework conditions, all instruments to establish and enforce security requirements must be used as only then the specific advantages of each instrument can be harnessed. This requires, however, co-ordination of the various instruments as otherwise the result would be duplicate provisions, which would put an unnecessary financial burden on the industry. It could also lead to contradictory requirements, which would affect legal security.

Civil liability and certification are two instruments which would seem to have the potential to be effective in enforcing security standards, which is why they will be discussed briefly in the following.

5.2 *Aspects of IT Security Through Liability*

5.2.1 Advantages of Steering Through Liability

As mentioned above, it is currently being discussed in Germany/Europe how IT security can be improved by introducing liability. This approach would seem to be laudable and rules on civil liability for breaches of duty have significant advantages as regulatory instruments.

Civil liability paves the way for the enforcement instrument of the civil lawsuit. This instrument is especially powerful, as the obligated party can be faced with a number of possible lawsuits if a breach of duty has occurred and, unlike in the case of regulatory oversight, the enforcement cannot be impaired by a lack of personnel.

5.2.2 Improved Regulation

The behavioural impact of liability provisions can only take effect if it is ensured that the addressee of the norm can expect civil lawsuits if he violates such a norm. This requires, as discussed, the existence of clear legal requirements, which hardly exist in the area of IT security.

There is still a great deal of research to be done. Clear conditions of liability can be attained in various, completely different, ways. One way, for example, is the introduction of strict liability, whereby liability no longer depends on proof of an error or a culpable breach of duty. In such cases, the proof of prerequisites for liability is made easier as compared to traditional fault-based liability, as a breach of duty no longer has to be proven.

A clarification of conditions for liability can also occur through technical standards. Standards can also be structured in different ways. Standards can be used as a kind of "safe harbor" whereby reaching the standard would automatically fulfil the relevant legal requirements.[104]

The result is the use of standards as a form of liability exemption and, indeed, the term "safe harbor" is used in the area of the liability of managing directors, and similarly in some other cases, such as in the context of the liability privilege for providers in US law, as well as in the context of liability for faulty prognoses in US capital markets law. Standards can also be established as mere minimum requirements in which case a failure to meet the standards would be deemed a breach of duty and lead to liability. However, even if the minimum standards are met, civil liability cannot be excluded.

A general preference for one regulatory model or another should not be categorically expressed. In general, provisions for strict liability as currently discussed in Europe for systems equipped with Artificial Intelligence[105] exist for very specific cases only. "Safe harbor" solutions do not seem to be suitable as basic principles for civil liability, as they would create serious liability gaps. However, minimum standards would seem to be of interest as they could, at least partially, solve the problem of legal uncertainty regarding the conditions for liability. This would require explicit legal regulation, which is still lacking in German as well as in European Law. Also, the current EU IT Security Legislation does not contain such a solution.

If IT security is to be improved through liability, then this will not only require clarification of the substantive conditions for liability but it will also be necessary to make it easier for claimants to demonstrate and prove the prerequisites for liability in a civil lawsuit. As mentioned, there are still great deficits in this area and therefore, legislative action needs to be taken.

A possible hypothesis could be that regulation of the burden of producing evidence for producers of IT products as well as for providers of IT services should be a minimum requirement for an effective enforcement of claims in civil lawsuits. The burden of presentation has the effect that the bearer of that burden, meaning the providers of IT services or products, must demonstrate in a civil lawsuit that he has fulfilled his duties regarding IT security.

Such a burden of producing evidence can be found in current law for certain cases. There is great legal uncertainty, however, regarding the extent of such burdens of producing evidence and such burden can lead to further problems, especially concerning the protection of commercial secrecy. The burden of producing evidence should not lead to a situation where the providers of IT services and producers of IT products can be forced to divulge commercial secrets and sensitive information in

---

[104] Borges (2016), p. 170; Spindler and Thorun (2016), p. 20 f.

[105] Cf. the Proposal of the European Parliament's committee on legal Affairs for a Regulation on liability for the operation of Artificial Intelligence-systems, 2020/2014(INL).

public civil proceedings. There are appropriate protective measures in place in civil proceedings in order to avoid this, but clarification of such cases seems necessary.

5.3 Certification

Inspection and certification would also seem to play an important role in improving IT security. Certification, meaning statements based on inspection, has significant advantages and is already widely implemented in many jurisdictions.

Inspection and certification for security purposes is already often required by law. A well known case is the legally mandated inspection of motor vehicles which exits in many countries. Under German law, for instance, Road traffic Law obliges every owner of a car to take their car to regular safety inspections[106] and the compliance with the safety requirements is confirmed at the end of the inspection by the inspecting authority by sticking an inspection badge on the car.[107]

Such an inspection badge has to be affixed on the car's number plate[108] where it is easily visible, so that the police and other authorities can see if the inspection was successfully completed.

Similar inspection and certification duties apply to many facilities. Such legal certification procedures are implemented by the legislator if a large number of facilities require inspection and the authorities would not have enough personnel to inspect all of them.

Voluntary certification would seem to present an even more interesting option and its relevance to IT security is currently being intensely discussed in Europe. The GDPR, in its provisions on certification (Art. 42 et seq) encourages Member States to support voluntary certification schemes and provides strong incentives for the use of certification by controllers and processors. In the case of voluntary certification, the provider of a service or the producer of a product is able to choose if he wants to be inspected and certified. The benefit of such inspection and certification would be that he can then use the certification to strengthen his reputation in the marketplace.[109] Voluntary certification aims to create trust.

This approach could indeed be of great interest for the area of the industrial internet. In the industrial internet, companies cooperate with many partners and offer them access to their electronic networks, which makes them vulnerable. In this situation, the possibility of being able to trust the IT security of the partners is of

[106] See e.g. Glinski and Rott (2018), p. 92.

[107] This obligation is contained in § 29 Abs. 1 S. 1 Road Traffic Permit Act (Straßenverkehrszulassungsverordnung—StVZO).

[108] § 29 Abs. 2 S. 2 StVZO reads: "Inspection stickers must be issued by the authority in charge under national law or by the persons authorized to carry out general inspections and must be affixed permanently to the rear registration plate in such a way that they cannot be misused."

[109] See e.g. in the context of voluntary certification within the GDPR, Bergt and Pesch (2020), Art. 42 at mn. 4; Maier et al. (2020), p. 446; Raschauer (2018), Art. 42 at mn. 2.

fundamental importance. It would seem reasonable to use certification to create trust. If it is assumed that the industrial internet means that, in the future, companies will be communicating with a multitude of cooperation partners through electronic networks, then it would seem clear that there may be no alternative to certification as a way of creating trust in IT security. Certification has the crucial benefit that the necessary inspection does not have to be performed by every cooperation partner, but only by one agency as the certification authority performs the inspection for all cooperation partners. Certifications equal efficiency regarding the inspection they are based on.

However, trust based on certification is only likely to be successfully built if the requirements for granting the certificate and the procedure, especially the requirements for a proper inspection, are clear and transparent. There are still large deficits in this regard as is highlighted by the internationally recognized ISO Standard 27001. Intensively used in certification, this standard is very flexible. Whilst this is not objectionable *per se*, it can lead to the situation that certifications according to ISO 27001 are not comparable for third parties. To create trust in a sufficient level of IT security, comparable standards are indispensable.

One possible way of removing this obstacle could be the establishment of binding certification standards. This approach is established in the area of data protection law. Today, the European General Data Protection Regulation (GDPR) encourages voluntary certification, which seems to be particularly promising in the field of IT security. The GDPR provides, for the first time in data protection regulation, a legal framework for certification including requirements for certification in its Art. 43 as well as legal effects of a certificate.[110]

In Germany, the mechanism of certification in the area of data protection has been set up earlier in a pilot project on behalf of the Federal Ministry for Economic Affairs and Energy.[111] In this project, a certification standard for cloud services, the so-called Trusted Cloud Data Protection Profile for Cloud Services (TCDP) based on ISO standards has been created using the instruments of the 27000 standard family to establish a binding minimum standard.[112]

6 Conclusion

The need for the regulation of IT security has been recognized by some national legislators, and the European legislator as well as some national legislators, e.g. Germany are working towards a satisfactory solution.

[110] E.g. Art. 24 para. 3, Art. 25 para. 3, Art. 28 para. 5, Art. 32 para. 3 of GDPR.

[111] See the project homepage, available at https://tcdp.de (accessed 1 September 2021); see also Borges (2017), p. 69 ff.

[112] Borges (2017), p. 70 f.

An interesting approach for the establishment of substantive security requirements would seem to be the combination of self-regulation with legal requirements. On the level of substantive requirements, standards are of interest. Certification could also be an interesting enforcement instrument.

The regulation of IT security requirements urgently needs further interdisciplinary research, especially to optimize the interaction of the individual regulatory instruments with each other.

References

Adler, D. B. (2014). Entdecke die Möglichkeiten – US discovery für ausländische Verfahren. *Zeitschrift für Deutsches und Amerikanisches Recht (ZDAR)*, 138–145.

Bareiß, A. (2014). *Pflichtenkollisionen im transnationalen Beweisverkehr. Offenbarungspflichten im Zivilprozessrecht der USA und Offenbarungsverbote nach deutschem und europäischem Recht*. Mohr Siebeck.

Beckmann, K., & Müller, U. (1999). Online übermittelte Informationen: Produkte i.S.d. Produkthaftungsgesetzes? *Multimedia und Recht (MMR), 1*, 14–19.

Bender, J. (2008). Aktuelle Entwicklung der Haftung bei Phishing. Zugleich Besprechung der Urteile LG Köln WM 2008, 354 und AG Wiesloch WM 2008, 1648. *Wertpapier-Mitteilungen (WM)*, 2049–2059.

Bergt, M., & Pesch, P. J. (2020). Art. 42 DSGVO. In J. Kühling & B. Buchner (Eds.), *Datenschutz-Grundverordnung, Bundesdatenschutzgesetz* (3rd ed.). C.H.Beck.

Beucher, K., & Utzerath, J. (2013). Cybersicherheit – Nationale und internationale Regulierungsinitiativen. Folgen für die IT-Compliance und die Haftungsmaßstäbe. *Multimedia und Recht (MMR), 6*, 362–367.

Beukelmann, S. (2017). Cyber-Attacken – Erscheinungsformen, Strafbarkeit und Prävention. *Neue Juristische Wochenschrift (NJW)*, Spezial 2017, 376.

Bitdefender. (2019). *Hacked Off! A comprehensive study into the cybersecurity attitudes of infosecurity professionals around the world*. Retrieved August 28, 2021, from https://www.bitdefender.com/files/News/CaseStudies/study/285/Bitedefender-Hacked-Off-Report.pdf

Borges, G. (2011). *Rechtsfragen der Haftung im Zusammenhang mit dem elektronischen Identitätsnachweis*. Nomos.

Borges, G. (2012). Haftung für Identitätsmissbrauch im Online-Banking. *Neue Juristische Wochenschrift (NJW), 65*(33), 2385–2389.

Borges, G. (2014). In G. Borges (Ed.), *Rechtfragen der Internet-Auktion* (2nd ed., pp. 387–414). Nomos.

Borges, G. (2016). Neue Ansätze zur Regulierung von IT-Sicherheit. In S. van Oostrom & S. Weth (Eds.), *Festschrift für Maximilian Herberger* (pp. 159–173). Juris.

Borges, G. (2017). In E. Schweighofer, F. Kummer, W. Hötzendorfer, C. Sorge (Eds.), *Trends und Communities der Rechtsinformatik - Trends and Communities of Legal Informatics, Tagungsband des 20. Internationalen Rechtsinformatik Symposions IRIS2017*. OCG.

Borges, G. (2020). § 3 ProdHaftG. In G. Borges, M. Hilber (Eds.), *Beck'scher Online-Kommentar IT-Recht* (1st ed. 1 Sep 2020). C.H.Beck.

Borges, G., Schwenk, J., Stuckenberg, C.-F., & Wegener, C. (2011). *Identitätsdiebstahl und Identitätsmissbrauch im Internet*. Springer.

Bundesamt für Sicherheit in der Informationstechnik (BSI)/Federal Office for Information Security. (2016). *Lagedosssier Ransomware*. Retrieved August 28, 2021, from https://www.bsi.bund.de/SharedDocs/Downloads/DE/BSI/Cyber-Sicherheit/Themen/Lagedossier_Ransomware.pdf?__blob=publicationFile&v=2

Bundesamt für Sicherheit in der Informationstechnik (BSI)/Federal Office for Information Security. (2018). *Die Lage der IT-Sicherheit in Deutschland 2018*. Retrieved August 28, 2021, from https://www.bmi.bund.de/SharedDocs/downloads/DE/publikationen/themen/it-digitalpolitik/bsi-lagebericht-2018.pdf?__blob=publicationFile&v=3

Bundesamt für Sicherheit in der Informationstechnik (BSI)/Federal Office for Information Security. (2019). *Cyber-Sicherheits-Umfrage – Cyber-Risiken & Schutzmaßnahmen in Unternehmen*. Betrachtungszeitraum 2018. Retrieved August 28, 2021, from https://www.allianz-fuer-cybersicherheit.de/SharedDocs/Downloads/Webs/ACS/DE/cyber-sicherheitsumfrage_2018.pdf?__blob=publicationFile&v=1

Bundesamt für Sicherheit in der Informationstechnik (BSI)/Federal Office for Information Security. (2021). *BSI-Standards*. Retrieved August 28, 2021, from https://www.bsi.bund.de/DE/Themen/Unternehmen-und-Organisationen/Standards-und-Zertifizierung/IT-Grundschutz/BSI-Standards/bsi-standards_node.html

Bundesanstalt für Finanzdienstleistungsaufsicht (Federal Financial Supervisory Authority – BaFin). (2017). *Annex 1: Annotated text of the Minimum Requirements for Risk Management (MaRisk) in the version of 27.10.2017*. Retrieved August 28, 2021, from https://www.bafin.de/SharedDocs/Downloads/EN/Rundschreiben/dl_rs0917_marisk_Endfassung_2017_pdf_ba_en.html;jsessionid=B1BCC1A3CE6BE5B1EC06FDE3BE3889BA.1_cid500?nn=9021442 (English version)

Bundesministeriums für Wirtschaft und Energie. (2016a). *Studie im Auftrag des Bundesministeriums für Wirtschaft und Energie*. IT-Sicherheit für die Industrie 4.0. Produktion, Produkte, Dienste von morgen im Zeichen globaler Wertschöpfungsketten. *Abschlussbericht*. Retrieved August 28, 2021, from https://www.bmwi.de/Redaktion/DE/Publikationen/Studien/itsicherheit-fuer-industrie-4-0.pdf?__blob=publicationFile&v=4

Bundesministeriums für Wirtschaft und Energie. (2016b). *Studie im Auftrag des Bundesministeriums für Wirtschaft und Energie*. IT-Sicherheit für die Industrie 4.0. Produktion, Produkte, Dienste von morgen im Zeichen globaler Wertschöpfungsketten. *Abschlussbericht – Kurzfassung*. Retrieved August 28, 2021, from https://www.bmwi.de/Redaktion/DE/Downloads/I/studie-it-sicherheit-fuer-industrie-4-0-kurzfassung.pdf?__blob=publicationFile&v=4

Cahn, A. (1996). Produkthaftung für verkörperte geistige Leistungen. *Neue Juristische Wochenschrift (NJW), 49*(44), 2899–2905.

Ceffinato, T. (2016). Vermögensstraftaten im und über das Internet. *Neue Zeitschrift für Wirtschafts-, Steuer- und Unternehmensstrafrecht (NZWiSt), 12*, 464–467.

Conrad, I. (2019). A survey on German IT security law. In A. Auer-Reinsdorff, I. Conrad (Eds.), *Handbuch IT- und Datenschutzrecht* (3rd ed., § 33, mn. 179–320). C.H.Beck.

Conrad, I., & Streitz, S. (2019). § 33 Compliance, IT-Sicherheit, Ordnungsmäßigkeit der Datenverarbeitung. II. Risikomanagement, Haftung der Geschäftsleitung, Compliance. In A. Auer-Reinsdorff & I. Conrad (Eds.), *Handbuch IT- und Datenschutzrecht* (3rd ed.). C.H. Beck.

De la Durantaye, K. (2020). § 823 BGB. In G. Borges, M. Hilber (Eds.), *Beck'scher Online-Kommentar IT-Recht* (1st ed. 1 Sep 2020). C.H.Beck.

Dombrowski, J. (2016). Discovery – auch im deutschen Gerichtsverfahren? *Gewerblicher Rechtsschutz und Urheberrecht Praxis im Immaterialgüterrecht und Wettbewerbsrecht (GRURPrax), 8*(15), 319–321.

Eckert, C. (2018). *IT-Sicherheit. Konzepte – Verfahren – Protokolle* (10th ed.). De Gruyter.

Fairgrieve, D., & Rajneri, E. (2019). Is software a product under the product liability directive? Member state and academic perspectives. *Zeitschrift für Internationales Wirtschaftsrecht (IWRZ), 4*(1), 24–27.

Fleischer, H. (2021). § 91 AktG. In M. Henssler, G. Spindler, E. Stilz (Eds.), *Beck'scher Online-Großkommentar zum Zivilrecht* (ed. 1 Feb 2021). C.H.Beck.

Fleming, J. G. (1992). *The Law of Torts* (8th ed.). The Law Book Company.

Förster, C. (2021a). § 823 BGB. In W. Hau, R. Poseck, *Beck'scher Online-Kommentar BGB* (57th ed. 1 Feb 2021). C.H.Beck.

Förster, C. (2021b). § 2 ProdHaftG. In R. Poseck & W. Hau (Eds.), *Beck-Online Kommentar zum BGB* (57th ed.). C.H.Beck.

Frisse, F., Glaßl, R., Baranowski, A., & Duwald, L. (2018). Unternehmenssicherheit bei Banken – IT-Sicherheit, Know-how Schutz, Datensicherheit und Datenschutz. *Zeitschrift für Bank- und Kapitalmarktrecht (BKR), 18*(5), 177–184.

Gesamtverband der Deutschen Versicherungswirtschaft e.V./Association of the German Insurance Industry. (2020). *Cyberrisiken im Mittelstand 2020.* Report Retrieved August 28, 2021, from https://www.gdv.de/resource/blob/61466/0456901217b39a5893bc6829b8d7d156/report-cyberrisiken-im-mittelstand-2020-data.pdf

Glinski, C., & Rott, P. (2018). The role and liability of certification organisations in transnational value chains. *Deakin Law Review, 23*, 83–117.

Hager, J. (2017). Vorbemerkung zu §§ 823 ff. In H.-P. Mansel (Ed.), BJ. v. Staudinger (foudner), *Staudingers Kommentar zum BGB* (ed. 2017). Sellier/de Gruyter.

Hamdan, M., & Günes, M. (2020). § 2 ProdHaftG. In M. Junker, R.M. Beckmann, H. Rüßmann (Eds.), *Juris PraxisKommentar BGB* (9th ed. Vol. 2). Juris.

Hazard, G. C., & Taruffo, M. (1993). *American civil procedure*. Yale University Press.

Henssler, M. (1993). Die zivil- und urheberrechtliche Behandlung von Software. *Monatsschrift für Deutsches Recht (MDR)*, 489–497.

Hornung, G. (2015). Neue Pflichten für Betreiber kritischer Infrastrukturen: Das IT-Sicherheitsgesetz des Bundes. *Neue Juristische Wochenschrift (NJW), 68*(46), 3334–3340.

Howarth, D., et al. (2015). *Hepple and Matthews' Tort law – cases and materials* (7th ed.). Hart Publishing.

Illies, G., & Lochter, M., & Stein, O. (2020). Part 15. Datensicherheit. 152 Kryptografie. In W. Kilian, B. Heussen (former Eds.), J. Taeger, J. Pohle (Eds.), *Computerrechts-Handbuch* (35th ed.). C.H.Beck.

Internet Organised Crime Threat Assessment (IOCTA). (2017). *Internet Organised Crime Threat Assessment 2017.* Retrieved August 28, 2021, from https://www.europol.europa.eu/iocta/2017/index.html

Kanevskaia, O. (2020). Governance of ICT standardization: Due process in technocratic decision-making. *North Carolina Journal of International Law, 45*(3), 549–618.

Kaplan, B. & Clermont, K. M. (1984). In Capelletti, M. (Ed.), *International Encyclopedia of Comparative Law, Vol. XVI: Civil Procedure*. Mohr Siebeck/Nijhoff.

Kern, C. (2010). Internationale Schiedsverfahren zwischen civil law und common law. *Zeitschrift für vergleichende Rechtswissenschaft (ZVglRWiss), 109*(1), 78–93.

Kipker, A.-K., Harner, A., & Müller, S. (2017). Der Mensch an der Schnittstelle zur Technik – Praxishilfe in der Umsetzung von Datensicherheit durch den IT-Security Navigator. In J. Taeger (Ed.), *Deutsche Stiftung für Recht und Informatik-Tagungsband (DSRITB)* (Vol. 8, pp. 651–665).

Kischel, U. (2019). *Comparative law*. Oxford University Press.

Klein, F. (2019). *Die Verwertbarkeit gem. 28 USC § 1782(a) erlangter Beweismittel im deutschen Zivilprozess*. Mohr Siebeck.

Klöhn, L., & Schmolke, K. U. (2015). Unternehmensreputation (Corporate Reputation). Ökonomische Erkenntnisse und ihre Bedeutung im Gesellschafts- und Kapitalmarktrecht. *Neue Zeitschrift für Gesellschaftsrecht (NZG), 18*(18), 689–697.

Kort, M. (1990). Produkteigenschaft medizinischer Software. Einordnung im deutschen und US-amerikanischen Produkthaftungsrecht. *Computer und Recht (CR)*, 171–176.

Krämer, N., & Dahmen, U. (2017). Trojaner im KIS. *Führen & Wirtschaften im Krankenhaus (f&w), 16*. Retrieved August 28, 2021, from https://www.bibliomedmanager.de/zeitschriften/fw/heftarchiv/ausgabe/artikel/fw-1-2017-crime/31425-trojaner-im-kis/

Landwehr, J. (2017). *Die Pretrial Discovery. Eine funktionale Analyse*. Peter Lang.

Lehmann, M. (1992). Produkt- und Produzentenhaftung für Software. *Neue Juristische Wochenschrift (NJW), 45*(28), 1721–1725.

Littbarski, S. (2020). Part 18. Produkthaftung In W. Kilian, B. Heussen (Former Eds.), J. Taeger, J. Pohle (Eds.), *Computerrechts-Handbuch* (35th ed., June 2020). C.H.Beck.

Magnus, U. (2016a). Germany. In P. Machnikowski (Ed.), *European product liability – An analysis of the state of the art in the era of new technologies* (pp. 237–274). Intersentia.

Magnus, U. (2016b). Vorbemerkungen zu §§ 249–255. In B. Dauner-Lieb, Heidel, T., & Ring, G. (Eds.), *NomosKommentar BGB* (3th ed., Vol. 2,1). Nomos.

Maier, N., Pawlowska, I. M., Lins, S., & Sunyaev, A. (2020). Die Zertifizierung nach der DS-GVO. Transparenz und Vertrauen für Nutzer digitaler Dienste? *Zeitschrift für Datenschutz (ZD), 9*, 445–450.

Marschollek, J. P. (2020). In T. Sassenberg & T. Faber (Eds.), *Rechtshandbuch Industrie 4.0 und Internet of Things* (2nd ed., pp. 272–276). C.H.Beck/Vahlen.

Meier, K., & Wehlau, A. (1998). Die zivilrechtliche Haftung für Datenlöschung. *Datenverlust und Datenzerstörung. Neue Juristische Wochenschrift (NJW), 51*(22), 1585–1591.

Meyer, A. (1997). Die Haftung für fehlerhafte Aussagen in wissenschaftlichen Werken. *Zeitschrift für Urheber- und Medienrecht (ZUM), 41*(1), 26–34.

Meyer, J. (2014). In G. Borges (Ed.), *Rechtfragen der Internet-Auktion* (2nd ed., pp. 36–60). Nomos.

Moriarty, B. (2015). *Evidence in civil law – Ireland. Maribor*. Institute for Local Self-Government and Public Procurement Maribor.

Morscher, L. (1999). Software-Überlassung nach Schweizer Recht. Urheberrecht, Gewährleistungspflichten, Produkthaftung. *Computer und Recht (CR)*, 262–274.

Müller-Hengstenberg, C. D. (1994). Computersoftware ist keine Sache. *Neue Juristische Wochenschrift (NJW), 47*(48), 3128–3134.

Müller-Stoy, T. (2005). Grundzüge des U.S.-amerikanischen Patentverletzungsverfahrens. *Gewerblicher Rechtsschutz und Urheberrecht International Journal of European and International IP Law (GRUR Int), 54*(7), 558–566.

Müllmann, D., Ebert, A., & Reissner, L. (2020). IT-Sicherheitsgesetz 2.0 und das Vorbild USA. Ein Rechtsvergleich. *Computer und Recht (CR)*, 584–592.

Neudörfer, A. (2014). *Konstruieren sicherheitsgerechter Produkte. Methoden und systematische Lösungssammlungen zur EG-Maschinenrichtlinie* (6th ed.). Springer Vieweg.

Nolte, N., & Becker, T. (2008). IT-Compliance. In *Betriebs-Berater (BB)* Beilage 2008, Nr. 005, 23–27.

Oechsler, J. (2018). § 2 ProdHaftG. In N. Habermann (Ed.), *Staudinger-Kommentar zum BGB*. Sellier/de Gruyter.

Oetker, H. (2019). § 249 BGB. In F. J. Säcker, R. Rixecker, H. Oetker, & B. Limperg (Eds.), *Münchener Kommentar zum Bürgerlichen Gesetzbuch* (Vol. 2, 8th ed.). C.H.Beck.

Plath, K.-U. (2001). Haftung für Datenverluste wegen fehlerhafter Software. Anspruchsgrundlagen, Anspruchsgegner und Konsequenzen für die Vertragsgestaltung. *Der IT-Rechtsberater (ITRB)*, 274–276.

Raschauer, N. (2018). Art. 42 DSGVO. In G. Sydow (Ed.), *Europäische Datenschutzgrundverordnung. Handkommentar* (2nd ed.). Nomos.

Rebin, I. (2021, 1 April). § 2 ProdHaftG. In B. Gsell, W. Krüger, S. Lorenz, C. Reymann (Eds.), *Beck'scher Online-Großkommentar zum Zivilrecht*. C.H.Beck.

Redeker, H. (2020). *IT-Recht* (7th ed.). C.H.Beck.

Reusch, P. (2019). Mobile updates – updatability, Update-Pflicht und produkthaftungsrechtlicher Rahmen. *Betriebs-Berater (BB), 16-17*, 904–909.

Riehm, T. (2014). Von Drohnen, Google-Cars und Software-Agenten. Rechtliche Herausforderungen autonomer Systeme. *Der IT-Rechtsberater (ITRB)*, 113–115.

Riehm, T., & Meier, S. (2020). Rechtliche Durchsetzung von Anforderungen an die IT-Sicherheit. Behörden, Private und Verbände in der Gesamtverantwortung. *Multimedia und Recht (MMR), 23*(9), 571–576.

Rosenberg, M. (1958). Sanctions to effectuate pretrial discovery. *Columbia Law Rev, 58*(4), 480–497.

Rothe, L. (1993). Produkthaftung. Rechtliche Grundlagen und ihre Auswirkungen auf die Industrie. *Computer und Recht (CR)*, 310–313.

Salomon, T. R. (2016). Cybercrime und Lösegeld – Strafbarkeit der Zahlung von Lösegeld als Reaktion auf Erpressungstrojaner. *MMR – Zeitschrift für IT-Recht und Recht der Digitalisierung, 9*, 575–578.

Sattler, A. (2018). Der Einfluss der Digitalisierung auf das Gesellschaftsrecht. *Betriebs-Berater (BB), 39*, 2243–2253.

Schack, H. (2020). *Einführung in das US-amerikanische Zivilprozessrecht* (5th ed.). C.H.Beck.

Schäfer, H.-B., & Ott, C. (2020). *Lehrbuch der ökonomischen Analyse des Zivilrechts* (6th ed.). Springer Gabler.

Schallbruch, M. (2017). IT-Sicherheitsrecht – Schutz kritischer Infrastrukturen und staatlicher IT-Systeme. Zur Entwicklung des IT-Sicherheitsrechts in der 18. Wahlperiode (Teil 1). *Computer und Recht (CR)*, 648–656.

Schlutz, J. H. (1994). Haftungstatbestände des Produkthaftungsrechts – Die Haftung des Herstellers fehlerhafter Produkte: Produkthaftungsgesetz. *Deutsches Steuerrcht (DStR), 22*, 791–794.

Schmidl, M. (2016). § 28. Recht der IT-Sicherheit. In C.E. Hauschka, K. Moosmayer, T. Lösler (Eds.), *Corporate Compliance. Handbuch der Haftungsvermeidung im Unternehmen* (3rd ed.). C.H.Beck.

Schmidt, B. (2009). Vertrauen ist gut, Compliance ist besser! – Anforderungen an die Datenverarbeitung im Rahmen der Compliance-Überwachung. *Betriebs-Berater (BB), 24*, 1295–1299.

Schmon, C. (2018). Product liability of emerging digital technologies – A fitness check of the 1985 product liability directive. *Zeitschrift für Internationales Wirtschaftsrecht (IWRZ), 6*, 254–259.

Schneider, J., & Witzel, M. (2015). In J. Schneider (Ed.), *Handbuch EDV-Recht* (5th ed.). Otto-Schmidt-Verlag.

Schönknecht, M. (2011). Beweisbeschaffung in den USA zur Verwendung in deutschen Verfahren. *Gewerblicher Rechtsschutz und Urheberrecht International Journal of European and International IP Law (GRUR Int), 60*(12), 1000–1009.

Schulte am Hülse, U., & Klabunde, S. (2010). Abgreifen von Bankzugangsdaten im Onlinebanking. Vorgehensweise der Täter und neue zivilrechtliche Haftungsfragen des BGB. *Multimedia und Recht (MMR), 13*(2), 84–90.

Schwarze, R. (2019). Vorbemerkungen zu §§ 280–285. In M. Löwisch (Ed.), J. v. Staudinger (founder), *Staudingers Kommentar zum BGB* (ed. 2019). Sellier/de Gruyter.

Shabbir, A., & Padgett, C. (2008). The UK code of corporate governance: Link between compliance and firm performance. In K. Grant (Ed.), *Proceedings of the 4th European Conference on Management Leadership and Governance (ECMLG 2008)* (pp. 167–177). Academic Conferences Ltd.

Sommerlad, K. (1997). Millenium Bug. *Betriebs-Berater (BB). Beilage, 18*, 5–10.

Spindler, G. (1998). Verschuldensunabhängige Produkthaftung im Internet. *Multimedia und Recht (MMR), 3*, 119–124.

Spindler, G. (2008). Rechtliche Defizite und rechtliche Differenzen. *Multimedia und Recht (MMR), 7*, 9–12.

Spindler, G. (2021). § 823 BGB. In B. Gsell, W. Krüger, S. Lorenz, C. Reymann (Eds.), *Beck'scher Online-Großkommentar zum Zivilrecht* (ed. 1 Feb 2021). C.H.Beck.

Spindler, G., & Thorun, C. (2016). Die Rolle der Ko-Regulierung in der Informationsgesellschaft. Handlungsempfehlung für eine digitale Ordnungspolitik. *Multimedia und Recht (MMR), Beilage, 2016*(6), 1–28.

Staudinger, A. (2019). Vorbemerkung zu §§ 823–853. In R. Schulze (Ed.), *Bürgerliches Gesetzbuch Handkommentar, NomosKommentar* (10th ed.). Nomos.

Stempfle, C. T. (2017). § 36 Internationale Versicherung. In K. Höra (Ed.), *Münchener Anwaltshandbuch Versicherungsrecht* (4th ed., pp. 2073–2092). C.H.Beck.

Stocklas, J. (2016). 3. Kommunale IT-Sicherheitskongress: Alle Wege führen nach Rom?, *Zeitschrift für Datenschutz* (ZD), 11, Aktuell, 05146.

Streatman, G., & Verhoeven, D. (2016). In P. Machnikowski (Ed.), *European product liability – An analysis of the state of the art in the era of new technologies* (pp. 17–110). Intersentia.

Sydow, L. (2019). *Das Spannungsverhältnis zwischen e-Discovery und Datenschutzrecht. Eine Analyse der e-Discovery nach US-Recht aus datenschutzrechtlicher Perspektive im Vergleich zur Informationsbeschaffung für Zivilprozesse in Deutschland und England*. Duncker & Humblot.

Taeger, J. (1996). Produkt- und Produzentenhaftung bei Schäden durch fehlerhafte Computerprogramme. *Computer und Recht (CR)*, 257–271.

Thiele, J. (2020). Starre und gleitende Verweise auf technische Normen im Recht. *Der Sachverständiger (DS), 47*(12), 308–312.

Tolman, L. (1958). Discovery under the federal rules: Production of documents and the work product of the lawyer. *Columbia Law Review, 58*(4), 498–515.

Trinks, M. (2016). Steuer(straf)rechtl. Konsequenzen für den Erpresser. *Neue Zeitschrift für Strafrecht (NStZ), 5*, 263–266.

Van Leeuwen, B. (2017). *European standardisation of services and its impact on private law. Paradoxes of convergence*. Bloomsbury Publishing.

Wagner, G. (2020a). § 2 ProdHaftG. In F. J. Säcker, R. Rixecker, H. Oetker, & B. Limperg (Eds.), *Münchener Kommentar zum Bürgerlichen Gesetzbuch* (Vol. 7, 8th ed.). C.H.Beck.

Wagner, G. (2020b). § 823 BGB. In F. J. Säcker, R. Rixecker, H. Oetker, & B. Limperg (Eds.), *Münchener Kommentar zum Bürgerlichen Gesetzbuch* (Vol. 7, 8th ed.). C.H.Beck.

Wagner, G. (2020c). Vor § 823 BGB. In F. J. Säcker, R. Rixecker, H. Oetker, & B. Limperg (Eds.), *Münchener Kommentar zum Bürgerlichen Gesetzbuch* (Vol. 7, 8th ed.). C.H.Beck.

Weidemann, M. (2019). StGB § 202a Ausspähen von Daten. In B. von Heintschel-Heinegg (Ed.), *Beck'scher Online-Kommentar StGB* (41st ed.). C.H.Beck.

Werner, D. (2010). *Verkehrspflichten privater IT-Nutzer in Bezug auf die Verbreitung von Schadsoftware*. Nomos.

Wilhelmi, R. (2020a). § 2 ProdhaftG. In H. P. Westermann, B. Grunewald, & G. Maier-Reimer (Eds.), *Erman-Kommentar zum BGB* (16th ed.). Otto-Schmidt-Verlag.

Wilhelmi, R. (2020b). § 823 BGB. In W. Erman (Founder), B. Grunewald, G. Maier-Reimer, H.P. Westermann (Eds.), *Bürgerliches Gesetzbuch Handkommentar* (16th ed., Vol. 2). Otto Schmidt.

Wilrich, T. (2015). Die neue Formel der Betriebssicherheitsverordnung: Sicherheit = Produktkonformität + Gefährdungsbeurteilung + Schutzmaßnahmen ≠ starrer Bestandsschutz. *Corporate Compliance Zeitschrift (CCZ), 8*(4), 175–180.

Zech, H. (2012). *Information als Schutzgegenstand*. Mohr Siebeck.

The Role of Criminal Law in Regulating Cybercrime and IT Security

Dominik Brodowski

Contents

Abstract It is all too common to call for an intervention by means of criminal law whenever a severe wrong occurs—such as when private personal data of politicians are leaked, when malicious websites use a special JavaScript applet to mine Bitcoin on visitors' computers (cryptojacking), or when a social bot spreads "fake news". In this article, I provide an overview over core concepts, models, and limitations of a regulation of IT security through criminal law. On the basis of the German and European Union legal orders, I discuss generic regulatory aspects of substantive

A Spanish version of this contribution, translated by Maria Belén Linares, is published in Brodowski (2021), pp. 85 ff.

D. Brodowski (✉)
Saarland University, Saarbrücken, Germany
e-mail: dominik.brodowski@uni-saarland.de

G. Borges, C. Sorge (eds.), *Law and Technology in a Global Digital Society*,
https://doi.org/10.1007/978-3-030-90513-2_12

criminal law (1) and provide an overview on the criminal law provisions on cybercrime (2). On this basis, I analyse the role criminal law already has in regulating IT security in the EU and in Germany, and how this role may expand in the future (3).

1 Criminal Law and Regulation: Core Concepts, Models, and Limitations

Everyone has some understanding of criminal law. For most non-specialists, that notion is heavily shaped by narrations of murderers being caught by the police, of public criminal trials where, when all seems lost, some turn of events leads to the conviction of the culprit or the acquittal of the mistakenly accused, and to stigmatising criminal punishment in overcrowded, harsh prisons. Contrary to this widespread myth, a broad range of behaviour is prohibited by an ever-expanding portfolio of criminal[1] law,[2] which may relate to subtleties such as incorrect labelling of beef products.[3] Most violations of criminal laws that authorities become aware of lead to alternative measures instead of a full-fledged public trial,[4] and (long) prison sentences are—at least within the EU—the rare exception. Therefore, it is important to be aware of the wide scope of modern criminal justice when analysing the role of (substantive) criminal law in general (Sect. 1.1), but also when determining which limits exist to incrimination (Sect. 1.2), and whether and how criminal law serves as a regulatory tool (Sect. 1.3).

1.1 The Role of Criminal Law

As already distinguished by *Binding* (1922),[5] any provision of substantive criminal law, firstly, prohibits the specific behaviour specified therein, and thereby sets out a norm on how to behave (behavioural norm component or primary norm,

[1] Except otherwise noted (see, in particular, *infra* Sect. 3.2), I will focus on criminal law in a formal sense in this article, and exclude administrative or other alternative, quasi-criminal sanctioning mechanisms.

[2] See, among others, Landau (2015a), pp. 525 ff.; Ignor (2016), pp. 380 ff.; Schlepper (2014), pp. 11 ff.

[3] See German Federal Constitutional Court, Decision of 21.09.2016 – 2 BvL 1/15 = ECLI:DE:BVerfG:2016:ls20160921.2bvl000115 = BVerfGE 143, 38.

[4] See Brodowski (2018), pp. 373 ff.

[5] Binding (1922), pp. 35 ff., 45.

"*Verhaltensnorm*"). Secondly, in case this norm was breached, the provision of substantive criminal law provides a legal basis for the state to punish the offender (sanctioning norm component or secondary norm, "*Sanktionsnorm*").[6] Despite limited empirical evidence on its effectiveness, punishment by means of criminal sanctions usually is considered to be a preventive measure (a prisoner has limited chances to commit new offences; he or she should be rehabilitated in prison or by means of the sanction; potential other offenders are deterred, and society overall can see that the rule of law trumps in the end), but also as a repressive or retributive measure (the guiltiness of the offender is negated by the punishment), and as a communicative function between (and within) society and an offender by means of the criminal sanction.[7] Beyond that, a certain harshness of the sanction may be required to pacify victims and to avoid vigilantism. However, such punishment by a criminal sanction in a strict sense (imprisonment or fine) is quite rare for non-severe forms of crime, but is oftentimes replaced by alternative forms of enforcement. These alternatives provide for greater flexibility, "with far more tailored legal consequences or – in a broad sense – sanctions being available to micro-manage the behaviour of an actual or presumed offender",[8] while still holding true at least to the preventive and, to a somewhat lesser extent, to the retributive and communicative functions outlined above.

At its core, the traditional role of criminal law may therefore be summarised as stipulating specific norms on (human and, only more recently, corporate)[9] behaviour and as providing an oftentimes harsh mechanism—criminal sanctioning and criminal enforcement—to assert that these norms are observed. Criminal justice is therefore a specific form of social control, addressing anti-social behaviour. Beyond intra-criminal law effects, provisions of substantive criminal law also serve as the basis for tort liability, for civil enforcement by private parties, and for police law-based intervention.[10] Moreover, the symbolism of (substantive) criminal law provides politics a useful tool to address a social problem—for a seemingly low cost—by adding a new or intensifying an existing criminal law provision.[11]

[6] On this distinction, see, among others, Kudlich (2003), p. 129; Renzikowski (2002), pp. 12 f.; Lagodny (1996), pp. 80 f.; Vogel (1993), pp. 49 ff. and critically Goeckenjan (2015), pp. 192 ff.

[7] See generally Hörnle (2011) and, on the communicative function, Feinberg (1965), p. 400 ff.; von Hirsch (2005), pp. 48 ff.; Duff (2001), p. 89.

[8] Brodowski (2018), p. 393.

[9] On the emergence of corporate criminal liability see, inter alia, Brodowski et al. (2014); Engelhart (2012); Laufer (2006); Pieth and Ivory (2011); Schmitt-Leonardy (2013).

[10] See, in further detail, Brodowski (2018), pp. 383 ff.

[11] See critically Hassemer (1989).

1.2 Generic Limitations to Criminal Law

Due to the severity of criminal sanctions (in a European context: up to imprisonment for life) and in light of what is regularly seen as overcriminalisation (see Husak, 2008), a regular topic of criminal law theory are the limitations to substantive criminal law. The extensive discussion on that matter cannot be repeated or even be summarised here, especially as such limits differ between legal systems and depend on their constitutional framing. Instead, I want to highlight some aspects relevant throughout the EU.

Criminal law is far less special than it seems: on the national level in Germany, criminal statutes are enacted by parliament under the same rules as other (normal) laws, without any additional procedural requirements or limitations (such as a qualified majority); contrary to other areas of law, criminal statutes generally do not even require a consent of the German *Länder* in the second chamber.[12] In a similar vein, EU directives on "minimum rules concerning the definition of criminal offences and sanctions" are "adopted in accordance with the ordinary legislative procedure" (Art. 83 [1] TFEU). However, there is a special "emergency brake" provided for in Art. 83 (3) TFEU to protect the "fundamental aspects of its criminal justice system" against changes imposed by such directives. While the specific extent of this safeguard, which no member state has yet triggered, remains doubtful,[13] it does not stipulate limitations to EU criminal law that a citizen can effectively rely upon. Instead, it is merely a modification of the legislative procedure (and its—at least temporary—delay) a member state may request. Usually it is up to the government of a member state whether to trigger the "emergency brake"; in Germany, however, based on a decision of the German Federal Constitutional Court,[14] parliament may require the government to do so (§ 9 IntVG). Yet this does not shift the "emergency brake" mechanism away from the political domain to become a safeguard of citizens' rights against (over-)criminalisation. It is and remains (mostly) a safeguard to protect political spheres of influence and decision-making.

Then, it is important to realise that whether the state may legally prohibit some conduct, or introduce a legal requirement to do something, is a matter of normal constitutional and human rights review, and does not differ between criminal law provisions and provisions of other areas of the law.[15] Therefore, the behavioural norm component of each provision of substantive criminal law is required to be in accordance with constitutional requirements and (national and/or European) human rights standards. Whether a breach of such a provision may—and should—lead to *criminal* sanctioning, to other forms of sanctioning, to liability under tort law, or to

[12] Cf. Jahn and Brodowski (2016), p. 972.

[13] See, inter alia, Eisele and Vogel (2018), mn. 98 f.; Satzger (2018), mn. 43 ff.; Suhr (2016), mn. 31 ff.; Summers et al. (2014), pp. 49 ff., 281.

[14] German Federal Constitutional Court, judgment of 30.6.2009 – 2 BvE 2/08 = ECLI:DE:BVerfG:2009:es20090630.2bve000208 mn. 358 ff., 365.

[15] Just see Frisch (2016); Gärditz (2016), p. 642; Jahn and Brodowski (2016), p. 970.

other forms of legal consequences is a different question, and usually framed under the label that criminal law is (or should be) the *ultima ratio*.[16]

Regarding this question of *ultima ratio* and further limitations to criminal law, it is far from clear what standard of review applies from the view of German[17] or European constitutional and human rights law.[18] The EU Council has adopted a (soft) guideline that "the European Union should as a general rule adhere to the principle of use of criminal law as a last resort", that any criminal provisions should "address clearly defined and delimited conduct, which cannot be addressed effectively by less severe measures" and "only prescribe penalties for acts which have been committed intentionally"; only exceptionally may "negligent conduct [...] be criminalised when a case-by-case assessment indicates that this is appropriate due to the particular relevance of the right or essential interest which is the object of protection, for example in cases of serious negligence which endangers human life or causes serious damage".[19] In quite similar terms, the EU Commission[20] and the European Parliament[21] have reflected upon their stance on criminal law.[22] And only recently, the ECJ stepped in when it considered a (national) sanction, enacted in the implementation of Union law, to be unproportionate and excessive to the breach of the behavioural norm.[23]

1.3 Criminal Law as a Regulatory Tool?

To what extent is criminal law therefore a tool which is available in the toolbox of the legislator, regulator and policy-maker? The answer to this question very much relates to one's opinion on the extent of the *ultima ratio* limitation to criminal law:[24] the more one stresses cultural, historical, or philosophical roots of criminal law, the more it is focused on its traditional "core", and the more static it becomes.[25] The options of the legislature to react to current trends (such as in technology and in anti-

---

[16] See Jahn and Brodowski (2016), p. 970 ff.; Jahn and Brodowski (2017), p. 366 ff.

[17] On different perspectives on *ultima ratio* see Jahn and Brodowski (2016), p. 974 ff.

[18] Cf. Asp et al. (2009); see furthermore the analysis by Meyer (2015), mn. 31.

[19] Council Document 16542/09 of 23.11.2009.

[20] COM [2011] 573 final of 20.09.2011.

[21] "An EU approach to criminal law. European Parliament resolution of 22 May 2012 on an EU approach to criminal law," 2010/2310[INI], OJ C 264E of 13.09.2013, p. 7.

[22] See also the comparison in Council Document 10137/15 of 24.06.2015, Annex 1.

[23] ECJ, decision of 12.07.2018 – C-707/17 (Pinzaru and Cirstinoiu) = ECLI:EU:C:2018:574, mn. 26 ff., 33, 35; see beforehand ECJ, judgment of 16.07.2015 – C-255/14 (Chmielewski) = ECLI:EU:C:2015:475, mn. 31, 35.

[24] See generally Wohlers (2018).

[25] See, for instance, Hassemer (1994), pp. 307 f.; Hamm (2016), pp. 1541 f.; Frisch (2016), p. 24.

social behaviour relating thereto) are limited to what may be embedded into the aforementioned roots,[26] and alternatives to criminal law—which may be as severe as criminal law, but offer less protections—become ever more promising.[27] If one stresses the democratic roots of (current) criminal law[28] and its features within the constitutional orders instead, the more flexible to legislative modification criminal law is and the more dynamic it may become, possibly at the cost of overcriminalization and blurring its *proprium*.

Using criminal law as a regulatory tool additionally presupposes that it, in its preventive function, *does* regulate behaviour by stipulating behavioural norms and (harsh) sanctions one has to bear in case the behavioural norm is breached. Indeed, the harshness of the criminal sanction makes it, under very most circumstances, rational *not* to commit a crime. However, humans oftentimes behave irrational, especially when they miscalculate risks (which lie, such as a criminal prosecution, in the far future) in light of gains seemingly certain and near at hand.[29] Such (at least seemingly) near gains may be economic—and most cybercrime is economically motivated[30]–, but they may as well be the satisfaction of curiosity, anger, or thrill. The cost associated with committing a crime—and particularly the likelihood of having to face a sanction in future—is oftentimes underestimated by potential criminals. They (especially highly skilled individuals) tend to think that they are too smart to get caught, and underestimate the skill set of their opponents (the police). And the actual knowledge of what exactly constitutes a crime, and especially the knowledge of what sanction a crime will lead to, is also quite low. Such limitations and features of human behaviour substantially limit the (general)[31] preventive function of criminal law,[32] and run counter to demands[33] and explanations oftentimes brought forward in favour of criminalization and/or of increasing punishment.

In any case, it is (not only politically) wise to first discuss what specific behaviour should be regulated by law—the aforementioned behavioural norm component.

---

[26] See critically Jahn and Brodowski (2016) on Landau (2015a), p. 533; Landau (2015b), p. 668.

[27] Cf. Mattes (1977), p. 41.

[28] Cf. Gärditz (2016), p. 641 ff.

[29] On time discounting, intertemporal choice and temporal myopia in criminals, see Mamayek et al. (2017); Nagin and Pogarsky (2004) (showing a high relevance for property offending); and already Wilson and Herrnstein (1985), p. 50. For these features of human-decision making from the perspective of general psychology and economic decision-making theory, see, inter alia, Ahlbrecht and Weber (1997); Kahneman and Tversky (1979) Loewenstein and Elster (1992); Öncüler (2010); Tversky and Kahneman (1992); Williams (2014).

[30] Cf. Kshetri (2010); Wall (2010).

[31] A different aspect, not to be addressed in this contribution, relates to *special prevention*, that is the aspect that a criminal cannot commit further crimes outside of prison while he or she is imprisoned, and that the sanction may teach him or her a lesson to not commit (further) crimes in future.

[32] See generally Bernasco et al. (2017) and Lüderssen (1979), pp. 70 f.; Bock (2013), pp. 310 f.; Meier (2016), pp. 281 ff.

[33] As an example, see the view on criminal law taken by Cárdenas et al. (2009).

Only once a (preliminary) decision on that aspect is reached, the question on how to assert compliance arises. There, various options ranging from criminal law over administrative sanctions to tort law need to be taken into account. At that step, the range of options for a specific intervention may turn out to be limited—at least—based on the proportionality principle.[34] Finally, if a criminal statute is lawfully chosen as mode of regulation, that may require more attention to the exact wording of the provision compared to other mechanisms of regulation (Art. 103 [2] GG).

2 Substantive Criminal Laws on Cybercrime in Europe and Germany: An Overview

2.1 Typology of Laws on Cybercrime

Under the umbrella term of cybercrime,[35] two major forms can be discerned[36]: firstly, and in a strict sense, criminal behaviour may affect IT (including the data stored therein) itself, and the criminal statutes refer—at least partly—to such effects. Typical examples here are criminal statutes specifically protecting the confidentiality, integrity, and/or availability of IT and/or the data stored therein, but also computer fraud, where it is not a human being who is deceived to transfer property, but instead a computer.[37]

Secondly, and in a broad sense, criminal behaviour may use IT as a means to commit a crime, without interfering with the technology as such. Typical examples for such forms are (criminal) hate speech, fraud, but also the dissemination of illegal content. To some approximation, criminal statutes are typically drafted in a technology-neutral manner, meaning that criminal liability exists the same whether the crime was committed "off-line" or "on-line".

Some crimes—and criminal statutes—lie, however, in between these categories. Their common root seems to be that a specific behaviour becomes exceedingly risky, likely, or problematic in light of the features of IT, in particular due to the heightened anonymity and the opportunity for automatization. For example, a recent addition to the German penal code aims at criminalising the operation of marketplaces for weapons and drugs, in particular in the "Darknet" (§ 127 StGB). This crime does not fit in the first category (as IT is not affected by itself) but also not into the second

[34] Just see Greco (2013), p. 23; Jahn and Brodowski (2016), p. 976; Jakobs (1991), 2/27 f.; Kühl (2008), p. 41; Swoboda (2010), p. 49 f.; Rengier (2018), § 3 at mn. 5; Roxin and Greco (2020), § 2 at mn. 97; Kaspar (2014), pp. 243 ff., 865.

[35] On the history and definition, just see Brodowski (2019a), p. 98; Chawki et al. (2015), p. 3; Gillespie (2016), pp. 1 ff.; Kshetri (2010), p. 3; Sieber (2008), p. 131; Schjolberg (2014); Vogel (2008); Wall (2007), p. 10.

[36] Just see Kshetri (2010), p. 3; Vogel (2008), p. 1 f. as well as UNODC (2013), pp. 11 f.

[37] Just see Brodowski (2013).

category (as IT is not merely a tool to commit a generic crime). In a similar vein, it is a crime under German criminal law to buy computer data which was "stolen" beforehand (§ 202d (1) StGB), and criminal provisions relating to the protection of personal data were, most generally speaking, limited to the processing of personal data by means of computer technology.[38]

2.2 *EU Criminal Law on Cybercrime*

Due to the current primary law limitation that the EU may only address "particularly serious crime with a cross-border dimension" (Art. 83 [1] TFEU) and similar limitations in previous founding treaties, the EU has only enacted "minimum rules concerning the definition of criminal offences and sanctions" (Art. 83 [1] TFEU) for a handful areas of criminal activities. Furthermore, in past decades the Council of Europe held a leading role in the harmonisation of substantive criminal law. Therefore, the first and still highly important landmark in European Criminal Law on Cybercrime is the Convention on Cybercrime (CCC, European Treaties Serie No. 185) tabled by the Council of Europe, which has also been endorsed both internally[39] and externally[40] by the European Union. Generally speaking,[41] this Convention calls upon the states which ratified or acceded to the convention to provide for, *inter alia*, criminal laws for the intentional accessing of computer systems without right (Art. 2 CCC), for illegal interception of data (Art. 3 CCC), for data (Art. 4 CCC) and system interference (Art. 5 CCC), for the handling of devices (including computer programs) "designed or adapted primarily for the committing" one of the aforementioned offences (Art. 6 CCC), for computer-related forgery (Art. 7 CCC) and fraud (Art. 8 CCC).

Nonetheless, the EU has followed suit[42] and enacted very similar requirements in its secondary law, firstly in a Council Framework Decision of 2005[43] and then in a Directive of 2013,[44] both concerning attacks against information systems. Member States must, under EU law,[45] criminalise (intentional) illegal access to information systems (Art. 3 Directive 2013/40/EU), illegal system (Art. 4 Directive 2013/40/EU)

---

[38] Cf. Art. 3 (1) Directive 95/46/EC. The scope of that Directive was limited to the "the processing of personal data wholly or partly by automatic means" or by filing systems. The GDPR (see *infra*), which replaced this Directive, does not contain a similar limitation.

[39] See Council Document 12711/17 of 02.10.2017, p. 8, 16 ff.

[40] See Council Document 10007/16 of 09.06.2016.

[41] For further details, see the Explanatory Report as well as Brodowski (2016), pp. 341 ff.; Schjolberg (2014), pp. 38 ff.

[42] See generally Haase (2017); Summers et al. (2014), pp. 231 ff.; Brodowski (2010).

[43] Cf. Council Framework Decision 2005/222/JHA, OJ 2005 L 69 of 15.03.2005, p. 67.

[44] Cf. Directive 2013/40/EU, OJ 2013 L 218 of 13.08.2013, p. 8.

[45] For further details, see Brodowski (2016), pp. 347 f.

and data interference (Art. 5 Directive 2013/40/EU), illegal interception [of data] (Art. 6 Directive 2013/40/EU) and the handling of tools used for committing offences, in particular of "computer programme[s], designed or adapted primarily for the purpose of committing any of the offences" mentioned before (Art. 7 Directive 2013/40/EU). The relevance of these provisions lies less in the specific additions compared to the CCC (which mostly relate to penalties). Instead, not all EU Member States have yet ratified or acceded to the CCC. Moreover, the EU Commission and the ECJ only have jurisdiction to monitor the correct implementation of requirements contained in EU law, but not of the CCC which stands beside the EU *acquis*. In addition, a common understanding on these computer crimes facilitates police and judicial cooperation in the EU area of freedom, security and justice.[46]

Concerning the use of IT in furtherance of a criminal activity, the most relevant EU secondary law concerns a minimum level of criminalising child pornography[47] as well as addressing fraud and counterfeiting of non-cash means of payment.[48] On a broader level, EU secondary law on terrorism[49] and money laundering[50] references, as an element of the crimes described therein, the commission of cybercrime directly;[51] other law—such as on organised crime[52]—may often be applicable in the course of cybercrime.

In contrast, the EU has abstained from introducing a *criminal* sanctioning mechanism when regulating the protection of personal data, in particular by means of the General Data Protection Regulation.[53] Instead, it relies on a specific administrative sanctioning mechanism (Art. 83 GDPR) to enforce and assert compliance with the GDPR.[54] Moreover, and easily explained by the limited competence of the EU in the field of criminal justice, EU criminal law does not provide for a minimum standard with regard to many "classic" crimes such as fraud, causing bodily harm (also through means of manipulating IT) or even death (for instance due to a malfunctioning self-driving vehicle). To summarise, the framework stipulated by EU criminal law is fragmentary, but has a particular focus on attacks against information systems.

---

[46] Cf. Brodowski (2016), p. 348.

[47] See, in particular, Art. 5 Directive 2011/93/EU, OJ 2011 L 335 of 16.12.2011, p. 1.

[48] See the Directive [EU] 2019/713, OJ 2019 L 123 of 10.05.2019, p. 18.

[49] Directive [EU] 2017/541, OJ 2017 L 88 of 31.03.2017, p. 6.

[50] Directive [EU] 2018/1673, OJ 2018 L 284 of 12.11.2018, p. 22.

[51] Art. 3 [1] [I] Directive (EU) 2017/541, Art.2 [1] [v] Directive [EU] 2018/1673.

[52] Council Framework Decision 2008/841/JHA, OJ 2008 L 300 of 10.11.2008, p. 42.

[53] Regulation [EU] 2016/679 [GDPR], OJ 2016 L 119 of 04.05.2016, p. 1. However, the GDPR leaves the option open to member states to sanction grave breaches by means of criminal sanctions, cf. Art. 84 GDPR.

[54] Just see Brodowski and Nowak (2020), mn. 4.

2.3 German Substantive Criminal Laws on Cybercrime

While current EU substantive criminal law—generally speaking—only prescribes a fragmentary minimum standard the member states have to implement, the criminal laws on the German (federal) level take an all-encompassing approach and are directly applicable to all natural (and, to some extent, legal) persons falling under its jurisdiction to prescribe. Therefore, the German Criminal Code contains a broad canon of "typical" criminal law offences such as murder (§ 211 [1], [2] StGB), negligent manslaughter (§ 222 StGB) and causing bodily harm (§§ 223 [1], 229 StGB), theft (§ 242 [1] StGB), and fraud (§ 263 [1] StGB).

To address the specific threats caused by the use of IT, specific statutes on computer fraud (§ 263a StGB), on illegal access of stored and specifically protected computer data (§ 202a [1] StGB), on illegal data manipulation (§ 303a [1] StGB), on illegal system interference (§ 303b StGB), and on data forgery (§ 269 StGB)[55] were added to the German Criminal Code already in the 1980s.[56] In the course of implementing the CCC and the EU Council Framework Decision in 2007,[57] Germany primarily amended the provision on system interference to criminalise network-based denial of service attacks (§ 303b (1) Nr. 2 StGB) and introduced the crime of handling "hacking tools" (§ 202c StGB). Beyond repeated modifications of the crimes relating to child pornography and to the (on-line) sexual exploitation of children,[58] legislative activity in the field of cybercrime has been rather limited in recent years, with the exception of the addition of a crime of "data laundering" (that is, the selling, transfer or distribution of "stolen" computer data, § 202d [1] StGB), a reform of the criminal law provisions relating to the protection of personal data (§ 42 BDSG), which in particular criminalises any processing of non-public personal data without authorisation, as long as the perpetrator obtains a (monetary) reward for the processing, aims to obtain a monetary gain, or aims to cause harm to another person (§ 42 [2] Nr. 1 BDSG), and the criminalisation of illegal online marketplaces, in particular in the "Darknet" (§ 127 StGB). Political attempts to criminalise *all* misuse of IT ("*Digitaler Hausfriedensbruch*")[59] have not been successful.

Yet despite this moderate rate of legislative turnover, the German Criminal Code is considered to be well-prepared to cover at least most forms of cybercrime seen in

---

[55] See also § 270 StGB.

[56] Zweites Gesetz zur Bekämpfung der Wirtschaftskriminalität, BGBl I 1986, p. 721.

[57] 41. StrÄndG, BGBl I 2007, p. 1786.

[58] Most recently in Siebenundfünfzigstes Gesetz zur Änderung des Strafgesetzbuches – Versuchsstrafbarkeit des Cybergroomings, BGBl I 2020, p. 431.

[59] See BR-Drs. 47/18.

practice. In particular, there are criminal law prohibitions which are applicable for most if not all forms of ransomware,[60] denial-of-service attacks,[61] phishing,[62] (online) identity theft,[63] and even side-channel attacks to obtain secretive computer data.[64] Most discussions on the extent of substantive criminal law protection relate to the question of how far the user has to be protected in case he or she is deceived to cause harm to data or an IT system his- or herself,[65] and how far users should be protected by means of criminal law even if they do not specifically protect their computer data by technical means.[66] Moreover, most experts agree that the procedural side of how to effectively investigate and prosecute cybercrimes needs more legislative (and fiscal) attention than substantive criminal law.

3 Cybercrime Law and Regulating IT Security

Let us link the question of regulation by means of criminal law (*supra* Sect. 1) and specifically the German and EU laws on cybercrime (*supra* Sect. 2) with the topic of IT security. Relevant in this context are the cybercrimes in the strict sense of the term (especially those related to the confidentiality, integrity and availability of IT systems and data), as the operation of IT is not affected as such if "neutral" IT is merely misused for the commission of a crime (such as for the transport of a message containing a death threat).[67]

3.1 Targeted Repression and Prevention of Attacks on IT Systems: Attacks and Attackers

When it comes to addressing information security or IT security by means of criminal law, the clear material focus of European and German criminal law—at least so far—rests on so-called *attacks,* that is on intentional breaches of the

[60] Just see Vogelgesang and Möllers (2016).

[61] Just see German Federal Court of Justice, decision of 11.01.2017 – 5 StR 164/16 = ECLI:DE:BGH:2017:110117B5STR164.16.0.

[62] See, though more critically, Popp (2006); Graf (2007).

[63] Just see Kammergericht Berlin, decision of 22.07.2009 – (4) 1 Ss 181/09 (130/09) = ECLI:DE:KG:2009:0722.4.1SS181.09.130.0.0A.

[64] Just see Brodowski (2019b).

[65] On this question, see Brodowski (2019b), p. 55 f. with further references.

[66] On this question, see Brodowski (2019b), p. 55 with further references.

[67] However, service providers are increasingly tasked with asserting the legality of the actions their users take on their platforms. Exemplary for this trend is the German *Gesetz zur Verbesserung der Rechtsdurchsetzung in sozialen Netzwerken* (NetzDG).

confidentiality, integrity, and/or availability of an information system and/or the data stored therein or transmitted between such systems. Linked thereto is the personal focus on the so-called *attacker*. This term is regularly used in the field of IT security to describe the person who intentionally and wilfully conducts an attack. The (primary) criminal liability in matters of IT security is therefore put on a person who seeks, for whatever reason (but in most cases his or her own economic gain),[68] to wilfully interfere with the operation of IT as foreseen by the owner, operator, and (legitimate) user of that IT.

It is obvious that the specific behaviour shown by attackers regularly falls under the criminal law provisions described above (in particular Articles 3 ff. Directive 2013/40/EU; §§ 202a ff., 303a f. StGB). However, each and every singular attack requires a thorough analysis whether it fulfils the specific objective and subjective elements of a criminal statute, and whether further elements of a crime such as jurisdiction is given, and no grounds for justifications and grounds for excuses are present. In this contribution, I will neither address these specifics, nor analyse some forms of attacks in detail. Instead, I want to address some aspects which limit the effectiveness of this approach in the repression and prevention of attacks, in light of the generic limitations of criminal law-based regulation explained above (see *supra* Sect. 1.3).

As it is not the (abstract) criminalisation, but the perceived threat of punishment which may have preventive effects, a swift and thorough investigation and prosecution of a crime is an important element for the preventive function of criminal law, but evidently also for its repressive function. In relation to cybercrime, the effectiveness of investigations and prosecutions is, however, hampered firstly as the codes on criminal procedure and specifically the legal framework for an effective investigation of cybercrime may lay behind: for example, the German Code on Criminal Procedure (StPO) still is largely focused on tangible evidence (such as a bloody knife or a pistol), but not on digital evidence,[69] and it does not allow to wiretap a server used in the commission of a large-scale attack on the confidentiality, availability or integrity of IT systems.[70] Secondly, even if a sufficient legal framework exists, it needs to be put in practice, requiring sufficient personal and material resources. Thirdly, it is not uncommon for cybercrime to cause minor individual damages, but—owing to the automatisation of crime—high overall damages. Yet due to the low individual effects, the incentive for the individual to report the crime and for police to follow through with the report may be severely limited.[71] Fourthly, encryption[72] and related privacy-enhancing technologies like TOR and the "Darknet" make it much more difficult—but rarely impossible, as criminals (same

[68] Cf. Kshetri (2010), Wall (2010).

[69] See Sieber and Brodowski (2018), mn. 7 ff. with further references.

[70] Sieber and Brodowski (2018), mn. 152.

[71] Sieber and Brodowski (2018), mn. 174. Legal requirements to report crimes exist, however, in the context of critical infrastructure and the GDPR (Art. 33, Art. 34 [1] GDPR).

[72] Just see Sieber and Brodowski (2018), mn. 108 ff. with further references.

as all humans) do make mistakes—to track down individuals implicated in a criminal activity. Fifthly, the typical transnational dimension of cybercrime and especially of high-profile, large-scale or advanced and persistent cybercrime activities requires that different jurisdictions speedily coordinate their investigations and prosecutions.[73] However, such investigations are oftentimes not conducted speedily and emphatically in cases where the damage only occurred abroad—if at all. Moreover, in transnational cases evidence and suspects may be present in different jurisdictions. Such transfer of evidence or the extradition of suspects generally requires tedious and resource-intensive procedures.

In light of these aspects, much effort is spent by actors such as the UNODC,[74] the G-8/G-7,[75] the Council of Europe[76] and the European Union[77] to harmonise a common set of investigation measures, to strengthen investigation skill sets, to streamline transnational cybercrime investigations and to enhance cooperation between the different criminal justice systems. However, the obstacles continue to limit the repressive and also the (general) preventive effects of criminal law. Moreover, in case (general) prevention failed and a cybercriminal launched an attack on an IT system, harm to IT security is already done – and such harm cannot be undone by a criminal prosecution.

3.2 *Mediated Repression and Prevention of Attacks on IT Systems: Criminal Law Requirements and Obligations on IT Security*

In contrast to the focus it spends on attacks and attackers, criminal law usually does not yet directly and specifically address manufacturers, programmers, engineers, owners, administrators and/or user of IT systems to provide for secure IT systems—

[73] Just see Brodowski (2016), p. 353 f. with further references.

[74] Just see UNODC (2013).

[75] Just see Schjolberg (2014), pp. 43 ff.

[76] Just see the CCC [*supra* Sect. 2.2] and recent discussions on a second additional protocol to this convention, T-CY(2017)3 (available at https://www.coe.int/en/web/cybercrime/t-cy-drafting-group. Accessed 27 Aug 2021).

[77] See, among others, Directive 2014/41/EU, OJ 2014 L 130 of 30.04.2014, p. 1, and a Proposal for a Regulation of the European Parliament and of the Council on European Production and Preservation Orders for electronic evidence in criminal matters, COM [2018] 225 final of 17.04.2018 as well as the [Commission] Recommendation for a Council Decision authorising the opening of negotiations in view of an agreement between the European Union and the United States of America on cross-border access to electronic evidence for judicial cooperation in criminal matters, COM [2019] 70 final of 05.02.2019.

at least if they act without specific knowledge of an attack and therefore cannot be implicated in its commission.[78]

As a first exception to this rule, some design elements of criminal liability place some (indirect) responsibility on the manufacturer, programmer, engineer, owner, administrator and/or user of IT systems. Such features do not[79] create legal requirements but only an obligation—if one does not meet this obligation, one "only" loses the protection offered by criminal law. For example, the German criminal statute on illegal access to data (§ 202a [1] StGB) is only applicable to data which was "especially protected against unauthorised access", meaning that protection by means of criminal law is only given as long as certain precautions are met;[80] and it is unclear whether the provision on data interference (§ 303a [1] StGB) is also applicable if users are tricked into deleting their own data.[81] Such limitations to criminal law are not only legitimate in view of the EU and CCC framework on cybercrime, they underline that the task of criminal law is not to provide an all-in-one package, but it is (at least, see *supra* Sect. 1.2 in light of the principle of proportionality) limited to address *severe* violations of the law. Moreover, they highlight that, same as in all areas of life, some responsibility for IT security rests with everyone[82]—and achieving a reasonable standard of IT security by technical and organisational measures can increase the costs for attackers, and make attacks impossible for less skilled attackers, and thereby effectively prevent at least some if not most attacks from occurring. But as such technical and organisational IT security is inherently limited, criminal law retains an important task if the precautions taken prove insufficient.[83]

More recently, the legislative focus has shifted to enact specific legal requirements on IT security *including* sanctioning provisions in case these requirements are breached. For example, the German legislature, in its law on IT security,[84] added in 2015 a provision requiring certain Internet Service Providers[85] to protect their services, by technical and organisational means, against violations of the protection

---

[78] While, for example, Art. 8 (1) Directive 2013/40/EU requires that incitement, or aiding and abetting in relation to one of the cybercrimes referenced in the Directive are punishable as a criminal offence, such incitement, aiding and abetting requires at least *some* knowledge of the specific crime, and/or *some* intent with regard to the commission of the crime.

[79] At least not on their own. It requires further research whether such obligations may become requirements in the interplay with contractual obligations, tort law, and (criminal or civil) compliance requirements.

[80] Brodowski (2019b), p. 55.

[81] Brodowski (2019b), p. 55 f.

[82] Just see Brodowski and Freiling (2011), pp. 187 f.

[83] Just see Brodowski and Freiling (2011), pp. 100, 189.

[84] Gesetz zur Erhöhung der Sicherheit informationstechnischer Systeme [IT-Sicherheitsgesetz], BGBl I 2015, 1324.

[85] In addition, it created a similar requirement for providers of critical infrastructure in § 14 (1) Nr. 1 i.V.m. § 8a (1) 1 BSI-G (Gesetz über das Bundesamt für Sicherheit in der Informationstechnik); see further Mansdörfer (2018), p. 166.

of personal data (§ 13 [7] 1 Nr. 2 [a] TMG), or otherwise face administrative (quasi-criminal) sanctioning,[86] which here bears a maximum fine of 50.000 €. A closely linked provision, which is not flanked with any sanctioning, requires service providers to protect against service disruptions, "even if they are caused by external attacks" (§ 13 [7] 1 Nr. 2 [b] TMG). Though external attacks are only mentioned specifically in this second provision, leading to some ambiguity, the first provision must be interpreted—also in light of the legislative reasoning[87]—to also require a protection against external (criminal) attacks, and not only to negligent or internal violations. More clearly, Art. 5 (1) (f) GDPR requires that processors of personal data, insofar as they are covered by its regulatory regime, must "ensure [...] appropriate security of the personal data, including protection against unauthorised or unlawful processing [...], using appropriate technical or organisational measures ('integrity and confidentiality')". External attacks are included in this reference to unlawful and unauthorised actions,[88] as is further underlined by Art. 32 (2) GDPR which clarifies the requirement and stipulates that it extends to the "*unlawful* destruction, loss, alteration, *unauthorised* disclosure of, or access to personal data" (emphasis added). Violations of the requirement to provide for appropriate (IT) security can be sanctioned, under the specific sanctioning regime provided for in the GDPR, with "administrative fines up to 20,000,000 EUR, or in the case of an undertaking, up to 4% of the total worldwide annual turnover of the preceding financial year, whichever is higher" (Art. 83 [5] [a] GDPR). Whether an (external) attack occurred which was enabled or facilitated by the lack of IT security, may, however, be taken into account when the supervisory authority decides whether to actually impose such a sanction (on its disputed discretion, see Brodowski & Nowak, 2020, mn. 41), and if so, in what height (Art. 83 [2] [a] GDPR).

Such an approach shifts the legislative focus, in particular areas of high concern for IT security, from the attacker to (especially: commercial) service providers. Though such provisions have not yet reached the domain of "classic" criminal law—and do not bear the punishment of imprisonment–, they still provide for potentially harsh sanctioning, albeit non-criminal in name. In contrast to the problematic preventive effects of criminal laws seen in relation to attackers, it may be reasoned here that the addressees of these norms—in particular corporations and other economic actors—are considered to be more rational and therefore more compliant in view of the sanction they otherwise would have to face ("criminal compliance").[89] Nonetheless, as famously coined by McBarnett,[90] (corporate) criminal compliance may not necessarily be committed, but also creative or even fraudulent. Moreover, it is questionable whether supervisory authorities actually

[86] Just see Sieber (2018), pp. 301 ff.; Wohlers (2018), pp. 238 f., 252 ff.

[87] BT-Drs. 18/4096, p. 34.

[88] Reimer (2018), mn. 48 f.

[89] See, among others, Bock (2011); Engelhart (2012); Fiorella (2012); Jahn et al. (2018); Laufer (2018); Rotsch (2014); Sieber and Engelhart (2014).

[90] Cf. McBarnett (2003), pp. 229 ff.

have the resources and a proper toolbox available to sufficiently investigate and sanction violations of generic IT security requirements; instead, it seems likely that they will focus on cases where IT security was breached by an external attacker, and then investigate whether an appropriate level of IT security existed. One could even bluntly say that the supervisory authorities then sanction the first victim of the crime for having been a too easy target. While such a statement would be factually inaccurate insofar as the "victim" him- or herself failed to obey the law and faces liability for providing insufficient IT security and not for having been "hacked", it expresses a risk of potential injustice: generally speaking, the gravity of an actual attack—where actual harm is intentionally caused by an attacker—is much higher than the gravity of a (mostly negligent) failure to provide for sufficient IT security—which, by itself, does not cause harm but merely enables others to cause harm. This difference in gravity and culpability must not only be expressed in levels of sanctioning, but also in the design of the criminal justice system and the allocation of resources. It would be inherently unjust to focus investigations and prosecutions too much on the providers of IT systems and—in the case of data protection—on data processors, and too few on those individuals who intentionally attack IT systems and violate the confidentiality of personal data. For these reasons, this legislative approach seems limited in scope, and unavailable to assert IT security in general but only in cases where insecurity of IT might cause a severe risk.

3.3 Indirect Repression and Prevention of Attacks on IT Systems: Negligent Criminal Liability

Last but not least, another form of (indirect) regulation of IT security by means of criminal law may gain importance in the future—a criminal liability which is based on the negligence of keeping IT secure. As mentioned above, criminal offences generally require intentional behaviour, in particular in relation to crimes against the confidentiality, integrity or availability of data. Negligent criminal offences are rare,[91] and usually focused to most severe harms such as negligent manslaughter (§ 222 StGB) and negligently causing bodily harm (§ 229 StGB). Yet modern IT can easily cause such severe harms, such as when self-driving cars wreak havoc. It is therefore a clear requirement to adhere to safety standards when designing, producing and operating such cyber-physical systems,[92] or otherwise face criminal liability.[93] Can a criminal liability, however, be also construed in relation to IT security standards? To exemplify: an attacker attacks the IT systems of a car and causes its breaks to fail,[94] resulting in the death of the passengers. Can the owner of the car be

[91] See also Council Document 16542/09 of 23.11.2009 on the policy question.
[92] On this term, just see Alur (2015), p. 1; Mansdörfer (2018), pp. 155 f.
[93] See, among others, Hilgendorf (2017); Wörner (2019).
[94] Keenlab (2016).

held criminally liable for negligent manslaughter if he/she failed to install a security update to the software of the car which would have precluded the attack? Or, to cite a recent case: the IT systems of a German hospital were attacked, and the emergency room had to be closed. A patient had to be re-routed to a hospital many miles away, and died as a result of that delay. Is only the attacker liable for negligent manslaughter, or also the IT operator at the hospital if he/she failed to apply security updates to the IT systems in time?[95]

Under German criminal law, such a criminal liability for negligent behaviour not only requires that harm has actually been done (death in the case of § 222 StGB, bodily injury in case of § 229 StGB) and that this harm was (also) caused by the behaviour of the person (in the example: the omission of installing the security update). Criminal liability for negligent behaviour also requires that the alleged perpetrator violated his objective duties, though it was objectively foreseeable that such harm might be caused due to the violation of the duty.[96] In relation to IT security, it is generally far from clear, however, what the "objective duties" of manufacturers, programmers, engineers, owners, administrators and especially of users of IT systems are. Except for specific fields (such as critical infrastructure) where industry standards and legal frameworks are slowly evolving, it is yet quite unclear what risks may be taken (such as by selling and buying a mobile phone which, at the time of sale, uses a year-old operating system with many known "security holes", or by delaying the installation of security updates) and are socially acceptable, and what behaviour can reasonably be expected.[97] Another level of complexity relates to failure to act (such as failing to create, distribute and/or install a security update to an IT system): in such circumstances, German jurisprudence states that anyone who has a source of risks ("*Gefahrenquelle*") under his or her control is legally responsible to undertake all just and reasonable steps that are available to him or her to mitigate that risk, or otherwise face (the risk of) criminal liability.[98] A major problem, however, lies in the lack of foreseeability: while it is evident that cyber-physical systems such as self-driving cars or WiFi-enabled heart pacemakers have the potential to *directly* cause harm to life and limb, most IT systems can only cause such effects *indirectly* (such as when manipulated medical documents trick a doctor to prescribe a harmful drug, or when hacked computers are used as stepping stones for an attack on the electrical grid). That largely precludes the foreseeability of the harm, and therefore limits the criminal liability for negligent violations of providing for IT security.

A further limitation to negligent criminal liability may stem from the aspect that the harm was caused by an *intentional* attack by a third person, the attacker.[99] In

---

[95] Cf. von Leitner (2020).

[96] Just see Vogel and Bülte (2020), mn. 164 ff. with further references.

[97] See, in further detail, Brodowski (2020), mn. 85; Mansdörfer (2018), pp. 160 f.

[98] See, in particular, German Federal Court of Justice, judgment of 06.07.1990 – 2 StR 549/89; on this judgment, see Timpe (2017).

[99] See, in further detail, Brodowski (2020), mn. 87.

German legal literature, some state that if an attacker intentionally causes the harm (and, for example, commits a murder), there is no legal basis for a "negligent participation" in causing the harm, and therefore criminal liability for negligence is precluded.[100] German jurisprudence differentiates whether the objective duty that was violated by the programmer/administrator/user specifically sought to protect against intentional misuse—such as the requirement to securely store lethal weapons away existing to preclude others to misuse these weapons.[101] More convincing to me is another standard suggested in German legal literature,[102] which—transferred to the question at hand—states that criminal liability for negligence is *not* precluded if one is specifically legally responsible to prevent attacks by third parties, *or* when there are concrete indications that an attack is imminent.[103] Such a specific legal responsibility may exist for the owner of a self-driving car and IT operators at hospitals, but it does not exist for end users who do not keep their laptops and mobile phones up to date.

To summarise: criminal liability for negligent manslaughter or negligent causing of bodily harm may therefore, under certain circumstances, be construed where one fails to provide for adequate IT security, and that failure was later utilised by an intentional attacker to cause bodily harm or the death of a person. This mode of liability is of specific interest in relation to self-driving cars and other cyber-physical systems, and to the *criminal compliance* relating to their design, operation, and maintenance, as well as to critical IT infrastructure. However, I am very sceptical towards tackling generic IT security in this roundabout way, as it addresses only the most severe—and fortunately: mostly hypothetical—consequences of a lack of IT security.

4 Conclusion

Regulators on the European as well as on the national level have provided, by means of criminal law provisions on cybercrime, for a framework which plays a significant role in asserting IT security and in particular in asserting the confidentiality, integrity and availability of IT systems and data. Criminal law statutes—such as those on illegal access, illegal system or data interference, or on interception of data—provide for a strong incentive *not* to attack IT systems and *not* to breach IT security; and they allow to address those by means of criminal justice (including its alternative enforcement mechanisms)[104] who fail to obey the law. However, the effectiveness of this *direct* approach of addressing (potential) attackers and their attacks is limited,

[100] In this direction, see Eisele (2019), mn. 101j.

[101] Cf. German Federal Court of Justice, decision of 22.03.2012 – 1 StR 359/11, mn. 35.

[102] See, in extensive detail, Frisch (1988), pp. 230 ff.

[103] Brodowski (2020), mn. 87.

[104] See Brodowski (2018), pp. 373 ff.

as preventive effects of criminalisation of cybercrime are reduced by practical *and* legal limitations to their investigation and prosecution.

Nonetheless, it would be futile and even counter-productive to focus, by means of criminal law, solely and generally on service providers, programmers, administrators and users instead, and require them, under threat of punishment, to provide for an adequate level of IT security. Though those persons may indeed be addressed more easily by criminal norms and recognise the incentive of committed criminal compliance, it would be inherently unjust to sanction them more harshly—for being an easy target—than the attacker who intentionally caused harm. Moreover, it would run counter to the constitutional requirement of proportionality and the notion that criminal law is only *ultima ratio*. For systems and constellations where a breach of IT security would be of extensive risk, however, specific requirements to provide for adequate IT security, flanked with administrative or minor criminal sanctions in case of violations, provide a useful addition to the regulator's toolbox and do not run counter to fundamental principles of criminal law.

In a similar vein, negligent criminal liability may be construed, under specific circumstances, in cases where an attacker caused harm to life and limb, and this attack was enabled, facilitated or not hindered owing to inadequate IT security. However, the concrete requirements on IT security following for whom (manufacturer, programmer, engineer, owner, administrator, user) for what specific actions or omissions, and the actual scope of such negligent criminal liability in light of an intentional attack are far from clear – and this model has not yet been tested in criminal court in the context of IT security. Therefore, the threat with such negligent criminal liability may provide regulators and policy-makers a useful additional tool in relation to high-risk cyber-physical systems (such as self-driving cars or drones), but it does not seem viable in relation to generic IT security.

In addition, these roles that criminal law plays—or may play—in regulating cybercrime and IT security must not be seen in isolation, but always in their interplay with other areas of law and regulation. There, criminal law not only plays the role of encroaching in rights most severely, by threatening with a criminal sanction. Paradoxically, criminal law can also provide for spaces of freedom. Instead of a preventive *ex ante* intervention (such as by setting out very specific requirements on how to design an IT system, and strict oversight by supervisory authorities whether these requirements are met), criminal justice intervenes only after the fact in case someone misuses this space of freedom and causes serious harm.[105] That freedom-preserving feature of criminal law underlines why it indeed deserves to have a primary role in the regulation of the fast-evolving world of IT.

[105] Tiedemann (1969), pp. 144 f. in note 22; Tiedemann (1975), p. 266 f.; Gärditz (2016), p. 646; Jahn and Brodowski (2016), p. 977; Wohlers (2018), pp. 258 f.

References

Ahlbrecht, M., & Weber, M. (1997). An empirical study on intertemporal decision making under risk. *Management Science, 43*(6), 813–826.

Alur, R. (2015). *Principles of cyber-physical systems*. MIT Press.

Asp, P., Bitzilekis, N., Bogdan, S., Elholm, T., Foffani, L., Frände, D., et al. (2009). Manifest zur Europäischen Kriminalpolitik. *Zeitschrift für Internationale Strafrechtsdogmatik, 4*(12), 697–706.

Bernasco, W., van Gelder, J.-L., & Elffers, H. (2017). *The Oxford handbook of offender decision making*. OUP.

Binding, K. (1922). *Die Normen und ihre Übertretung I* (4th ed.). Engelmann.

Bock, D. (2011). *Criminal compliance*. Nomos.

Bock, M. (2013). *Kriminologie* (4th ed.). Vahlen.

Brodowski, D. (2010). EU actions on cybercrime and cybercrime investigations. In M. Bellini, P. Brunst, & J. Jähnke (Eds.), *Current issues in IT security. Proceedings of the interdisciplinary conference in Freiburg i. Br./Germany, May 12–14, 2009* (pp. 145–161). Max-Planck-Institut für ausländisches und internationales Strafrecht/Duncker & Humblot.

Brodowski, D. (2013). Preparatory colloquium for the XIXth international congress of penal law (Moscow, 24–27 April 2013). *Russian Law Journal, 1*(1), 94–103.

Brodowski, D. (2016). Transnational organised crime and cybercrime. In P. Hauck & S. Peterke (Eds.), *International law and transnational organised crime* (pp. 334–360). OUP.

Brodowski, D. (2018). Alternative enforcement mechanisms in Germany. In M. Dyson & B. Vogel (Eds.), *The limits of criminal law* (pp. 365–396). Intersentia.

Brodowski, D. (2019a). Cybercrime, human rights and digital politics. In B. Wagner, K. Vieth, & M. C. Kettemann (Eds.), *Research handbook on human rights and digital technology. Global politics, law and international relations* (pp. 98–112). Edward Elgar.

Brodowski, D. (2019b). Hacking 4.0: Seitenkanalangriffe auf informationstechnische Systeme; Zugleich ein Beitrag zur Theorie und Dogmatik des IT-Strafrechts. *Zeitschrift für Internationale Strafrechtsdogmatik, 14*(1), 49–61.

Brodowski, D. (2020). Gefahrenabwehr und Sanktionierung. In D.-K. Kipker (Ed.), *Cybersecurity: Rechtshandbuch*.

Brodowski, D. (2021). *Cibercrimen y protección de la seguridad informática*. AdHoc.

Brodowski, D., Espinoza de los Monterros de la Parra, M., Tiedemann, K., & Vogel, J. (2014). *Regulating corporate criminal liability*. Springer.

Brodowski, D., & Freiling, F. C. (2011). *Cyberkriminalität, Computerstrafrecht und die digitale Schattenwirtschaft*. Forschungsforum Öffentliche Sicherheit.

Brodowski, D., & Nowak, D. (2020). § 41 BDSG. In H. A. Wolff & S. Brink (Eds.), *Beck'scher Onlinekommentar Datenschutzrecht* (32th ed.). Beck.

Cárdenas, A. A., Amin, S., Sinopoli, B., Giani, A., Perrig, A., & Sastry, S. (2009). Challenges for securing cyber physical systems. *Workshop on Future Directions in Cyber-physical Systems Security, 5*(1).

Chawki, M., Darwish, A., Khan, M. A., & Tyagi, S. (2015). *Cybercrime, digital forensics and jurisdiction*. Springer.

Duff, A. (2001). *Punishment, communication, and community*. OUP.

Eisele, J. (2019). Vor §§ 13ff. In: A. Schönke & H. Schröder (founders), *Strafgesetzbuch: Kommentar* (30th edn). Beck.

Eisele, J., & Vogel, J. (2018). Art. 83 AEUV. In E. Grabitz, M. Hilf, & M. Nettesheim (Eds.), *Das Recht der Europäischen Union* (65th ed. suppl). Beck.

Engelhart, M. (2012). *Sanktionierung von Unternehmen und compliance* (2nd ed.). Duncker & Humblot.

Feinberg, J. (1965). The expressive function of punishment. *The Monist, 49*(3), 397–423.

Fiorella, A. (2012). *Corporate criminal liability and compliance programs: Toward a common model in the European Union*. Jovene.

Frisch, W. (1988). *Tatbestandsmäßiges Verhalten und Zurechnung des Erfolgs*. Müller.
Frisch, W. (2016). Voraussetzungen und grenzen staatlichen Strafens. *Neue Zeitschrift für Strafrecht, 36*(1), 16–25.
Gärditz, K. F. (2016). Demokratizität des Strafrechts und Ultima Ratio-Grundsatz. *Juristenzeitung, 71*(13), 641–650.
Gillespie, A. A. (2016). *Cybercrime: Key issues and debates*. Routledge.
Goeckenjan, I. (2015). Überprüfung von Straftatbeständen anhand des Verhältnismäßigkeitsgrundsatzes: überfällige Inventur oder Irrweg? In M. Jestaedt & O. Lepsius (Eds.), *Verhältnismäßigkeit* (pp. 184–209). Mohr Siebeck.
Graf, J.-P. (2007). "Phishing" derzeit nicht generell strafbar! *Neue Zeitschrift für Strafrecht, 27*(3), 129–132.
Greco, L. (2013). Verfassungskonformes oder legitimes Strafrecht? Zu den Grenzen einer verfassungsrechtlichen Orientierung der Strafrechtswissenschaft. In B. Brunhöber, K. Höffler, J. Kaspar, T. Reinbacher, & M. Vormbaum (Eds.), *Strafrecht und Verfassung* (pp. 13–36). Nomos.
Haase, A. (2017). *Computerkriminalität im Europäischen Strafrecht*. Mohr Siebeck.
Hamm, R. (2016). Richten mit und über Strafrecht. *Neue Juristische Wochenschrift, 69*(22), 1537–1542.
Hassemer, W. (1989). Symbolisches Strafrecht und Rechtsgüterschutz. *Neue Zeitschrift für Strafrecht, 9*(12), 553–559.
Hassemer, W. (1994). Strafrechtswissenschaft in der Bundesrepublik Deutschland. In D. Simon (Ed.), *Rechtswissenschaft in der Bonner Republik* (pp. 259–310). Suhrkamp.
Hilgendorf, E. (Ed.). (2017). *Autonome Systeme und neue Mobilität*. Nomos.
Hörnle, T. (2011). *Straftheorien*. Mohr Siebeck.
Husak, D. N. (2008). *Overcriminalization*. OUP.
Ignor, A. (2016). Gedanken zur Krise des Strafrechts. In V. Bouffier, H.-D. Horn, R. Poseck, H. Radtke, & C. Safferling (Eds.), *Grundgesetz und Europa. Liber Amicorum für Herbert Landau zum Ausscheiden aus dem Bundesverfassungsgericht* (pp. 375–392). Tübingen.
Jahn, M., & Brodowski, D. (2016). Krise und Neuaufbau eines strafverfassungsrechtlichen Ultima Ratio-Prinzips. *Juristenzeitung, 71*(20), 969–980.
Jahn, M., & Brodowski, D. (2017). Das Ultima Ratio-Prinzip als strafverfassungsrechtliche Vorgabe zur Frage der Entbehrlichkeit von Straftatbeständen. *Zeitschrift für die gesamte Strafrechtswissenschaft, 129*(2), 363–381.
Jahn, M., Schmitt-Leonardy, C., & Schnoop, C. (2018). Unternehmensverantwortung für Unternehmenskriminalität – "Frankfurter Thesen". *Wistra, 37*(1), 27–31.
Jakobs, G. (1991). *Strafrecht Allgemeiner Teil* (2nd ed.). de Gruyter.
Kahneman, D., & Tversky, A. (1979). Prospect theory: An analysis of decision under risk. *Econometrica, 47*(2), 263–291.
Kaspar, J. (2014). *Verhältnismäßigkeit und Grundrechtsschutz im Präventionsstrafrecht*. Nomos.
Keenlab (2016). *Car Hacking Research: Remote Attack Tesla Motors*. Retrieved August 27, 2021, from https://keenlab.tencent.com/en/2016/09/19/Keen-Security-Lab-of-Tencent-Car-Hacking-Research-Remote-Attack-to-Tesla-Cars/
Kshetri, N. (2010). *The global cybercrime industry*. Springer.
Kudlich, H. (2003). Grundrechtsorientierte Auslegung im Strafrecht. *Juristenzeitung, 58*(3), 127–133.
Kühl, K. (2008). Fragmentarisches und subsidiäres Strafrecht. In U. Sieber & G. Dannecker (Eds.), *Strafrecht und Wirtschaftsstrafrecht* (pp. 29–46). Heymann.
Lagodny, O. (1996). *Strafrecht vor den Schranken der Grundrechte*. Mohr Siebeck.
Landau, H. (2015a). Das ultima-ratio-Prinzip im Strafrecht. In W. Hadding & U. Herrmann (Eds.), *Festschrift für Wolfgang Schlick* (pp. 523–533). Heymann.
Landau, H. (2015b). Die jüngere Rechtsprechung des Bundesverfassungsgerichts zu Strafrecht und Strafverfahrensrecht. *Neue Zeitschrift für Strafrecht, 35*(12), 665–671.
Laufer, W. S. (2006). *Corporate bodies and guilty minds*. University of Chicago Press.

Laufer, W. S. (2018). The compliance game. *Revista dos Tribunais, 107*(2), 67–80.
Loewenstein, G., & Elster, J. (1992). *Choice over time*. Russell Sage.
Lüderssen, K. (1979). Die generalpräventive Funktion des Deliktssystems. In W. Hassemer, K. Lüderssen, & W. Naucke (Eds.), *Hauptprobleme der Generalprävention* (pp. 54–80). Metzner.
Mamayek, C., Paternoster, R., & Loughran, T. A. (2017). Temporal discounting, present orientation, and criminal deterrence. In W. Bernasco, J.-L. van Gelder, & H. Elffers (Eds.), *The Oxford handbook of offender decision making* (pp. 209–227). OUP.
Mansdörfer, M. (2018). Über die strafrechtliche Verantwortlichkeit beim Einsatz maschinell-autonomer cyberphysischer Systeme – oder: Prolegomena zu einem Allgemeinen Teil des Cyberstrafrechts. In S. Barton, R. Eschenbach, M. Hettinger, E. Kempf, C. Krehl, & F. Salditt (Eds.), *Festschrift für Thomas Fischer* (pp. 155–170). Beck.
Mattes, H. (1977). *Untersuchungen zur Lehre von den Ordnungswidrigkeiten. Erster Halbband: Geschichte und Rechtsvergleichung*. Duncker & Humblot.
McBarnett, D. (2003). When compliance is not the solution but the problem: From changes in law to changes in attitude. In V. Braithwaite (Ed.), *Taxing democracy: Understanding tax avoidance and evasion* (pp. 229–243). Ashgate.
Meier, B. (2016). *Kriminologie* (5th ed.). Beck.
Meyer, F. (2015). Art. 83 AEUV. In H. von der Groeben, J. Schwarze, & A. Hatje (Eds.), *Europäisches Unionsrecht* (7th ed.). Nomos.
Nagin, D. S., & Pogarsky, G. (2004). Time and punishment: Delayed consequences and criminal behavior. *Journal of Quantitative Criminology, 20*(4), 295–317.
Öncüler, A. (2010). How do we manage an uncertain future? Ambiguity today is not ambiguity tomorrow. In E. Michel-Kerjan & P. Slovic (Eds.), *The irrational economist. Making decisions in a dangerous world* (pp. 107–115). PublicAffairs.
Pieth, M., & Ivory, R. (2011). *Corporate criminal liability*. Springer.
Popp, A. (2006). "Phishing", "Pharming" und das Strafrecht. *Multimedia und Recht, 9*(2), 84–86.
Reimer, P. (2018). Art. 5. In G. Sydow (Ed.), *Europäische Datenschutzgrundverordnung* (2nd ed.). Nomos.
Rengier, R. (2018). *Strafrecht Allgemeiner Teil* (10th ed.). Beck.
Renzikowski, J. (2002). Die Unterscheidung von primären Verhaltens- und sekundären Sanktionsnormen in der analytischen Rechtstheorie. In D. Dölling (Ed.), *Festschrift für Karl Heinz Gössel zum 70. Geburtstag* (pp. 3–13). Müller.
Rotsch, T. (Ed.). (2014). *Criminal compliance*. Nomos.
Roxin, C., & Greco, L. (2020). *Strafrecht Allgemeiner Teil, Band 1* (5th ed.). Beck.
Satzger, H. (2018). Art. 83 AEUV. In R. Streinz (Ed.), *EUV/AEUV* (3rd ed.). Beck.
Schjolberg, S. (2014). *The history of cybercrime, 1976–2014*. Books on Demand.
Schlepper, C. (2014). *Strafgesetzgebung in der Spätmoderne*. Springer.
Schmitt-Leonardy, C. (2013). *Unternehmenskriminalität ohne Strafrecht?* Müller.
Sieber, U. (2008). Mastering complexity in the global cyberspace. In M. Delmas-Marty, M. Pieth, & U. Sieber (Eds.), *Les chemins de l'harmonisation pénale* (pp. 127–202). Société de Législation Comparée.
Sieber, U. (2018). Administrative sanction law in Germany. In M. Dyson & B. Vogel (Eds.), *The limits of criminal law* (pp. 301–332). Intersentia.
Sieber, U., & Brodowski, D. (2018). Teil 19.3: Strafprozessrecht. In T. Hoeren, U. Sieber, & B. Holznagel (Eds.), *Handbuch Multimedia-Recht* (47th ed. suppl). Beck.
Sieber, U., & Engelhart, M. (2014). *Compliance programs for the prevention of economic crimes: An empirical survey of German companies*. Duncker & Humblot.
Suhr, O. (2016). Art. 83 AEUV. In C. Calliess & M. Ruffert (Eds.), *EUV/AEUV* (5th ed.). Beck.
Summers, S., Schwarzenegger, C., Ege, G., & Young, F. (2014). *The emergence of EU criminal law*. Hart.
Swoboda, S. (2010). Die Lehre vom Rechtsgut und ihre Alternativen. *Zeitschrift für die gesamte Strafrechtswissenschaft, 122*(1), 24–50.

Tiedemann, K. (1969). *Tatbestandsfunktionen im Nebenstrafrecht*. Mohr Siebeck.
Tiedemann, K. (1975). Der Entwurf eines Ersten Gesetzes zur Bekämpfung der Wirtschaftskriminalität. *Zeitschrift für die gesamte Strafrechtswissenschaft, 87*(2), 253–328.
Timpe, G. (2017). Die strafrechtliche Produzentenhaftung. *Höchstrichterliche Rechtsprechung im Strafrecht, 18*(6), 272–281.
Tversky, A., & Kahneman, D. (1992). Advances in prospect theory: Cumulative representation of uncertainty. *Journal of Risk and Uncertainty, 5*(4), 297–323.
UNODC. (2013). *Comprehensive study on cybercrime*. United Nations.
Vogel, J. (1993). *Norm und Pflicht bei den unechten Unterlassungsdelikten*. Duncker & Humblot.
Vogel, J. (2008). *Towards a global Convention against cybercrime. First World Conference of Penal Law*. Revue électronique de l'AIDP C-07:1 Retrieved August 27, 2021, from http://www.penal.org/sites/default/files/files/Guadalajara-Vogel.pdf
Vogel, J., & Bülte, J. (2020). § 15. In H. W. Laufhütte, R. Rissing-van Saan, & K. Tiedemann (Eds.), *Leipziger Kommentar* (Vol. 1, 12th ed.). De Gruyter.
Vogelgesang, S., & Möllers, F. (2016). Ransomware als moderne Piraterie: Erpressung in Zeiten digitaler Kriminalität. *Juris – Die Monatszeitschrift, 3*(10), 381–387.
von Hirsch, A. (2005). *Fairness, Verbrechen und Strafe*. BWV.
von Leitner, F. (2020). *Fefes blog*. Retrieved August 27, 2021, from https://blog.fefe.de/?ts=a19dfb18
Wall, D. S. (2007). *Cybercrime: The transformation of crime in the information age*. Polity Press.
Wall, D. S. (2010). The organization of cybercrime and organized cybercrime. In M. Bellini, P. Brunst, & J. Jähnke (Eds.), *Current Issues in IT Security. Proceedings of the interdisciplinary conference in Freiburg i. Br./Germany, May 12–14, 2009* (pp. 51–66). Max-Planck-Institut für ausländisches und internationales Strafrecht/Duncker & Humblot.
Williams, S. H. (2014). Probability errors: Overoptimism, ambiguity aversion, and the certainty effect. In E. Zamir & D. Teichmann (Eds.), *The Oxford handbook of behavioral economics and the law* (pp. 335–353). OUP.
Wilson, J. Q., & Herrnstein, R. J. (1985). *Crime and human nature*. Free Press.
Wohlers, W. (2018). Criminal law as a regulatory tool. In M. Dyson & B. Vogel (Eds.), *The limits of criminal law* (pp. 235–262). Intersentia.
Wörner, L. (2019). Der Weichensteller 4.0: Zur strafrechtlichen Verantwortlichkeit des Programmierers im Notstand für Vorgaben an autonome Fahrzeuge. *Zeitschrift für Internationale Strafrechtsdogmatik, 14*(1), 41–48.

Phishing in Online Banking – An Overview of the Development and the European and German Legal Positions

Sonja Oleownik

Contents

Abstract Conducting banking transactions via Online Banking is well established in today's society. It is therefore not surprising that it is subject to frequent criminal attacks which lead to high economic damage. The so-called phishing attacks, which have been occurring in Germany since about 2005 are a particular example of this. The investigation of phishing cases is interesting from both a technical and a legal perspective. This article gives a basic overview of the development of phishing in recent years, different attack methods and various Online Banking procedures with which attempts are made to protect against phishing attacks. Furthermore, this contribution explains the basic European and German legal rules which apply in case of a phishing attack. Since the attacker cannot usually be identified and held accountable, it must be determined who is liable for the damage caused by the phishing attack. Phishing in Online Banking has received very specific legal

S. Oleownik (✉)
Saarland University, Saarbrücken, Germany
e-mail: sonja.oleownik@uni-saarland.de

G. Borges, C. Sorge (eds.), *Law and Technology in a Global Digital Society*,
https://doi.org/10.1007/978-3-030-90513-2_13

regulation, initially European and subsequently at German level. This paper will therefore examine the apportionment of risk under German law and according to the currently applicable European standards introduced by the First and Second Payment Services Directives. In addition, the most important innovations introduced by the Second Payment Services Directive are considered, in particular the so-called Two-Factor authentication.

1 Introduction

The subject of phishing in Online Banking is interesting from both a legal and a technical perspective. On the technical side, there are many different methods by which attacks can be made and many different processes in Online Banking have been put in place to try to protect against them. On the legal side, it must be determined who is liable for the damage caused by a phishing attack.

This paper will start with a definition of phishing illustrated by examples (see Sect. 2.1). Then the development of phishing over the last few years will be analysed (see Sect. 2.2). This will be followed by the listing of a number of technical points about phishing methods and Online Banking processes (see Sect. 3) and an overview of the legal key issues (see Sect. 4). Phishing in online banking has received very specific legal regulation, initially European and subsequently at German level. The basic legal situation is first explained in the form of an overview of European legislation through the Second Payment Services Directive (PSD 2) which came into effect in January 2018 (see Sect. 4.1.1) and then with regards to the German legislation (see Sect. 4.1.2). From a legal perspective, it is important to clarify who should bear the risk of a phishing attack in principle (see Sect. 4.2). Related to this, the obligations of those participating in Online Banking and how liability is to be allocated must be considered (see Sect. 5). This will be explained by reference to the relevant German legislation, with particular regard to European law and the changes brought about by the PSD 2. Finally, there will be a summary of the basic standards of the complex regulatory system and the significance of the changes brought about by PSD 2 (see Sect. 6).

2 Terminology and Development

2.1 Definition and Examples

Firstly, it is necessary to clarify what is meant by the term ‘phishing’. According to the Oxford English dictionary, phishing is

> the fraudulent practice of sending emails purporting to be from reputable companies in order to induce individuals to reveal personal information, such as passwords and credit card numbers.[1]

In a broad sense, phishing is an attack via the internet, which helps the attacker to obtain authorisation data from another party or to misuse such data.[2]

So, for example in a classical case of phishing in Online Banking, the attacker sends the victim an email which looks like an email from the victim's own bank and asks the victim to follow a link to a website and log into his bank's website using his username, password and TAN. However, the website is fake and is controlled by the attacker. This gives the attacker direct access to the victim's transaction details.[3]

Two current examples of phishing cases are as follows:

(1) Trick with a mobile phone invoice[4]

In this method, the attacker sends the victim an email which looks like it comes from the victim's own mobile phone provider. The email has an attachment which supposedly contains the mobile phone invoice or there is a button at the end of the email. The email refers an unusually high invoice amount or is a payment reminder. In the first case the victim opens the attachment immediately to look for the reason for this high amount. In the second case, the email states additionally, that the victim's account is blocked, so the victim clicks the button to try to remedy this situation. In both cases damaging software is put on the victim's computer by this attachment and this software enables the attacker to obtain the victim's personal contact details and to misuse them.

(2) Trick with a card reader, sent by post[5]

This phishing scam has been widely reported in the British press. In this method, the attacker sends an elderly victim a card reader by post. The accompanying letter looks exactly like an official letter from the Bank of Scotland. The attacker then calls the victim and asks if they need the card reader. When the victim says no, the attacker asks them for their personal account details, saying that they need them to deactivate the card reader. The details are willingly given and these details give the attackers access to the victim's account.

---

[1] See https://en.oxforddictionaries.com/definition/phishing (accessed 27 Nov 2021).

[2] Borges (2017), mn. 257.

[3] Borges (2017), mn. 263; Borges (2005), pp. 3313 f.; Erfurth (2006), p. 2200; Gajek et al. (2005), pp. 640 f.; Karper (2006), p. 215; Maihold (2017b), mn. 30; Recknagel (2005), p. 51; Schmidt and Pruß (2019), mn. 274; Zahrte (2019a), mn. 130c.

[4] See for example https://www.t-online.de/digital/sicherheit/id_88490166/telekom-betrug-vorsicht-bei-dieser-e-mail-droht-die-internet-abzocke.html; https://www.netzwelt.de/betrugswarnungen/151383-deutsche-telekom-gefaelschte-rechnungsmails-verbreiten-schadsoftware.html (accessed 27 Aug 2021).

[5] See for example Robinson (2017).

2.2 Development of Phishing

Phishing first appeared in the USA and has spread worldwide. It has been possible to trace the development of phishing in Germany since around 2005.[6]

In Germany, the *Federal Authority for Criminality* (*"Bundeskriminalamt"*, abbreviated to *BKA*), keeps statistics about the annual number of phishing cases. In 2012, there were 3440 reported cases of phishing in Germany. The figures initially rose to a peak of 6984 cases in 2014 and then descended to 2175 cases in 2016.[7]

One must however keep in mind that there are an unknown number of cases which go unreported. Not everyone wants to admit that they have been scammed. However, the numbers do seem to be in decline since the clear peak reached in 2014. In 2018 there were only 723 cases reported, a decrease of almost 50% compared to the previous year; and in 2017, the decline in the number of cases compared to the previous year was approximately 35%.[8]

This decline in cases is probably due to the fact that the new Online Banking processes provide better protection against attack.[9]

The numbers also provide data about the financial losses caused by phishing and it is clear that such losses are enormous—in 2014 the loss was in the region of 27.9 million Euro. In 2016, the amount of damage had fallen to 8.7 million Euro which, whilst obviously much less than the 2014 figure, is still a substantial loss.[10]

The fact that the figures fluctuate can also be seen from the statistics of the *Anti-Phishing Working Group (APWG)* which covers phishing cases worldwide.[11]

Here, the peak was in 2015. The statistics also show that the figures vary greatly even during a three-month period: In the 4th quarter of 2014 the figure was between approximately 60,000 and 70,000,[12] in the 4th quarter of 2015 even between approximately 80,000 and 195,000[13] and in the 4th quarter of 2016 between approximately 50,000 and 95,500.[14] In 2019 *Greg Aaron*, *APWG* Senior Research Fellow and President of Illumintel Inc. stated: *"The year 2019 turned out to be a roller-coaster ride for phishing. July though October was the worst period for phishing that the APWG had seen in three years, and then phishing levels settled back down to more normal levels"*.[15]

---

[6] Borges (2017), mn. 259; Borges (2005), p. 3313; BSI (2006), p. 28.

[7] BKA (2016), p. 21.

[8] BKA (2019a), p. 17.

[9] BKA (2019b), p. 2.

[10] BKA (2016), p. 22.

[11] The work of the Anti-Phishing Working Group can be seen on https://www.apwg.org (accessed 27 Nov 2021).

[12] APWG (2015), p. 4.

[13] APWG (2016), p. 4.

[14] APWG (2017), p. 4.

[15] APWG (2020), p. 4.

The *BKA* emphasises the fact that the attackers will, in the future, specialise on attacks which compromise mobile devices such as smart phones or tablets.[16]

It is therefore hard to predict how phishing will develop in the future. In view of the high potential for damage, compliance with technical protective measures, such as those now specified by the PSD 2 (see Sect. 5.2.3.3), is in any case of great importance. In any event, the legal and technical questions remain highly relevant for phishing cases.

3 Phishing Methods and Online Banking Processes

The legal classification of phishing always depends on the phishing method and the Online Banking process used in each case. Therefore, a brief overview of the most common methods and procedures follows.

3.1 Phishing Methods

Firstly, there is classical phishing. As mentioned above (see Sect. 2.1), in this method, the victim receives an email requesting that the username, password and TAN of his Online Banking account be entered via a fake website. Typical for classical phishing is the mass mailing of emails.[17]

A special form of classical phishing is spear-phishing. Here, the attacker focuses specifically on particular targets, uses personal information about an individual or organisation and pretends to be a business associate, friend or service provider known to the user.[18] In contrast to classical phishing, the victims are not part of a mass mailing of emails.[19]

There are many other common methods of attacks, which can fall under the definition of phishing in its broader sense, mentioned in Sect. 2.1. Some examples are:

Key loggers keep track of the entering of PINs and TANs via the keyboard and send this data to the offender.[20]

[16] BKA (2016), p. 22; BKA (2019a), p. 18.

[17] Cornelius (2013), mn. 275.

[18] Borges (2017), mn. 263; BKA (2020), p. 44 (footnote 35); Mohsin (2016); Malin et al. (2017), p. 153.

[19] Malin et al. (2017), p. 153.

[20] LG Bonn, Court Decision f. 07.07.2009 – 7 KLs 1/09 = BeckRS 2009, 28058; Borges (2017), mn. 264; Jungmann (2020a), mn. 35; Pruß and Sarre (2019), keyword “Spyware”; Zahrte (2019a), mn. 130h.

Through Trojans on a victim's computer which manipulate the host-file or the Domain Name Servers (DNS), attackers can entice the victim to a fake website controlled by the offender. This website would look very similar to the bank's own website. This is known as pharming.[21]

In the case of so-called 'man in the middle' attacks, the attacker spies on communications between two internet users and either misuses the data he obtains this way or gets involved in the communication directly.[22]

Phishing also has a social aspect. In the case of social engineering or social hacking, offenders spy on the social and business contacts of the victim and use the knowledge they gain to get the victim to behave in a certain way; for example, to pass on confidential information.[23] Social engineering is, as a rule, always a part of phishing.[24]

3.2 *Online Banking-Processes*

Which phishing method an attacker chooses to use depends on what type of Online Banking process he wants to attack.

The Online Banking processes which are currently being used are listed below in a brief overview:

Most processes are based on the use of a PIN and TAN.[25] A PIN is the personal identification number[26] which, in many cases, is used in connection with the username to log in to Online Banking; TAN is a transaction number[27] which is used to authenticate a declaration given by the customer, for example a mandate to transfer money.[28]

The customer can obtain TANs in a number of different ways. A few classic examples are:

In a simple PIN/TAN process, the customer receives a TAN-list from his bank and can choose any TAN from this list for the transaction.[29]

---

[21] BGH, Court Decision f. 24 Apr 2012 – XI ZR 96/11 = NJW 2012, 2422, 2424; Bär (2020), mn. 32; Borges (2017), mn. 265; Köbrich (2015), p. 10; Maihold (2017b), mn. 31; Zahrte (2019a), mn. 130i.

[22] Borges (2017), mn. 266; Bundeskartellamt (BKartA) (4. Beschlussabteilung), Resolution f. 29 June 2016 – B 4-71/10, BeckRS 2016, 107337, mn. 53 (note 39); BSI, Glossar der Cyber-Sicherheit; Graf (2017), mn. 94; Maihold (2017b), mn. 32; Zahrte (2019a), mn. 130k.

[23] BSI (2017), p. 33; Gilch (2016), mn. 39 (note 70); Zahrte (2019a), mn. 130l.

[24] Borges (2017), mn. 269; Borges (2012), p. 2388; Schulte am Hülse and Kraus (2016), p. 436.

[25] Maihold (2017b), mn. 7.

[26] Aljawarneh (2017), p. 16; Zahrte (2019a), mn. 42.

[27] Zahrte (2019a), mn. 42a.

[28] Borges (2017), mn. 34.

[29] Borges (2017), mn. 35; Maihold (2017b), mn. 10.

In the case of an iTAN-process, each TAN on the TAN-list is linked to a sequence number and the bank requests a specific sequence number for each transaction made.[30]

In chipTAN and eTAN+ processes, the customer generates the TAN himself using a token.[31] This works for example with a TAN generator into which the customer inserts his chip card.[32] These processes are characterised by the fact that the TAN is linked to a specific transaction; the TAN is generated for specific transaction details and is only valid for this one single transaction.[33]

In the smsTAN processes, the bank sends the customer the TAN to his mobile phone and the customer has to give the bank his mobile phone number in advance; this process is also transaction-specific.[34]

Relatively new processes, such as photoTAN or pushTAN, use apps on the user's smart phone.[35]

4 Basic Apportionment of Risk in Online Banking

Phishing is, of course, a criminal act and the offender also faces civil liability for the damage which he causes. However, in order to take the offender to court, it is first necessary to establish who the offender actually is. In most cases, it is simply not possible to find out the offender's identity.

In these cases, it is important to know whether the bank or the bank's customer must bear the risk of phishing.

4.1 Legal Basis

To understand the apportionment of risk in Online Banking according to European and German law, one must first look at the legal basis for the payment transaction in Online Banking.

[30] Borges (2017), mn. 36; Maihold (2017b), mn. 12.

[31] Borges (2017), mn. 38 ff.

[32] See for example: https://www.sparkasse.de/service/sicherheit-im-internet/tan-verfahren.html (accessed 27 Nov 2021).

[33] Borges (2017), mn. 38; Maihold (2017b), mn. 17 ff.

[34] Borges (2017), mn. 41; see also: https://www.sparkasse.de/service/sicherheit-im-internet/tan-verfahren.html (accessed 27 Nov 2021).

[35] Borges (2017), mn.42 f.; see also: https://www.paycenter.de/servicecenter/aktuelles/pressemitteilungen/189-phototan-chiptan-smstan-pushtan-welches-verfahren-ist-sicher (accessed 27 Nov 2021).

4.1.1 Payment Services Directive (PSD 2)

The aim of the European legislator was to unify European payment traffic. This aim was achieved in particular by the enactment of the European payment services directive (PSD)[36] in 2007.

A directive is a legal act of the European Union and individual member states must implement directives into their national law.[37] The PSD introduced a Single Euro Payments Area (SEPA) to unify European wide payment transactions.[38] In January 2016, the Second Payment Services Directive (PSD 2) came into effect and replaced the PSD of 2007.[39] The PSD 2 had to be substantially incorporated into national law by 13th January 2018.[40] With regards to phishing, the rules were not substantially changed by the PSD 2.[41] However, there are some modifications in the details, which will be explained later (see the explanations on the liability provisions under Sect. 5.2).

The main rules of the PSD 2 in connection with phishing in Online Banking can be found in Art. 64, 73 and 74.

Art. 64 section 1 sentence 1 PSD 2 states that a payment transaction can only be considered to be authorised if the payment services user consented to it. This is not the case in the event of a phishing attack.

The consequence of an unauthorised payment transaction is according to Art. 73 section 1 sentence 1 of the PSD 2 in principle that the bank has to refund the customer the amount of the unauthorised transfer already made.

Therefore, the bank initially bears the risk of a phishing attack.

However, Art. 74 PSD 2 codifies some exceptions which result in the customer being liable for unauthorised payment transactions.

The first subparagraph of section 1 provides that the payer may be obliged by Member States to bear the loss relating to any unauthorised payment transactions, up to a maximum of 50 Euro, resulting from the use of a lost or stolen payment instrument or from the misappropriation of a payment instrument. The second subparagraph states two exceptions to this rule: The payer is not liable if the loss, theft or misappropriation was not detectable or if the causation of the loss is attributable to the Bank.

---

[36] Directive 2007/64/EC of the European Parliament and of the Council on payment services in the internal market amending Directives 97/7/EC, 2002/65/EC, 2005/60/EC and 2006/48/EC and repealing Directive 97/5/EC, Abl. EU Nr. L 319/1 v. 5.12.2007.

[37] Nettesheim (2009), mn. 124.

[38] Kolassa (2017), mn. 134.

[39] Directive 2015/2366 of the of the European Parliament and of the Council on payment services in the internal market amending Directives 97/7/EC, 2002/65/EC, 2005/60/EC and 2006/48/EC and repealing Directive 97/5/EC, Abl. EU Nr. L 337/35 v. 23.12.2015.

[40] The incorporation into the BGB happened because of Art. 2, 15 Abs. 4 S. 1 "Gesetz zur Umsetzung der Zweiten Zahlungsdiensterichtlinie" (BGBl. I 2017, p. 2446).

[41] Borges (2017), mn. 287a; Borges (2016), pp. 257 f.

Subparagraph 2 of section 1 prescribes the full liability of the payer for all of the loss relating to any unauthorised payment transactions if he has intentionally or negligently breached his obligations under Art. 69 PSD 2. These obligations are the use of the payment instrument according to the terms governing its issue and the notification of the payment service provider of the misuse of a payment instrument without undue delay of the customer becoming aware of it.

Exceptions to the client's liability exist if there is an exclusion of liability.

In case of a merely negligent breach of obligations Art. 74 section 1 subparagraph 3 allows Member States to reduce liability.

Section 2 prescribes an exclusion of liability where the payment service provider does not require strong customer authentication. A further exclusion of liability applies in accordance with section 3 as soon as the customer has reported the loss of the payment instrument to the bank.

As a reverse-exception of this Art. 74 PSD 2 provides in several places that the payer is fully liable in any case if he has acted fraudulently.

4.1.2 German Regulation

These PSD 2 rules have been implemented into German law in various places.

The following description of the German legal situation refers to the standards implemented by the First and Second Payment Services Directives at the most relevant points.

The legal basis for a payment transaction in Online Banking in Germany is regulated in the 12th Title of the German Civil Code (Bürgerliches Gesetzbuch (BGB)), the statute governing German civil law. This Title deals with mandates, contracts for the management of the affairs of another and payment services (§§ 662 – 676c BGB). Paragraphs 675c et seq. BGB explicitly deal with the law relating to payment services. §§ 675c to 676c were inserted into the BGB as a result of the 2007 European payment services directive (PSD) and changed as a result of the PSD 2.

The main provision in German Law dealing with payment services is § 675 BGB (management of the affairs of another).[42] A payment transfer is, in effect, a contract for the management of the affairs of another the basis of which is the delivery of a payment service. Pursuant to §675 c section 1 BGB, certain provisions relating to mandates are applicable to payment transfers. Therefore, § 670 provides that if the bank, for the purpose of performing the mandate, incurs expenses that could be considered to be necessary in the circumstances, then the bank can seek reimbursement of these expenses.

[42] Begründung zum Gesetzesentwurf zur Umsetzung der Verbraucherkreditrichtlinie, des zivilrechtlichen Teils der Zahlungsdiensterichtlinie sowie zur Neuordnung der Vorschriften über das Widerrufs- und Rückgaberecht, Bundestags-Drucksache (BT-Drs.) 16/11643, 98; Heermann (2020), mn. 54 f.

This means the following: when a bank customer undertakes a bank transfer; he gives the bank a mandate. Due to the payment services contract, the bank is obliged to carry out this transaction. The bank will incur expenses to the value of the amount which it transfers. Pursuant to § 670 BGB, the bank can ask the customer to indemnify it against these expenses, but only, and this is key, if it was necessary for the bank to incur such expenses in the first place.

In particular, it would be necessary for the bank to incur the expenses if the mandate given to the bank was valid.[43] Pursuant to § 675j section 1 BGB a mandate is only valid, if it was authorised.[44]

According to the legal definition in § 675j para 1 BGB, authorisation exists if the payer consented to the payment transaction.

§ 675j BGB implemented Art. 54 section 1 sentence 1 PSD. This section of the PSD also states that a payment transaction can only be considered to be authorised if the payment services user consented to it. In the PSD 2 the same rule is set out in Art. 64 section 1 sentence 1.

From the above it follows that the expenses incurred by the bank are not deemed to have been necessary if the payment was not authorised by the bank's customer. In this case, the bank then has no right to ask the customer for indemnification.[45] This result is expressly set out in § 675u sentence 1 BGB.

The logical consequence of this is that the bank has to reimburse the customer for an unauthorised transfer already made. This is codified in § 675u sentence 2 BGB. This rule implemented Art. 60 section 1 of the PSD which makes the same provision. In the PSD 2 the same rule is set out in Art. 73 section 1 sentence 1.

4.2 Basic Principle: Payment Services Provider Bears the Risk

With regards to the apportionment of risk in Online Banking, the result of the rules explained above is clear.

If a payment transaction is not authorised, the bank cannot be indemnified for the expenses which it incurs in carrying out the transaction.[46] In addition, according to §

[43] BGH, Court Decision f. 11 Dec 1994 – XI ZR 238/93 = NJW 1994, p. 3345; AG Wiesloch, Court Decision f. 20 June 2008 – 4 C 57/08 = MMR 2008, p. 627; Borges et al. (2011), p. 256; Fischer (2020), mn. 59; Schäfer (2020a), mn. 24.

[44] Borges (2017), mn. 285.

[45] Fischer (2020), mn. 59.

[46] BGH, Court Decision f. 17 July 2001 – XI ZR 325/00 = NJW 2001, p. 2969; BGH, Court Decision f. 16 June 2015 – XI ZR 243/13 = BKR 2015, p. 473; BGH, Court Decision f. 26 Jan 2016 – XI ZR 91/14 = NJW 2016, p. 2025, ECLI:DE:BGH:2016:260116UXIZR91.14.0; LG Düsseldorf, Court Decision f. 26 Oct 2018 – 6 O 72/17 = BKR 2019, p. 155; Borges et al. (2011), p. 256; Fischer (2020), mn. 59; Häuser (2019), mn. 338; Maihold (2017b), mn. 67.

675u sentence 2 BGB, the bank is obliged to reimburse the customer's account with the amount of the non-authorised payment.[47]

The conclusion which can be drawn is that, as a basic legal principle, the bank bears the risk of misuse in Online Banking.[48]

5 Liability in Online Banking

Irrespective of this legal principle of risk bearing, in many situations the user is actually liable in the case of misuse of his online-account.

We can broadly distinguish between two scenarios:

1. The existence of authorisation is assumed as a result of a legal fiction.
2. There is no authorisation and the customer is liable himself.

5.1 Assumption of Authorisation Through Legal Fiction

The existence of the legal fiction that authorisation exists can be established by using either a substantive law or a procedural law approach. In the case of substantive law, liability for apparent authority must be considered. The procedural law approach uses idea of so-called prima facie evidence.

5.1.1 Liability for Apparent Authority

According to the principles of apparent authority, the customer will be liable for a fraudulent payment transaction if the payment transaction undertaken by the offender can be attributed to him.[49]

The concept of apparent authority should, in German law, be considered in connection with the principles of agency law.[50] In order to make a valid declaration of intent in the name of another person, the basic principle is that one must have

[47] BGH, Court Decision f. 16 June 2015 – XI ZR 243/13 = BKR 2015, p. 473; LG Düsseldorf, Court Decision f. 26 Oct 2018 – 6 O 72/17 = BKR 2019, p. 155.

[48] BGH, Court Decision f. 17 July 2001 – XI ZR 325/00 = NJW 2001, p. 2969; AG Wiesloch, Court Decision f. 20 June 2008 – 4 C 57/08 = MMR 2008, p. 627; Borges (2017), mn. 287; Borges (2005), p. 3317; Casper (2009), mn. 17; Fischer (2020), mn. 59; Häuser (2019), mn. 338; Hossenfelder (2009), p. 791; Jungmann (2020b), mn. 16; Linardatos (2015), p. 96.

[49] OLG Schleswig, Resolution f. 19 July 2010 – 3 W 47/10 = CR 2011, p. 52; LG Darmstadt, Court Decision f. 28 Aug 2014 – 28 O 36/14 = BKR 2014, p. 481; Borges (2017), mn. 291; Schmalenbach (2020a), mn. 4.

[50] BGH, Court Decision f. 11 May 2011 – VIII ZR 289/09 = NJW 2011, p. 2423; Borges (2017), mn. 292; Kaufhold (2017), mn. 32; Schäfer (2020b), mn. 14; Schubert (2018b), mn. 15.

consent from the person being represented.[51] However, under certain circumstances, it can suffice if it appears, from the perspective of the person receiving the declaration, that there was consent and this would be a case of apparent authority.[52]

If one assumes that this type of apparent authority is also possible in the case of authorisation pursuant to § 675j BGB,[53] one must examine what the requirements for apparent authority actually are. These are the existence of an apparent authority situation and a ground for attribution.[54]

5.1.1.1 Apparent Authority Situation

There must be an apparent authority situation. This means that there must be an objective basis for the recipient of the declaration to be able to rely on it.[55] In Online Banking, this objective basis could be use of the authentication media, such as a bank card, password or TAN.[56] Therefore, if a customer's authentication media is used to undertake a transaction, there is an objective basis for the bank to be able to rely on there being authorisation for such transaction. The requirement for an apparent authority situation is the existence of an apparent authority worthy of protection, based on justified trust.[57] The terms 'strong' or 'weak' apparent authority are sometimes used in this context: strong apparent authority exists if the authentication media offers a great deal of protection against misuse; in insecure Online Banking processes which have been the subject of frequent successful attacks in the past, there is only weak apparent authority.[58] A bank thus has good reasons to use strong (secure) processes, because it is more likely that a transfer can be attributed to the account holder in the end (for the requirement for attribution see below Sect. 5.1.1.2).[59] Some courts have affirmed the liability of the user for apparent authority

---

[51] Mansel (2021), mn. 1; Schubert (2018a), mn. 1.

[52] BGH, Court Decision f. 11 May 2011 – VIII ZR 289/09 = NJW 2011, p. 2422; Benning and Oberrath (2015), Glossar/Definitionen – Rechtsschein; Köhler (2020), mn. 35.

[53] This is contentious. For it e.g.: OLG Schleswig, Resolution f. 19 July 2010 – 3 W 47/10 = CR 2011, p. 52; LG Darmstadt, Court Decision f. 28 Aug 2014 – 28 O 36/14 = BKR 2014, p. 482; Borges (2017), mn. 295 ff.; Borges et al. (2011), p. 256; Zimmermann (2020), mn. 17; against it e.g.: BGH, Court Decision f. 16 June 2015 – XI ZR 243/13 = BKR 2015, p. 473 (mn. 22); Linardatos (2015), p. 98; Maihold (2017b), mn. 68; Wösthoff (2015), p. 2068; open, but with arguments against it: BGH Court Decision f. 26 Jan 2016 – XI ZR 91/14 = NJW 2016, p. 2029, ECLI:DE:BGH:2016:260116UXIZR91.14.0.

[54] BGH, Court Decision f. 11 May 2011 – VIII ZR 289/09 = NJW 2011, pp. 2422f.; Borges (2017), mn. 296; Stadler (2020), mn. 43 and 47.

[55] BGH, Court Decision f. 11 May 2011 – VIII ZR 289/09 = NJW 2011, p. 2422; Borges (2017), mn. 297; Borges (2011a), p. 2401; Stadler (2020), mn. 43 and 47.

[56] LG Darmstadt, Court Decision f. 28 Aug 2014 – 28 O 36/14 = BKR 2014, pp. 482 f.; Borges (2017), mn. 298.

[57] Schäfer (2020b), mn. 14; Schubert (2018a), mn. 129.

[58] Borges (2017), mn. 298 ff.; Borges (2011a), p. 2402; Borges (2007), pp. 390 ff.

[59] Borges (2017), mn. 299; Borges (2011a), p. 2402; Borges (2007), p. 391.

only in the case of a strong apparent authority.[60] However, this is contentious.[61] According to the—in my view correct—opposite opinion, an insecure process may be sufficient to affirm an apparent authority situation and thus protect the bank, because the customer would still have good reason to behave in a prudent and careful manner,[62] as, if he fails to do so, he could be liable as a result of apparent authority.

5.1.1.2 Attribution

The apparent authority must also be attributable to the bank customer.[63] This is the case if it would have been possible and reasonable for the customer to have prevented the fraudulent declaration from having been given.[64] This can at least be assumed if the customer consciously takes a risk—for example through passing on his TAN list—or stops himself from being able to become aware of the risk because, for example, he does not double-check the details of the transaction before submitting them.[65] Case law also requires that there be a certain length and frequency of the unauthorised use of the authentication media. Thus a one-off unauthorised use would not suffice.[66]

5.1.2 Prima Facie Evidence

From a procedural perspective, it can sometimes be assumed on an evidentiary level that authorisation exists. This is achieved through so-called prima facie evidence. If the authorisation is contentious, then the bank must prove that a transaction was authorised (§ 675w S. 1 BGB). Pursuant to § 675w S. 2 BGB, authorisation occurs when the bank has verified the use of a particular payment authentication instrument,

---

[60] See next footnote ("For the requirement of strong apparent authority").

[61] For the requirement of strong apparent authority: BGH, Court Decision f. 11 May 2011 – VIII ZR 289/09 = NJW 2011, pp. 2422 f.; OLG Köln, Court Decision f. 13 Jan 2006 – 19 U 120/05 = NJW 2006, p. 1677; OLG Hamm, Court Decision f. 16 Nov 2006 – 28 U 84/06 = NJW 2007, p. 611; LG Bonn, Court Decision f. 7 Aug 2001 – 2 O 450/00 = MMR 2002, p. 257; LG Bonn, Court Decision f. 19 Dec 2003 – 2 O 472/03 = MMR 2004, p. 181; against the requirement of strong apparent authority: Borges (2017), mn. 298 ff.; Borges (2011a), p. 2402; Borges (2007), pp. 390 f.; Schubert (2018a), mn. 129; Stöber (2011), p. 552.

[62] Borges (2017), mn. 298 ff.; Borges (2011a), p. 2402; Borges (2007), pp. 390 f.

[63] BGH, Court Decision f. 11 May 2011 – VIII ZR 289/09 = NJW 2011, pp. 2422 f.; Borges (2017), mn. 296; Stadler (2020), mn. 43 and 47.

[64] Borges (2017), mn. 302 f.; Schubert (2018a), mn. 119; Stadler (2020), mn. 43 and 47.

[65] Borges (2017), mn. 303; Borges (2011a), p. 2402.

[66] BGH, Court Decision f. 16 Mar 2006 – III ZR 152/05 = NJW 2006, p. 1972; BGH, Court Decision f. 10 Jan 2007 – VIII ZR 380/04 = NJW 2007, p. 989; BGH, Court Decision f. 11 May 2011 – VIII ZR 289/09 = NJW 2011, p. 2422; LG Darmstadt, Court Decision f. 28 Aug 2014 – 28 O 36/14 = BKR 2014, p. 483; LG Köln, Court Decision f. 13 Aug 2020 – 14 O 77/19 = GRUR-RS 2020, 21212 (mn. 72).

including the personalised security elements. In this case, prima facie evidence can exist for the premise that the customer authorised the bank transfer. The applicability of prima facie evidence to constellations of payment services was highly controversial,[67] but has ultimately been affirmed by the German Federal Court.[68,69]

A prerequisite for the provision of prima facie evidence is the existence of a prima facie case. This is given, if there is a typical sequence of events that, according to life experience, points to a certain cause or consequence.[70]

In Online Banking, a typical event which could be prima facie evidence for the fact that a customer authorised a transaction can only be found in processes which have been shown to be resistant to attack.[71] The German Federal Court has recently introduced the requirement that a process be "practical insurmountable" before prima facie evidence can be used.[72] Given the tangible threat of a phishing attack, prima facie evidence should no longer suffice in the classic PIN/TAN situations.[73]

The prima facie case that the authorisation is a result of the bank customer's own actions can be rebutted if the customer can prove an alternative course of events.[74]

5.1.3 Legal Consequence of the Legal Fiction

As stated above (see Sect. 4.1.2), in the case of an authorised transfer, the bank may demand reimbursement for its expenses from the customer.

If there is legal fiction for the existence of an authorisation on the basis of an apparent authority or a prima facie evidence, an authorised transfer is deemed to exist.

---

[67] See an overview at Maihold (2017a), mn. 108 or at Zetzsche (2020b), mn. 22 ff.; Omlor (2019), p. 105.

[68] BGH, Court Decision f. 26 Jan 2016 – XI ZR 91/14 = NJW 2016, p. 2026, ECLI:DE:BGH:2016:260116UXIZR91.14.0; Linardatos (2017), p. 2145.

[69] The new provision of Art. 72 section 2 sentence 2 PSD 2 (implemented in § 675w sentence 4 BGB) specifies the rules of evidence, but does not affect the applicability of prima facie evidence: Linardatos (2017), p. 2150, Zetzsche (2020b), mn. 27.

[70] BGH, Court Decision f. 17 Apr 1951 – I ZR 28/50 = NJW 1951, p. 654; BGH, Court Decision f. 3 Mar 1990 – VI ZR 239/89 = NJW 1991, p. 231; BGH, Court Decision f. 4 Dec 2000 – II ZR 293/99 = NJW 2001, p. 1141; BGH, Court Decision f. 19 Jan 2010 – VI ZR 33/09 = NJW 2010, p. 1072; BGH, Court Decision f. 11 Dec 2018 – KZR 26/17 = NJW 2019, p. 664; Bacher (2020), mn. 94; Saenger (2019), mn. 38; Schmidt (2020), mn. 43; Schulz-Merkel and Quarch (2021), mn. 1.

[71] Borges (2007), p. 41; Borges (2009), pp. 86 f.; Hofmann (2021d), mn. 63; Kind and Werner (2006), p. 359; Köbrich (2015), p. 12; Mühlenbrock and Dienstbach (2008), p. 630.

[72] BGH, Court Decision f. 26 Jan 2016 – XI ZR 91/14 = NJW 2016, p. 2027, ECLI:DE:BGH:2016:260116UXIZR91.14.0.

[73] See e.g. penultimate footnote.

[74] See for the rebut of prima facie evidence: BGH, Court Decision f. 3 Mar 1990 – VI ZR 239/89 = NJW 1991, p. 231; BGH, Resolution f. 6 July 2010 – XI ZR 224/09 = VuR 2010, p. 386; BGH, Court Decision f. 26 Jan 2016 – XI ZR 91/14 = NJW 2016, p. 2026; Bacher (2020), mn. 98; Saenger (2019), mn. 40.

5.2 Liability of the Customer Due to Non-Existence of an Authorisation

In cases where there is no authorisation—either actual or as a result of a legal fiction—the customer may be liable for the damage. The customer's liability to pay compensation was set out first in Art 61 PSD. German law implemented Art. 61 PSD in § 675v BGB and this has now been modified as a result of Art. 74 PSD 2.

If the customer is liable for damages in whole or in part under § 675v BGB, the bank may set off this claim (in whole or in part) against the customer's claim under § 675u BGB, so that the customer's claim is extinguished (in whole or in part) and in the end the customer bears the loss (in whole or in part).[75]

§ 675v BGB describes three situations (see below Sects. 5.2.1 and 5.2.2) in which a customer will be liable to pay damages.

5.2.1 Misplaced Payment Instrument (Situation 1)

The first situation is governed by § 675v section 1 sentence 1 variant 1 BGB. This states that the bank has a right to claim for damages—however only to a maximum of 50 Euro (under PSD 2007 it was 150 Euro)—if the non-authorised transaction was based on the use of a stolen or otherwise misplaced payment instrument. A payment instrument (previously "payment authentication instrument") is a personalised instrument or process which is agreed between the customer and the bank and which the customer uses to give a mandate to the bank.[76] This includes physical things such as chip cards or TAN generators as well as information such as passwords.[77]

In the case of § 675v section 1 sentence 1 variant 1 BGB, a physical payment instrument must have been misplaced.[78] This situation is hardly relevant in the case of phishing in Online Banking.

---

[75] OLG München, Court Decision f. 23 Jan 2012 – 17 U 3527/11 = BKR 2012, pp. 476 f.; LG Köln, Court Decision f. 10 Sep 2019 – 21 O 116/19 = BKR 2020, p. 311; Köbrich (2015), p. 13; Maier (2018), p. 171; Mehrbrey and Schreibauer (2016), p. 78, Zahrte (2016), p. 317.

[76] See the legal definition in Art. 4 Nr. 14 PSD 2 (before Art. 4 Nr. 23 PSD) and in § 1 section 20 Zahlungsdiensteaufsichtsgesetz (ZAG) (before § 1 section 5 ZAG); Herresthal (2020), mn. 9; Zetzsche (2020a), mn. 17.

[77] Borges (2017), mn. 309; Borges (2020), mn. 4; Scheibengruber (2010), p. 17; Schmalenbach (2020a), mn. 13a; Zahrte, 2.2. ("Zahlungsinstrumente").

[78] Hofmann (2021c), mn. 26; Maihold (2017b), mn. 95; Omlor (2020b), mn. 10; Scheibengruber (2010), p. 17; Schmalenbach (2020c), mn. 3.

5.2.2 Misuse or Breach of Duty (Situations 2 and 3)

Usually, in phishing cases in Online Banking, the other 2 situations described in § 675v BGB are relevant.[79]

According to § 675v section 1 sentence 1 variant 2 BGB, the customer will be liable—up to a maximum of 50 Euro—if the payment instrument has been misused. Until the implementation of the PSD 2, it was additionally required that the customer at least negligently failed to safeguard his personal security features (for the term see below regarding the requirements of § 675l BGB).[80] Since the PDS 2 came into effect, liability under both alternatives of § 675v section 1 sentence 1 BGB is strict, i.e. liability is independent of fault.[81]

According to § 675v section 3 BGB, the customer is obliged to replace the total amount of the loss. The requirements for this are either a fraudulent intention of the customer or a deliberate or grossly negligent breach of duty. For the latter, the customer must have breached his duties which arise from § 675l section 1 BGB or contractual agreements regarding the handing out and use of the payment instrument.

The 3rd situation of grossly negligent breaches of duties is particularly relevant in phishing cases.[82]

5.2.2.1 Possible Breaches of Duty

Such breaches refer to the breach of duties arising out of § 675l BGB. As a rule, these are identical to the contractually agreed duties or even are determined or supplemented by them.[83] A standard example are the "Special conditions of the Association of German Banks".[84,85]

According to § 675l sentence 1 BGB, the customer is obliged to take all reasonable steps to safeguard his personal security features from unauthorised access. Personal security features are intangible things with which it is possible to identify the customer, such as, for example, PIN and TAN.[86]

The concrete measures to be taken depend on the individual case. However, there are categories of obligations that must be always be considered. Of particular

[79] Borges (2017), mn. 312; Keßler (2020b), mn. 4, 14; Zetzsche (2020a), mn. 22, 53 ff.

[80] Maihold (2017b), mn. 102 ff.; Scheibengruber (2010), pp. 16 f.; Zetzsche (2017), mn. 27.

[81] Keßler (2020b), mn. 3; Omlor (2019), p. 112; Spitzer (2018), p. 565; Zetzsche (2020a), mn. 13.

[82] Borges (2017), mn. 312; Keßler (2020b), mn. 14; Zetzsche (2020a), mn. 53 ff.

[83] Borges (2017), mn. 315; Hofmann (2021a), mn. 14; Omlor (2020a), mn. 11 ff.; Schmalenbach (2020b), mn. 7; Zetzsche (2020a), mn. 45.

[84] Sonderbedingungen des Bundesverbands deutscher Banken (BdB), printed by *Bunte/Zahrte*.

[85] Borges (2017), mn. 315; Omlor (2020a), mn. 11 ff.; Zetzsche (2020a), mn. 45.

[86] BGH, Court Decision f. 25 July 2017 – XI ZR 260/15 = NJW 2017, p. 3224; Borges (2017), mn. 312; Borges (2020), mn. 5; Schmalenbach (2020b), mn. 3.

importance in this context are the duty to keep the personal security features confidential, the duty to respond appropriately to suspicious circumstances as well as the duty to secure one's own IT-infrastructure.[87]

5.2.2.1.1 Duty of Confidentiality

It is clear that the customer must protect his personal security features from unauthorised attack and must keep such features confidential.[88] This does not necessarily mean that he has to keep intangible features in his head and is not allowed to write them down anywhere, but he must ensure that he keeps the notes somewhere safe.[89] This also includes the fact that notes must not be kept together with another payment instrument such as the chip card.[90] For the same reason, in the case of the smsTAN process, the receiving device for the TAN, for example a smart phone cannot simultaneously be used for a transaction.[91]

5.2.2.1.2 Appropriate Reactions to Suspicious Circumstances

Part of protection against unauthorised attack is the fact that the customer must react appropriately to any suspicious circumstances.[92] This means that if there are suspicious circumstances, which indicate that someone is trying to commit fraud, then the customer must recognise and react to them and cannot simply ignore them.[93] Fraud attempts can, for example, be recognised by spelling and grammar mistakes in an email or on a website.[94] The request to enter numerous TANs from a list could also

---

[87] Borges (2017), mn. 315 ff.; Omlor (2020a), mn. 11 ff.

[88] See Nr. 7.2. (1) of "Sonderbedingungen für das Online Banking" (see footnote 84); BGH, Court Decision f. 26 Jan 2016 – XI ZR 91/14 = NJW 2016, p. 2029 (mn. 58), ECLI:DE:BGH:2016:260116UXIZR91.14.0; Borges (2017), mn. 316; Borges (2005), p. 3314; Knupfer (2004), p. 642; Linardatos (2019), mn. 119; Omlor (2019), pp. 107 f., 113.

[89] Borges (2017), mn. 316; Hofmann (2021c) mn. 76 f.; Hofmann (2021a), mn. 45; Maihold (2017b), mn. 115; Schmalenbach (2020b), mn. 4.

[90] See Nr. 7.2. (2) of "Sonderbedingungen für das Online Banking" (see footnote 84); Hofmann (2021a), mn. 87; Linardatos (2019), mn. 120; Schmalenbach (2020b), mn. 4; Zetzsche (2020a), mn. 45.

[91] See Nr. 7.2. (2) of "Sonderbedingungen für das Online Banking" (see footnote 84); Borges (2017), mn. 317

[92] Borges (2017), mn. 318; Borges (2005), pp. 3314 f.; Jungmann (2020b), mn. 42; Omlor (2020a), mn. 12; Schulte am Hülse and Klabunde (2010), p. 84.

[93] BGH, Court Decision f. 24 Apr 2012 – XI ZR 96/11 = NJW 2012, p. 2424; LG Köln, Court Decision f. 30 July 2015 – 15 O 505/14 = BeckRS (2015), 15385; Borges (2012), p. 2386; Hofmann (2021a), mn. 94 ff.; Schwintowski (2020), mn. 6.

[94] Borges (2017), mn. 319; Borges (2012), p. 2386; Maihold (2017b), mn. 30.

be a suspicious circumstance.[95] Of course, the customer must check the details of the bank transfer before he confirms the transaction.[96]

5.2.2.1.3 Duties Relating to the Securing of One's Own Infrastructure

Further measures to protect the personal security elements from unauthorised attack result from the customer's duty to secure his own IT-infrastructure.[97] The minimum requirement is the installation and keeping up to date of anti-virus software.[98] The customer must also keep his operating system and internet browser up to date.[99] There are different views as to whether there is also an obligation to use a firewall.[100] A preinstalled firewall is usually sufficient, but must not be removed.[101]

5.2.2.1.4 Notification Obligations

Pursuant to section 675l section 1 sentence 2 BGB, the customer is obliged to notify the payment service provider or an entity designated by the payment service provider of the loss, theft, misuse or other unauthorised use of a ZAI without undue delay after becoming aware of it.

5.2.2.2 Gross Negligence

If there is a breach of duty of the type mentioned above, whether the customer has been at least grossly negligent must be considered. The liability under § 675v section 1 Sentence 1 BGB is, as mentioned before (see above Sect. 5.2.2), strict. Therefore, not even negligence is required. The bank only has a right to claim damages of maximum 50 Euro.

---

[95] BGH, Court Decision f. 24 Apr 2012 – XI ZR 96/, NJW 2012, 2422, 2424; Borges (2017), mn. 320; Hofmann (2021a), mn. 95; Omlor (2020a), mn. 12; however, today there is usually an exclusion of liability for the customer according to § 675v section 4 BGB if the bank still uses a procedure with lists (see below Sect. 5.2.3.3.2).

[96] See Nr. 7.4. of „Sonderbedingungen für das Online Banking" (see footnote 84); LG Köln, Court Decision f. 26 Aug 2014 – 3 O 390/13 = NJW 2014, 3735, 3736; Borges (2017), mn. 321.

[97] See Nr. 7.3. of „Sonderbedingungen für das Online Banking" (see footnote 84); LG Köln, Court Decision f. 5 Dec 2007 – 9 S 195/07 = MMR 2008, 259, 261; Borges (2017), mn. 326; Hofmann (2021a), mn. 98; Maihold (2017b), mn. 50, 134 ff.; Omlor (2020a), mn. 13; Omlor (2019), p. 111; Zetzsche (2020a), mn. 54.

[98] LG Köln, Court Decision f. 5 Dec 2007 – 9 S 195/07 = MMR 2008, 259, 261; Borges (2017), mn. 328; Hofmann (2021a), mn. 98; Hossenfelder (2009), p. 794; Omlor (2020a), mn. 13.

[99] LG Köln, Court Decision f. 5 Dec 2007 – 9 S 195/07 = MMR 2008, 259, 261; Borges (2017), mn. 330; Omlor (2020a), mn. 13; Zetzsche (2020a), mn. 54.

[100] See the state of opinion Borges (2017), mn. 329.

[101] In this way for example Omlor (2019), p. 111; Werner (2010), p. 163.

Unlimited liability under § 675v section 2 BGB, in contrast, requires there to be at least gross negligence.

Gross negligence is defined as a significant breach of the required duty of care, which is considered if the person acting had ignored fundamental considerations or did not notice something blindingly obvious.[102]

One must decide in each individual case whether or not actions can be considered to be grossly negligent, in particular because the subjective abilities of the person acting are relevant here.[103] German case law is reluctant to find in favour of gross negligence.[104] However, for instance in cases where the customer is requested to enter numerous TANs, courts are increasingly deciding that the customer was grossly negligent.[105] This is also because society is becoming more aware of the digital age in which we live and the average internet user is far more aware of the dangers of Online Banking than they used to be.[106]

5.2.3 No Exclusion of Liability

The introduction of the PSD 2007 already established an exclusion of liability of the customer if he fulfilled his duty of notification according to § 675l BGB in time, i.e. in particular before the damage occurred. This has not been changed by PSD 2 (see immediately Sect. 5.2.3.1).

Until PSD 2 came into effect, the customer could only assert further possible counter rights in the form of offsetting due to potential contributory negligence on the part of the bank (see below Sect. 5.2.5).

By implementing Art. 74 section 1 subparagraph 2 PSD 2 and Art. 74 section 2 PSD 2 into § 675v sections 2 and 4 BGB, the requirements for possible further

[102] BGH, Court Decision f. 11 May 1953 – IV ZR 170/52 = NJW 1953, 1139; BGH, Court Decision f. 5 Dec 1983 – II ZR 252/82 = NJW 1984, 789, 791; Hofmann (2021c), mn. 64 f.; Maihold (2017a), mn. 70; Zetzsche (2020a), mn. 42.

[103] BGH, Court Decision f. 11 May 1953 – IV ZR 170/52 = NJW 1953, 1139; Hofmann (2021c), mn. 65; Maihold (2017a), mn. 70; Omlor (2020b), mn. 24 ff.; Schmalenbach (2020c), mn. 10; Zetzsche (2020a), mn. 42.

[104] Borges (2017), mn. 336 (with some respective sources from the case law); Schmalenbach (2020c), mn. 10.

[105] See for example OLG München, Court Decision f. 23 Jan 2012 – 17 U 3527/11 = MMR 2013, 163; LG Berlin, Court Decision f. 11 Aug 2009 – 37 O 4/09 = BeckRS 2009, 28142; LG Berlin, Court Decision f. 8 Nov 2011 – 21 O 80/11 = NJW-RR 2012, 570; LG Düsseldorf, Court Decision f. 27 Mar 2014 – 21 S 211/13 = BeckRS 2015, 5137; AG Krefeld, Court Decision f. 6 July 2012 – 7 C 605/11 = BKR 2012, 480, 481.

[106] LG Düsseldorf, Court Decision f. 27 Mar 2014. – 21 S 211/13 = BeckRS 2015, 5137; LG Essen, Court Decision f. 4 Dec 2014 – 6 O 339/14 = BeckRS 2015, 1693; Borges (2017), mn. 336; Knupfer (2004), p. 642; Omlor (2020b), mn. 26.

exclusions of liability of the customer have been legally regulated (see immediately Sects. 5.2.3.2 and 5.2.3.3).[107]

5.2.3.1 § 675v Section 5 BGB

Pursuant to § 675v section 5 sentence 1 BGB, the customer is released from further liability as soon as he has submitted a notification pursuant to § 675l section 1 sentence 2 BGB (see above Sect. 5.2.2.1.4), as the bank is then able to take the necessary blocking measures.[108]

This applies to all cases of liability under sections 1 and 3 except in the event of the customer acting fraudulently.[109]

5.2.3.2 § 675v Section 2 BGB

As explained above, since PSD 2 the requirement of fault has in principle been omitted in the case of misuse of the payment instrument (see above Sect. 5.2.2), so there is no need for the customer to excuse his behaviour. However, this is replaced by the possibility of an exclusion of liability under § 675v section 2 BGB and unlike before the requirement of fault, it now extends to all claims under § 675v section 1 BGB.[110]

§ 675v section 2 BGB provides for two cases of exclusion of liability:

The customer shall not be liable if the loss, theft or misappropriation of a payment instrument was not detectable to the payer prior to a payment (no. 1).

The customer shall not be liable if the loss was within the responsibility of the payment service provider (caused by acts of an employee, agent or branch of a payment service provider or of an entity to which its activities were outsourced) (no. 2).

5.2.3.3 § 675v Section 4 BGB

The most prominent innovation by the PDS 2 is the introduction of Two-Factor Authentication.[111] This has the following effect on a possible exclusion of liability:

According to § 675v section 4 sentence 1number 1 BGB, the payment service provider shall not have a claim for compensation against the payer if the payer's

---

[107] See Gesetzesentwurf der Bundesregierung: Bundestags-Drucksache (RegE, BT-Drs.) 18/11495, 165 ff.

[108] Hofmann (2021c), mn. 139; Schmalenbach (2020c), mn. 15; Zetzsche (2020a), mn. 69.

[109] Schmalenbach (2020c), mn. 15.

[110] See RegE, BT-Drs. 18/11495, 165ff. (see footnote 107); Zetzsche (2020a), mn. 32.

[111] Zahrte (2019c), p. 485.

payment service provider does not require strong customer authentication within the meaning of § 1 section 24 ZAG[112]. The definition in § 1 section 24 implements the definition in Art. 4 section 30 PDS 2.

This results in the following main requirements for the customer's exclusion of liability: Two-Factor Authentication and independence of the security elements used.[113]

5.2.3.3.1 Two-Factor Authentication

The client's authentication must consist of at least two elements, which must be from the following three categories: knowledge, possession, inherence.[114] It is not necessary for the elements to come from more than one category.[115]

The category 'possession' refers to something that only the user possesses, e.g. a payment card or mobile phone.[116]

The category 'knowledge' includes the personalised security features, e.g. a password,[117] PIN or TAN,[118] as these fulfil the requirement that they must be elements known only to the authorised person.[119]

The category 'inherence' refers to something, that the user is and covers the biometric features as fingerprint and facial recognition.[120]

---

[112] § 1 section 24 ZAG reads: *Starke Kundenauthentifizierung ist eine Authentifizierung, die so ausgestaltet ist, dass die Vertraulichkeit der Authentifizierungsdaten geschützt ist und die unter Heranziehung von mindestens zwei der folgenden, in dem Sinne voneinander unabhängigen Elementen geschieht, dass die Nichterfüllung eines Kriteriums die Zuverlässigkeit der anderen nicht in Frage stellt:*

1. *Kategorie Wissen, also etwas, das nur der Nutzer weiß,*
2. *Kategorie Besitz, also etwas, das nur der Nutzer besitzt oder*
3. *Kategorie Inhärenz, also etwas, das der Nutzer ist.*

[113] Hofmann (2021c), mn. 100; Keßler (2020a), mn. 7; Omlor (2020b), mn. 38; Schmalenbach (2020c), mn. 13.

[114] Borges (2017), mn. 29c; Hofmann (2021c), mn. 100; Keßler (2020a), mn. 7; Zetzsche (2020a), mn. 62.

[115] Borges (2017), mn. 29e; Borges (2016), p. 256; Hoffmann (2016), p. 249; Hofmann (2021c), mn. 100; different opinion Hingst and Lösing (2014), p. 321.

[116] Borges (2017), mn. 29c; Hoffmann (2016), p. 248; Keßler (2020a), mn. 7; Omlor (2020b), mn. 39; Schmalenbach (2020c), mn. 13.

[117] Borges (2017), mn. 29c; Keßler (2020a), mn. 7; Omlor (2020b), mn. 39; Schmalenbach (2020c), mn. 13.

[118] Borges (2017), mn. 29c; Keßler (2020a), mn. 7; Schmalenbach (2020c), mn. 13; Zetzsche (2020a), mn. 62.

[119] Hofmann (2021c), mn.102; Zetzsche (2020a), mn.

[120] Hoffmann (2016), p. 248 f.; Hofmann (2021c), mn. 103; Omlor (2020b), mn. 39; Schmalenbach (2020c), mn. 13.

5.2.3.3.2 Independence of the Security Elements Used

For the existence of strong customer authentication, it is also necessary that the two security elements used are independent of each other. According to the wording of the law (Art. 4 section 30 PSD 2), this requires "that the breach of one does not compromise the reliability of the others".

This is predominantly interpreted to mean that independence is lacking if the compromise of one element also calls into question the security of the second element, i.e. it is possible to circumvent the security procedure by means of a uniform attack.[121]

In particular, the requirement of independence means that simple PIN/TAN processes as well as the iTAN process do not provide strong customer authentication.[122]

This is because with both processes, there is no specific reference between the TAN and the transaction,[123] so that a single attack, e.g. a man in the middle-attack (see above Sect. 3.1), on the terminal device used for Online Banking is sufficient to manipulate the data transmitted to the bank.[124]

The newer eTAN+ or chip TAN processes can fulfil the requirements of a strong customer authentication.[125] They are characterised by the fact that the TAN is generated for a specific transaction (see above Sect. 3.2).

5.2.4 Reverse Exception: Fraud

In all variants of an exclusion of liability of the customer regulated in § 675v BGB, there are reverse exceptions for the case that the customer was acting fraudulently.

The reverse exception to the exclusion of liability in section 2 is found in section 3 no. 1. According to this, the customer acting fraudulently is even liable without any limitation, so that the exclusion of liability with regard to the liability limited to 50 Euro according to section 2 is no longer relevant.

In Art. 74 PSD 2, this reverse exception is directly regulated to the loss or misuse of the payment instrument in subparagraph 2 letter a 2nd half-sentence. However, since this does not have its own scope of application due to the unlimited liability according to subparagraph 4 (corresponds to § 675v section 3 BGB), it was not separately implemented into German law.[126]

The reverse exception to the exclusion of liability in section 4 sentence 1 is found in section 4 sentence 2.

---

[121] Hoffmann (2016), p. 249; Hofmann (2021c), mn. 100; Omlor (2020b), mn. 38.

[122] Borges (2017), mn. 29j; Jungmann (2020c), mn. 21; *Omlor* (2019), p. 113; Schneider and Atzler (2020); Zahrte (2019c), p. 485.

[123] Borges (2017), mn. 29j; Borges (2016), p. 256; Linardatos (2019), mn. 31; Klemm (2019).

[124] Borges et al. (2011), p. 28 f.; Maihold (2017b), mn. 13.

[125] Maihold (2017b), mn. 85; Omlor (2019), p. 113.

[126] See RegE, BT-Drs. 18/11495, 166 (see footnote 107).

The reverse exception to the exclusion of liability in section 5 sentence 1 is found in section 5 sentence 3.

5.2.5 No Contributory Negligence of the Bank

If the bank breaches its duties of care, it may be contributorily negligent and its claim under § 675v BGB must be reduced accordingly.[127]

The bank's duties of care include, above all, duties of information and clarification. This requires the banks in particular to inform their customers about the general risks involved in Online Banking.[128] One often finds notifications of such risks on the start page or in current notifications on an Online Banking portal.[129]

Whilst the bank does not have a fundamental obligation to check all customer transactions for suspicious circumstances, if it is aware of suspicious circumstances, then it must, of course, inform the customer of this.[130]

In certain cases, the breach of information and clarification obligations by the bank can also lead to a claim for damages by the customer; this can be the case, for example, if he authorised the transfer himself but was induced to do so by a fraud which he would have been able to notice if he had been sufficiently informed by the bank.[131]

Before the introduction of PDS 2, the use of an insufficiently secure Online Banking procedure also led to contributory negligence on the part of the bank,[132] whereby what constitutes a sufficient system has been the subject of lively discussions.[133] This should only be relevant for old cases.[134]

Today, the lack of a system for strong customer authentication already leads to the complete exclusion of liability of the customer (see above Sect. 5.2.3.3). Currently, however, there is probably no system that fulfils the requirements of strong customer

[127] Borges (2017), mn. 338; Hofmann (2021b), mn. 1; Mehrbrey and Schreibauer (2016), p. 78; Werner (2020), mn. 73.

[128] Borges (2017), mn. 340; Borges (2012), p. 2388; Maihold (2017b), mn. 55.

[129] See e.g. https://www.targobank.de/de/identification/authentification.html#Anker_Sicherheit (accessed 27 Nov 2021); https://www.sparkasse.de/service/sicherheit-im-internet.html (accessed 27 Nov 2021).

[130] BGH Court Decision f. 24 Apr 2012 – XI ZR 96/11 = NJW 2012, 2422 (mn. 32); BGH, Court Decision f. 6 May 2008 – XI ZR 56/07 = NJW 2008, 2245 (mn. 14); Borges (2017), mn. 341; Maihold (2017b), mn. 55; Schmalenbach (2020c), mn. 14; Zahrte (2019b), p. 131.

[131] Borges (2012), pp. 2387 f.

[132] KG Berlin, Court Decision f. 29 Nov 2010 – 26 U 159/09 = MMR 2011, 338, 339, ECLI:DE:KG:2010:1129.26U159.09.0A; Borges (2011b), p. 336; Maihold (2017b), mn. 11; Mehrbrey and Schreibauer (2016), p. 78; Zahrte (2019a), mn. 130e.

[133] Borges (2017), mn. 343; Borges (2012), p. 2388; description and evaluation of different possible processes: Kind and Werner (2006), p. 357 ff.; Köbrich (2015), p. 9 ff.

[134] Zahrte (2019a), mn. 130e.

authentication, yet is nevertheless so weak that it causes contributory negligence on the part of the bank.

6 Conclusion

In conclusion, it can be stated that under German law on the basis of the European regulation, in principle, the banks bear the risk of a phishing attack. However, there are many constellations in which the customer will bear the risk because he is liable for the attack. This is the case if either a legal fiction of authorisation is created or if the bank has a claim for damages against the customer as a result of a breach of obligation. However, the risk reverts to the bank—either partially or fully—, if there is an exclusion of liability or if the bank is contributorily negligent in respect of providing clarification to the customer.

As this contribution should show, the regulation of liability issues in Online Banking in the case of phishing consists of a complex structure of rules, exceptions and reverse exceptions.

These are summarised again in the overview below.

(1) In the event of the authorisation of a transaction (actual or fictitious), the following applies:
    The bank has a claim for compensation, the customer has no claim for reimbursement (see Sect. 4.1).
(2) Without the authorisation of a transaction, the following applies in principle:
    The bank has no claim for compensation, the customer has a claim for reimbursement (see Sects. 4.1 and 4.2).

If other factors exist, such as a breach of duty by the client, this result may change. In this case, the customer may be liable with the consequence that the bank receives a claim for damages with which it can set off against the customer's claim for reimbursement, leading to the extinguishing of the customer's claim.

(3) If the customer is liable under § 675v section 1 or section 3 BGB, the following applies:
    The bank has a claim for compensation (partial or full), the customer has no claim for reimbursement (partial or full) (see Sect. 5.2).

Exceptions exist if there is an exclusion of liability according to § 675v section 2, 4 or 5 BGB or if there is contributory negligence on the part of the bank.

(4) If there is an exclusion of liability or contributory negligence on the part of the bank, then the following applies:
    The bank has no claim for compensation (partial or full), the customer has a claim for reimbursement (partial or full) (see Sects. 5.2.3 and 5.2.5).
(5) There is an exception in the case of fraudulent intent on the part of the customer. Then the following applies again:

The bank has a claim for compensation, the customer has no claim for reimbursement (see Sect. 5.2.4).

One of the most important changes brought by PSD 2 is that the obligations of the banks and the legal consequences of their violation are now explicitly standardised: § 675v section 2 No. 2 and 4 (see Sects. 5.2.3.2 and 5.2.3.3). Particularly prominent is the provision in section 4, which requires banks to introduce strong customer authentication for Online Banking. With this, the European legislator has also clarified a question that had previously been the subject of lively discussions in case law and literature, namely which older procedures are no longer sufficient for Online Banking (see Sect. 5.2.5). This has been done by setting precise technical specifications.

The regulation of technology at the highest possible level, namely the regulation by law, at this point speaks for the great importance to be attached to the security of Online Banking and protection against phishing and other criminal attacks.

References

Aljawarneh, S. A. (2017). *Online banking security measures and data protection*. IGI Global. https://doi.org/10.4018/978-1-5225-0864-9

Anti-Phishing Working Group (APWG). Retrieved November 27, 2021, from https://www.apwg.org

APWG (Anti-Phishing Working Group). (2015). *Phishing activity trends report, 4th quarter 2014*. Retrieved November 27, 2021, from https://docs.apwg.org//reports/apwg_trends_report_q4_2014.pdf

APWG (Anti-Phishing Working Group). (2016). *Phishing activity trends report, 4th quarter 2015*. Retrieved November 27, 2021, from https://docs.apwg.org//reports/apwg_trends_report_q4_2015.pdf

APWG (Anti-Phishing Working Group). (2017). *Phishing activity trends report, 4th quarter 2016*. Retrieved November 27, 2021, from https://docs.apwg.org//reports/apwg_trends_report_q4_2016.pdf

APWG (Anti-Phishing Working Group). (2020). *Phishing activity trends report, 4th quarter 2019*. Retrieved November 27, 2021, from https://docs.apwg.org/reports/apwg_trends_report_q4_2019.pdf

Bacher, K. (2020, 1 Sept). § 284. In O. K. Beck, V. Vorwerk, & C. Wolf (Eds.), *Beck'scher Online-Kommentar ZPO* (38th ed.). C.H. Beck.

Bär, W. (2020). 15th Chapter. In H.-B. Wabnitz, T. Janovsky, & L. Schmitt (Eds.), *Handbuch Wirtschafts- und Steuerstrafrecht*. C.H. Beck.

Benning, A., & Oberrath, J.-D. (2015). *Bürgerliches Recht I*. Boorberg.

BKA (Bundeskriminalamt). (2016). *Cybercrime: Bundeslagebild 2016*. BKA. Retrieved November 27, 2021, from https://www.bka.de/SharedDocs/Downloads/DE/Publikationen/JahresberichteUndLagebilder/Cybercrime/cybercrimeBundeslagebild2016.pdf?__blob=publicationFile&v=5

BKA (Bundeskriminalamt). (2019a). *Cybercrime: Bundeslagebild 2018*. BKA. Retrieved November 27, 2021, from https://www.bka.de/SharedDocs/Downloads/DE/Publikationen/JahresberichteUndLagebilder/Cybercrime/cybercrimeBundeslagebild2018.html

BKA (Bundeskriminalamt). (2019b). *Cybercrime – BKA stellt Bundeslagebild 2018 auf Pressekonferenz vor. Press release 11 Nov 2019*. BKA. Retrieved November 27, 2021, from

https://www.bka.de/DE/Presse/Listenseite_Pressemitteilungen/2019/Presse2019/191111_PMBLBCybercrime.html
BKA (Bundeskriminalamt). (2020). *Cybercrime: Bundeslagebild 2019*. BKA. Retrieved November 27, 2021, from https://www.bka.de/SharedDocs/Downloads/DE/Publikationen/JahresberichteUndLagebilder/Cybercrime/cybercrimeBundeslagebild2019.html?nn=28110
Borges, G. (2005). Rechtsfragen des Phishing - Ein Überblick. *Neue Juristische Wochenschrift (NJW), 46*, 3313–3317.
Borges, G. (2007). *Rechtsfragen der Internet-Auktion*. Nomos. https://doi.org/10.5771/9783845200057
Borges, G. (2009). Comment on LG Mannheim, Court Decisioin f. 16 May 2008 – 1 S 189/07. *Zeitschrift für Bank- und Kapitalmarktrecht (BKR), 2*, 84, 85–84, 87.
Borges, G. (2011a). Rechtsscheinhaftung im Internet. *Neue Juristische Wochenschrift (NJW), 33*, 2400–2403.
Borges, G. (2011b). Sorgfaltspflichten beim Online-Banking zur Vermeidung von Phishing-Attacken. Comment on KG Berlin, Court Decision f. 29 Nov 2010 – 26 U 159/09. *Entscheidungen zum Wirtschaftsrecht (EWiR), 11*, 335–336.
Borges, G. (2012). Haftung für Identitätsmissbrauch im Online-Banking. *Neue Juristische Wochenschrift (NJW), 33*, 2385–2389.
Borges, G. (2016). Identitätsmissbrauch im Online-Banking und die neue Zahlungsdiensterichtlinie (PSD2). *Zeitschrift für Bankrecht und Bankwirtschaft (ZBB), 4*, 249–259. https://doi.org/10.15375/zbb-2016-0406
Borges, G. (2017). § 11 Electronic Banking. In P. Derleder, K.-O. Knops, & H. G. Bamberger (Eds.), *Deutsches und europäisches Bank- und Kapitalmarktrecht*. Springer. https://doi.org/10.1007/978-3-662-52807-5
Borges, G. (2020). 8th Chap. § 675j. In K. Langenbucher, D. H. Bliesener, & G. Spindler (Eds.), *Bankrechts-Kommentar*.
Borges, G., Schwenk, J., Stuckenberg, C.-F., & Wegener, C. (2011). *Identitätsdiebstahl und Identitätsmissbrauch im Internet*. Springer. https://doi.org/10.1007/978-3-642-15833-9
BSI (Bundesamt für Sicherheit in der Informationstechnik). (2006). *Jahresbericht 2005*. BSI. Retrieved November 27, 2021, from https://www.ipih.de/system/files/upload/ipih-archive/2005%20BSI%20Jahresbericht2005.pdf
BSI (Bundesamt für Sicherheit in der Informationstechnik). (2017). *Die Lage der IT-Sicherheit in Deutschland 2017*. BSI. Retrieved November 2021, from https://www.bsi.bund.de/SharedDocs/Downloads/DE/BSI/Publikationen/Lageberichte/Lagebericht2017.pdf?__blob=publicationFile&v=4
BSI (Bundesamt für Sicherheit in der Informationstechnik). *Glossar der Cybersicherheit*. Retrieved November 27, 2021, from https://www.bsi.bund.de/SharedDocs/Glossareintraege/DE/M/Man-In-The-Middle-Angriff.html
Bunte, H.-J., & Zahrte, K. *AGB-Banken, AGB-Sparkassen, Sonderbedingungen*. C.H. Beck.
Casper, M. (2009). § 676a. In: Münchener Kommentar. In F. J. Säcker, R. Rixecker, H. Oetker, & B. Limperg (Eds.), *Münchener Kommentar zum Bürgerlichen Gesetzbuch* (Vol. 6). C.H. Beck.
Cornelius, K. (2013). Teil 10: Besonderheiten des Straf- und Strafprozessrechts. In A. Leupold & S. Glossner (Eds.), *Münchener Anwaltshandbuch IT-Recht*. C.H. Beck.
Erfurth, R. (2006). Haftung für Missbrauch von Legitimationsdaten durch Dritte beim Online-Banking. *Wertpapiermitteilungen (WM), 47*, 2198–2207.
Fischer, D. (2020). § 675. In: BeckOK. In W. Hau & R. Poseck (Eds.), *Beck'scher Online-Kommentar BGB. 56. Ed., 1 Nov 2020*. C.H. Beck.
Gajek, S., Schwenk, J., & Wegener, C. (2005). Identitätsmissbrauch im Onlinebanking. *Datenschutz und Datensicherheit (DuD), 29*, 639–642.
Gilch, A. (2016). § 41 Compliance und neue Medien. In C. E. Hauschka, K. Moosmayer, & T. Lösler (Eds.), *Corporate Compliance: Handbuch der Haftungsvermeidung im Unternehmen*. C.H. Beck.

Graf, J. (2017). § 202a. Münchener Kommentar. In W. Joecks & K. Miebach (Eds.), *Münchener Kommentar zum StGB* (Vol. 4). C.H. Beck.
Häuser, F. (2019). Part 1. B. "Überweisungsverkehr". Münchener Kommentar. In K. Schmidt & W.-F. Ebke (Eds.), *Münchener Kommentar zum Handelsgesetzbuch* (Vol. 6). C.H. Beck.
Heermann, P. (2020). § 675. Münchener Kommentar. In F. J. Säcker, R. Rixecker, H. Oetker, & B. Limperg (Eds.), *Münchener Kommentar zum Bürgerlichen Gesetzbuch* (Vol. 6). C.H. Beck.
Herresthal, C. (2020). 3rd Chap., § 675j. In K. Langenbucher, D. H. Bliesener, & G. Spindler (Eds.), *Bankrechts-Kommentar*. C.H.Beck.
Hingst, K.-M., & Lösing, C. (2014). Die geplante Fortentwicklung des europäischen Zahlungsdiensteaufsichtsrechts durch die Zweite Zahlungsdienste-Richtlinie. *Zeitschrift für Bank- und Kapitalmarktrecht (BKR), 8*, 315–324.
Hoffmann, J. (2016). Kundenhaftung unter der Neufassung der Zahlungsdiensterichtlinie. *Verbraucher und Recht (VuR), 7*, 243–254.
Hofmann, C. (2021a). § 675l. BeckOGK. In B. Gsell, W. Krüger, S. Lorenz, & C. Reymann (Eds.), *Beck-online.GROSSKOMMENTAR BGB. Version 1 Nov 2020*. C.H.Beck.
Hofmann, C. (2021b). § 675m. BeckOGK. In B. Gsell, W. Krüger, S. Lorenz, & C. Reymann (Eds.), *Beck-online.GROSSKOMMENTAR BGB. Version 1 Nov 2020*. C.H.Beck.
Hofmann, C. (2021c). § 675v. BeckOGK. In B. Gsell, W. Krüger, S. Lorenz, & C. Reymann (Eds.), *Beck-online.GROSSKOMMENTAR BGB. Version 1 Nov 2020*. C.H.Beck.
Hofmann, C. (2021d). § 675w. BeckOGK. In B. Gsell, W. Krüger, S. Lorenz, & C. Reymann (Eds.), *Beck-online.GROSSKOMMENTAR BGB. Version 1 Nov 2020*. C.H.Beck.
Hossenfelder, M. (2009). Onlinebanking und Haftung: Zu den Sorgfaltspflichten des Bankkunden im Lichte des neuen Zahlungsdiensterechts. *Computer und Recht (CR), 12*, 790–794. https://doi.org/10.9785/ovs-cr-2009-790
Jungmann, C. (2020a). § 675j. Münchener Kommentar. In F. J. Säcker, R. Rixecker, H. Oetker, & B. Limperg (Eds.), *Münchener Kommentar zum Bürgerlichen Gesetzbuch* (Vol. 6). C.H. Beck.
Jungmann, C. (2020b). § 675l. Münchener Kommentar. In F. J. Säcker, R. Rixecker, H. Oetker, & B. Limperg (Eds.), *Münchener Kommentar zum Bürgerlichen Gesetzbuch* (Vol. 6). C.H. Beck.
Jungmann, C. (2020c). § 675m. Münchener Kommentar. In F. J. Säcker, R. Rixecker, H. Oetker, & B. Limperg (Eds.), *Münchener Kommentar zum Bürgerlichen Gesetzbuch* (Vol. 6). C.H. Beck.
Karper, I. (2006). Sorgfaltspflichten beim Online-Banking – Der Bankkunde als Netzwerkprofil?: Zur möglichen Neubewertung des Haftungsmaßstabs. *Datenschutz und Datensicherheit (DuD), 30*, 215–219. https://doi.org/10.1007/s11623-006-0062-0
Kaufhold, S. (2017). Part 3. B. In P. Bräutigam & D. Rücker (Eds.), *E-Commerce*. C.H. Beck.
Keßler, O. (2020a). § 675m. In: Ebenroth/Boujong/Joost/Strohn. In D. Joost & L. Strohn (Eds.), *Handelsgesetzbuch* (Vol. 2). C.H. Beck/Vahlen.
Keßler, O. (2020b). § 675v. In: Ebenroth/Boujong/Joost/Strohn. In D. Joost & L. Strohn (Eds.), *Handelsgesetzbuch* (Vol. 2). C.H. Beck/Vahlen.
Kind, M., & Werner, D. (2006). Rechte und Pflichten im Umgang mit PIN und TAN. *Computer und Recht (CR), 05*, 353–360. https://doi.org/10.9785/ovs-cr-2006-353
Klemm, T. (2019). Abschied von der Tan-Liste. Resource document. *Frankfurter Allgemeine Zeitung*. Retrieved November 27, 2021, from https://www.faz.net/aktuell/finanzen/digital-bezahlen/online-banking-wird-komplizierter-abschied-von-der-tan-liste-16160793.html
Knupfer, J. (2004). Phishing for Money. *MultiMedia Recht (MMR), 10*, 641–642.
Köbrich, T. (2015). Pishing 2.0 – Ein Überblick über die zivilrechtlichen Streitstände. *Verbraucher und Recht (VuR), 1*, 9–14.
Köhler, H. (2020). *BGB Allgemeiner Teil*. C.H. Beck.
Kolassa, D. (2017). § 137. In: H. Schimansky, H.-J. Bunte, & H.-J. Lwowski (Eds.), *Bankrechts-Handbuch*. C.H. Beck.
Linardatos, D. (2015). Die Rechtsscheinhaftung im Zahlungsdiensterecht: Zugleich eine Anm. zu LG Darmstadt, Urt. v. 28.8.2014 - 28 O 36/14. *Zeitschrift für Bank- und Kapitalmarktrecht (BKR), 3*, 96–100.

Linardatos, D. (2017). Von Anscheinsbeweisen im Zahlungsdiensterecht und fehlgeleiteten Gesetzgebern. *Neue Juristische Wochenschrift (NJW), 30*, 2145–2150.
Linardatos, D. (2019). Vol. 6, Part 1. K. "Online-Banking". In: Münchener Kommentar – K. Schmidt, & W-F. Ebke (Eds.), *Münchener Kommentar zum Handelsgesetzbuch* (Vol. 6). C.H. Beck.
Maier, A. (2018). Bankrecht 2017. *Verbraucher und Recht (VuR), 5*, 163–174.
Maihold, D. (2017a). § 54. In H. Schimansky, H.-J. Bunte, & H.-J. Lwowski (Eds.), *Bankrechts-Handbuch*. C.H. Beck.
Maihold, D. (2017b). § 55. In H. Schimansky, H.-J. Bunte, & H.-J. Lwowski (Eds.), *Bankrechts-Handbuch*. C.H. Beck.
Malin, C., Gudaitis, T., Holt, T., & Kilger, M. (2017). 5 – phishing, watering holes, and scareware. In *Deception in the Digital Age* (pp. 149–166). Elsevier (Academic Press). https://doi.org/10.1016/B978-0-12-411630-6.00005-0.
Mansel, H. (2021). § 164. In: Jauernig BGB – R. Stürner, (Ed.), *Bürgerliches Gesetzbuch*. C.H. Beck.
Mehrbrey, K. L., & Schreibauer, M. (2016). Haftungsverhältnisse bei Cyber-Angriffen: Ansprüche und Haftungsrisiken von Unternehmen und Organen. *MulitMedia Recht (MMR), 2*, 75–82.
Mohsin, T. (2016). *What is spear-phishing? Resource document.* InfoSec Institute. Retrieved November 27, 2021, from https://resources.infosecinstitute.com/topic/what-is-spear-phishing
Mühlenbrock, T., & Dienstbach P. H. (2008). Comment on AG Wiesloch, Court Decisioin f. 20 June 2008 – 4 C 57/08. *MultiMedia und Recht (MMR), 9,* 626 (630)–631.
Nettesheim, M. (2009). Art. 249. In E. Grabitz & M. Hilf (Eds.), *Das Recht der Europäischen Union*. C.H. Beck.
Omlor, S. (2019). Online-Banking unter Geltung der Zweiten Zahlungsdiensterichtlinie (PSD II). *Zeitschrift für Bank- und Kapitalmarktrecht (BKR)., 3*, 105–114.
Omlor, S. (2020a). § 675l. In: Staudinger. *BGB*. Neubearbeitung 2020. de Gruyter.
Omlor, S. (2020b). § 675v. In: Staudinger. *BGB*. Neubearbeitung 2020. de Gruyter.
Oxford English and Spanish Dictionary, Thesaurus, and Spanish to English Translator. Retrieved November 27, 2021, from https://www.lexico.com
Pruß, M. & Sarre, F. (2019). Technisches Glossar. In: A. Auer-Reinsdorff, & I. Conrad (Eds.), *Handbuch IT- und Datenschutzrecht*. C.H. Beck.
Recknagel, E. (2005). *Vertrag und Haftung beim Internet-Banking*. C.H.Beck.
Robinson, M. (2017). *Dailymail.* Retrieved November 27, 2021, from http://www.dailymail.co.uk/news/article-4696242/Grandmother-bankaccount-emptied-fraudsters.html
Saenger, I. (2019). § 286. In: *Zivilprozessordnung*. Nomos.
Schäfer, F. (2020a). § 670. Münchener Kommentar. In F. J. Säcker, R. Rixecker, H. Oetker, & B. Limperg (Eds.), *Münchener Kommentar zum Bürgerlichen Gesetzbuch* (Vol. 6). C.H. Beck.
Schäfer, H. (2020b). § 167. BeckOK. In W. Hau, & R. Poseck (Eds.), *Beck'scher Online-Kommentar BGB*. 56. Ed., 1 Nov 2020. C.H. Beck.
Scheibengruber, C. (2010). Zur Zulässigkeit und Sinnhaftigkeit der Verlagerung des Missbrauchsrisikos bei Zahlungsdiensten auf die Nutzer: Ein Beitrag zur Analyse der Umsetzung der Zahlungsdiensterichtlinie in das BGB und die AGB der Banken. *Zeitschrift für Bank- und Kapitalmarktrecht (BKR), 1*, 15–23.
Schmalenbach, D. (2020a). § 675j. BeckOK. In W. Hau, & R. Poseck (Eds.), *Beck'scher Online-Kommentar BGB*. 56. Ed., 1 Nov 2020. C.H. Beck.
Schmalenbach, D. (2020b). § 675l. BeckOK. In W. Hau, & R. Poseck (Eds.), *Beck'scher Online-Kommentar BGB*. 56. Ed., 1 Nov 2020. C.H. Beck.
Schmalenbach, D. (2020c). § 675v. BeckOK. In: W. Hau, & R. Poseck (Eds.), *Beck'scher Online-Kommentar BGB*. 56. Ed., 1 Nov 2020. C.H. Beck.
Schmidt, K. (2020). Chap. 36. Geigel. In K Haag (Ed.), *Der Haftpflichtprozess*. C.H. Beck.
Schmidt, M. & Pruß, M. (2019). § 3. In: A. Auer-Reinsdorff, & I. Conrad (Eds.), *Handbuch IT- und Datenschutzrecht*. C.H. Beck.

Schneider, K., & Atzler, E. (2020). *Banken und Sparkassen ändern ihre Sicherheitsverfahren fürs Onlinebanking*. Resource document. Retrieved November 27, 2021, from https://www.handelsblatt.com/finanzen/banken-versicherungen/onlinebanking-banken-und-sparkassenaendern-ihre-sicherheitsverfahren-fuers-onlinebanking/24573918.html?ticket=ST-20041664-WsmmbmPRciLYzF23B3sW-ap2

Schubert, C. (2018a). § 167. Münchener Kommentar. In F. J. Säcker, R. Rixecker, H. Oetker, & B. Limperg (Eds.), *Münchener Kommentar zum Bürgerlichen Gesetzbuch* (Vol. 1, 8th ed.). C.H. Beck.

Schubert, C. (2018b). § 179. Münchener Kommentar. In F. J. Säcker, R. Rixecker, H. Oetker, & B. Limperg (Eds.), *Münchener Kommentar zum Bürgerlichen Gesetzbuch* (Vol. 1, 8th ed.). C.H. Beck.

Schulte am Hülse, U., & Klabunde, S. (2010). Abgreifen von Bankzugangsdaten im Onlinebanking: Vorgehensweise der Täter und neue zivilrechtliche Haftungsfragen des BGB. *Multimedia und Recht (MMR), 2*, 84–90.

Schulte am Hülse, U., & Kraus, M. (2016). Das Abgreifen von Zugangsdaten zum Online-Banking: Ausgeklügelte technische Angriffsformen und zivilrechtliche Haftungsfragen. *Multimedia und Recht (MMR), 7*, 435–440.

Schulz-Merkel, P., & Quarch, M. (2021). § 7. In R. Balke, O. Just, G. Reisert, & P. Schulz-Merkel (Eds.), *Regulierung von Verkehrsunfällen*. Nomos.

Schwintowski, H.-P. (2020). § 675l. JurisPK. In M. Herberger, M. Martinek, H. Rüßmann, S. Weth, & M. Würdinger (Eds.), *Juris PraxisKommentar BGB* (Vol. 2). Juris GmbH.

Spitzer, A. (2018). Gesetz zur Umsetzung der Zweiten Zahlungsdiensterichtlinie. *Monatsschrift für Deutsches Recht (MDR), 10*, 561–568. https://doi.org/10.9785/mdtr-2018-721003

Stadler, A. (2020). *Allgemeiner Teil des BGB*. C.H.BECK.

Stöber, M. (2011). Comment on BGH Court Decision f. 11 Mai 2011 – VIII ZR 289/09. *Entscheidungen zum Wirtschaftsrecht (EWiR)., 17*, 551–552.

Werner, D. (2010). *Verkehrspflichten privater IT-Nutzer in Bezug auf die Verbreitung von Schadsoftware*. Nomos.

Werner, S. (2020). Part 13.5. In T. Hoeren, U. Sieber, & B. Holznagel (Eds.), *Handbuch Multimedia-Recht*. C.H. Beck.

Wösthoff, P. (2015). Comment on BGH, Court Decisioin f. 16 June 2015 – XI ZR 243/13. *Betriebs-Berater (BB), 35*, 2065 (2068)–2068.

Zahrte, K. (2016). Angriffe auf das Onlinebanking im Jahr 2016: Zugleich eine Besprechung von LG Köln (30 O 330/14) und OLG Köln (13 U 223/15). *Zeitschrift für Bank- und Kapitalmarktrecht (BKR), 8*, 315–320.

Zahrte, K. (2019a). In H.-J. Bunte & K. Zahrte (Eds.), *AGB-Banken, AGB-Sparkassen, Sonderbedingungen, 4th part "Sonderbedingungen der Banken und Sparkassen", VI. "Sonderbedingungen für das Online-Banking"*. C.H.Beck.

Zahrte, K. (2019b). Haftungsverteilung im Zahlungsdiensterecht beim CEO-Fraud: zugleich eine Besprechung von LG Düsseldorf BKR 2019, 154 und LG Karlsruhe BKR 2019, 151. *Zeitschrift für Bank- und Kapitalmarktrecht (BKR), 3*, 126–131.

Zahrte, K. (2019c). Die "zweite Stufe" der PSD-2-Umsetzung: Änderungen der Sonderbedingungen für Online- und Kartenzahlungen zum 14.9.2019. *Zeitschrift für Bank- und Kapitalmarktrecht (BKR), 10*, 484–490.

Zetzsche, D. (2017). § 675v. Münchener Kommentar. In F. J. Säcker, R. Rixecker, H. Oetker, & B. Limperg (Eds.), *Münchener Kommentar zum Bürgerlichen Gesetzbuch* (Vol. 6, 7th ed.). C.H. Beck.
Zetzsche, D. (2020a). § 675v. Münchener Kommentar. In F. J. Säcker, R. Rixecker, H. Oetker, & B. Limperg (Eds.), *Münchener Kommentar zum Bürgerlichen Gesetzbuch* (Vol. 6, 8th ed.). C.H. Beck.
Zetzsche, D. (2020b). § 675w. Münchener Kommentar. In F. J. Säcker, R. Rixecker, H. Oetker, & B. Limperg (Eds.), *Münchener Kommentar zum Bürgerlichen Gesetzbuch* (Vol. 6, 8th ed.). C.H. Beck.
Zimmermann, M. (2020). § 675u. BeckOGK. In B. Gsell, W. Krüger, S. Lorenz, & C. Reymann (Eds.), *Beck-online.GROSSKOMMENTAR BGB. Version 1 Nov 2020*. C.H. Beck.

Internet of Things and Consumers' Privacy in a Brazilian Perspective: Digital Vulnerability and Dialogue of Sources

Guilherme Mucelin

Contents

Abstract The Internet of Things (IoT) will revolutionize the way we interact with physical objects, connecting each other and also connecting them to the Internet to make everyday life easier, with an intense flow of data and information from all consumers. The purpose of this article is to verify the functioning of the Internet of Things, as well as analyze the right to privacy in this new technology in a smart network. For that, the method used was the deductive method. It is concluded that the principle of transparency is the guideline for legal conduct on the capture and processing of data and information in the IoT structure.

1 Introduction

Society is computerizing all aspects of human experience, such as affective relationships, commerce, culture, leisure, and all manifestations that were eminently off-line. It goes into the virtual world, guided by images, sounds and texts at instant speed, where everything and everyone are connected.

In addition to social networks that interconnect *all*, the novelty is the connection of *everything*, the things that surround people. With the philosophy of rapidity of postmodernity and connectivity, a new technology has been developed: the Internet of Things, which aims to connect physical objects among themselves and with the Internet to create interfaces with the user to facilitate everyday life.

G. Mucelin (✉)
Federal University of Rio Grande do Sul, Porto Alegre, Brazil

G. Borges, C. Sorge (eds.), *Law and Technology in a Global Digital Society*,
https://doi.org/10.1007/978-3-030-90513-2_14

For this technology to function properly, a multitude of personal data and information from consumers who purchase products and services integrated into the IoT are collected and processed regardless of their consent or even without them knowing about such collection. At this point, the issue of data and information privacy becomes relevant, since they are, not rarely, distributed in the supply chain—not necessarily with the related branch, but for a variety of uses that the consumer does not have control.

The objective is, in this sense, to demonstrate that the technological innovation brought by the Internet of Things brings severe risks to the consumer privacy, analyzing the principle of transparency as a possible regulatory framework for future and eventual regulation of the subject. For this reason, the deductive approach is used, starting from a broad and general analysis until reaching its main problem by examining its peculiarities. The article was structured in two parts, the first one will analyze the functioning and specificities of the Internet of Things and the second part is to address the issue regarding consumer privacy and the theory of dialogue of sources.

2 Internet of Things and Data Collection

The expression "Internet of Things" (IoT) has its origin in the English language and first appeared in the literature in the year 1995 in the productions of Kevin Ashton, one of the founders of the MIT Auto ID Center.[1] The definition of its concept is still not unanimous:[2] among several interpretations of this technological phenomenon, one can usually establish that it is a matter of connecting physical objects to the Internet and also the connection among themselves, through wireless sensors and technologies capable of interacting with the environment and with the users,[3] transmitting information,[4] creating a proper ecosystem[5] called Smart Environment.[6] For Tidor, the Internet of Things connects humans and intelligent machines in a new, incredible and often frightening way. It deals with the movement and interaction between various areas such as people, animals, vehicles, air drafts, viruses and many other things.

Specifically regarding the consumer market, when suppliers possess such technology and offer it to the consumers through products and services or a mixture of both, this may allow IoT to recognize relationships and predict patterns that are often complex to the human mind, as well as discover the conditions of a bridge, changes

[1] Rifkin (2016), p. 94.

[2] Helberger (2016), p. 2.

[3] Dutton (2013), p. 15.

[4] Thierer (2014), p. 11.

[5] Evans (2011), p. 9.

[6] About this subject, see more in: Atzori et al. (2010), p. 10.

in the atmosphere, for instance. The Internet of Things can still operate independently of humans and become more and more intelligent over time using adaptive algorithms.[7]

IoT is already widely used in the industrial sector, because it has the ability to establish connections between machines so that they can operate without the need for large-scale human intervention in the operations for which they are programmed.[8] A different note, regarding this article, concerns the use of the smart environment by consumers, who perceive the IoT as something revolutionary that offers and will offer numerous benefits,[9] without necessarily taking into account the risks of using the information collected.

Despite the innovative and promising aspect IoT brings to consumers, both with regard to experiences and the facilitation of life in general, it is true that this new technology also revolutionizes how suppliers learn about who their consumers are and their habits,[10] in order to provide a hyper-personalized product or service.[11] For this, the popularization of technology is a condition.

According to Rifkin, in 2030 it is estimated that the quantity will be 100 trillion devices connected to the IoT and in developed countries the *per capita* quantity of connected items could reach 5 thousand.[12] The result will be an intense array of wireless sensors that will surround people and, specifically regarding the consumer market, it will enable suppliers to know *every* aspect of consumer life,[13] through the ability to transform data into information and combine it with those already available on the Internet.

The information collected on devices that are not components of the IoT structure, such as cellphones and computers, differ from those coming from the smart environment. In other words, the quality of information differs. While the consumer, when entering an application or site, provides data to the platform only when using the device, in IoT the data will be collected 24 h a day and 7 days a week,[14] being able to create a complete profile of a particular person. An interesting study by Youyou, Kosinski and Stillwell points out that only 150 to 200 'likes' on social networks allow the device to have enough information about the user so that he knows the person more than his or her cohabitant.[15]

For these reasons, data collected from smart things have high potential of use by part of product and services suppliers, may they be direct or invisible in the

[7]Tidor *apud* Kadow and Camargo (2016), p. 154.

[8]United States Federal Trade Commission (2015), p. 5.

[9]Davies (2015).

[10]Helberger (2016), p. 3.

[11]Stolpe (2016), p. 22.

[12]Rifkin (2016), p. 93.

[13]Helberger (2016), p. 3.

[14]Helberger (2016), p. 3.

[15]Youyou (2015), p. 138.

electronic commerce architecture, such as insurers and credit companies.[16] Thus, even if data collection is not restricted to the IoT structure, it is noticeable how the information collected by it is capable of issuing alerts regarding more personalized products and services, according to the preference and individual habits of each person, since the consumer who interacts directly with the device cause suppliers to switch mass advertising to targeted advertising with great accuracy for certain situations or phases.[17] The use of such data, however, is not restricted to advertising.

Cognizant has developed a study that has identified the ability to create a realistic profile and to direct advertising as one of IoT's development and application trends to the business world. Through interviews with two hundred companies that develop smart products and services, the author concluded that almost one-third (28%) of the respondents indicated that they will use the data of the consumers collected to personalize it *[products and services]*, while 39% will treat the data to analyze the use of the product or service and their life cycle, and 32% will share information with the supply chain for improvements in products and services.[18]

It means to say the relationship between supplier and consumer has changed. That is, the traditional consumption relationship consisted of the individual going to the store or accessing the site, making a purchase and, if the product or service did not present a fault or defect, it would terminate the supplier-consumer relationship. In Internet of Things, the temporal duration of the relationship with the supply chain is abruptly changed, since it establishes a long and dynamic interaction of the consumer with the suppliers because their data and personal information are circulating in the smart environment[19] with a durability equivalent to the useful life of the good.

At the same point, consumers, by operating and interacting frequently in the structure of the IoT, will at most of the times enter into transactions and contracts remunerating a plurality[20] of visible and invisible actors in the supply chain, either with money or with data, showing even more complexity and contractual connectivity[21] in this technology, which cannot be used to put into question the consumer rights. According to the brazilian Consumer Protection Code, for instance, it is considered as a part of the supply chain every and each part that handles personal data (if concerning service provision), and it may be held responsible as a supplier strictu sensu if not in compliance with the legislation. This is possible because the

[16]Marques (2017), p. 255.

[17]Helberger (2016), p. 4.

[18]Cognizant (2015), p. 8.

[19]Cognizant (2015), p. 9.

[20]In Jayme's doctrine, postmodern culture has four elements: (a) the plurality of forms and styles, according to which each person chooses his or her own way of life, marking the right to difference; (b) communication without frontiers, driven not only by technology, but also by the will and the desire to communicate, emerging from a common value; (c) narration, insofar as communicating is also describing, telling and narrating; and (d) return to the feelings, i.e., in relation to cultural identity, the defense of cultural identity, religion and other expressions of individualism (cf. Jayme, 2003).

[21]About this subject: Bergstein (2017).

Brazilian Civil Rights Framework didn't limit the hypotheses of civil liability for the Consumer Protection Code in case of consumer relations.[22]

In this sense, the result will be a growing personalization of products and services, allowing not only the advertisement targeting, but also a differentiated pricing by consumer groups and, mainly, the elaboration of individual contractual clauses[23] based specifically on all the activity captured by IoT. In a next step, the analysis of data and information will allow suppliers even to modify the habits and behaviors of the consumer.[24]

It is important to note that IoT products and services are smart precisely because their functionality feeds a continuous stream of personal data that translates personal habits and uses to provide real-time feedback performance. In other words, in order to achieve the full functionality of the IoT, consumers need to provide *data*, to the point the payment will no longer be confined to the cash consideration, including also all the information collected.

"Payment with data"[25] will become part of almost all transactions in the future consumer market (which is already happening, albeit with less intensity), since data and information are the new commodity with incredible economic value.[26] Meglena Kuneva, in 2009, warned that "personal data are the new oil of the Internet and the new currency of the digital world",[27] which makes it difficult to price information rather than money, considering there is a lack of transparency and information in establishing the prices of a given product or service in the IoT based on the data collected.

This extreme valuation of information, consequence of the connection of everything and everyone in a single network, "takes humanity from the age of privacy, a dominant feature of modernity"[28] to an era of unnoticed surveillance, classification and control.[29] In other words, instead of moving to a transparent consumer society, as it should ideally be, we will move to a veiled surveillance consumer society, since

[22] Important to point out that the relations between applications on the Internet (that, according to the Brazilian legislation, are any web pages or social networks, e-commerce or not) and the lay user and natural person are almost in all cases consumer relations, because they fit into the legal definition of supplier, ex vi art. 3°, CDC, when present the consumer/user. This understanding has its origin in one of Brazil's Supreme Court (Superior tribunal de Justiça), when debates were held on whether the consumer legislation should apply to the relation between Google's research mechanism and the user. The Court's judgement was positive, arguing that the Google company explores this tool commercially, establishing a consumer relation, present all the other legal requirements. (BRAZIL. Superior Court of Justice. Recuro Especial 1316921/RJ, Rel. Ministra NANCY ANDRIGHI, 26/06/2012).

[23] Professor Bruno Miragem brings an insurance contract example. See: Miragem (2017).

[24] Helberger (2016), p. 12.

[25] Helberger (2016), p. 7.

[26] See: Galvâo (1999).

[27] Kuneva (2009).

[28] Rifkin (2016), p. 96.

[29] Rodotà (2008), pp. 145 ff.

suppliers will have access to every consumer's information without them even knowing or agreeing in this sense. Privacy now gains a new definition, making it the "right to keep control over one's own information.".[30]

With the contemporary lifestyle, all one's activities are constantly recorded, whether virtual or physical. According to a survey by Rudd, a Briton has his/her image captured more than 300 times a day, whether on the street or in private places. With the Internet of Things, this surveillance will reach the private space of the consumer: in early 2017, there was data leakage, "including conversations between parents and children, by the invasion of children's software such as the CloudPets or the Cayla doll, leading to, including, in relation to the doll, the prohibition of its commercialization in some countries".[31] Bundesnetzagentur, in 2017, responsible for telecommunications in Germany, warned about possible failures on privacy in the Cayla doll, as well as security problems, such as the fact, for instance, that a user in the same network as the toy might have access and connect to it and transmit/receive data without the parents' consent, turning the doll into a spy, enabling access to the child or teenager by malicious people. Under the German Law, not only its commercialization was forbidden, but also its possession is considered a criminal offence, as it classified the doll as an illegal spying device.[32]

A similar problem was the brand Samsung with its smartTVs, which captured and transmitted everything people said in front of the devices in case the voice command option was activated, which was confirmed by the Korean manufacturer,[33] and even with the robot vacuum cleaners rom the Roomba brand, integrated into the smart environment, which, by vacuuming the places*, mapped and filmed the images and transmitted them to the supplier.[34] Note, all this information has as a background of improvement in the quality of the product and the provision of services, but there is no information about what is actually done with the collection of personal data, resulting in a serious violation of informational self-determination, which is diminished along with the right to privacy.

Although the IoT is a revolutionary promise that life will be made easier for everyone, Consumer Law must be alert to technological change, especially when dealing with personal information from consumers who do not always *know* or *want* their data collected. In this scenario, because of the strong overlap also in contract theory—especially in terms of adhesion—, to freedom of choice, right to information and to competition,[35] consumer privacy in the use of IoT should be preserved.

[30] Rodotà (2008), p. 92.

[31] Miragem (2017).

[32] Bolger (2017).

[33] O Globo (2015).

[34] Xataka (2016).

[35] Helberger (2016), p. 5.

3 Consumer Privacy in the Iot and Transparency

Postmodern society is marked by the rapidity and simultaneity in which things occur, revealing the fragility of interpersonal relationships.[36] At the same time, technological development accompanies this same speed, giving people an endless amount of information, which they are not able to absorb and process, causing, among other factors, the ill-being of the individual.[37] Technology, as a way of remedying the situation, creates mechanisms and devices that are increasingly interactive and connected to the Internet, in order to make daily life easier, both in business and in private life, reinforcing the fragility of interpersonal relations and increasing the relations with *things*.

For Luño, the development of technology and Internet can not be suppressed, since it means an unannounced progress and a sign of progress of the current time, but should not take a passive position regarding the risks immanent in the virtual environment.[38] In this sense, what should be clear is that with the "treatment of personal information by computer and telematic means, the intimacy of people is more exposed to vulnerability, since, as stated above, these means greatly facilitate the treatment of these information".[39]

In the Internet of Things it is no different. It is developed to facilitate life in postmodernity by collecting and processing data from the people who use the devices that are part of the smart environment in order to create an accurate user profile. Through the use of this smart environment, the transmission of data and information becomes increasingly common and facilitated, which contributes to the rupture of consumers' privacy rights when they do not know who manipulates their data or the destiny given to them.

Recently, on 3rd and 4th of July 2017, the issue of data was addressed at the Second Meeting of the Group of Experts on Consumer Law, promoted by UNCTAD (United Nations Conference on Trade and Development) in Geneva, Switzerland. There, "in a number of demonstrations, emphasis was placed on the protection of personal data in the context of a digital economy, especially considering the expectation that in the next three years it is expected to reach the number of approximately 20 million objects connected among themselves. Thus, in addition to adequate regulation, the challenge is to educate consumers about the level of risks involved in Internet-mediated consumption in relation to the availability of their own data.".[40]

It is understood that the use of this information should be restricted only to what was collected, but the data owners use it in an indiscriminate manner, mainly for advertising purposes and for its sale or assignment to third parties, in general, other

[36] Bauman (2003), p. 12.

[37] Although Freud analyzes from modernity, his teachings are also valuable for postmodernity: Freud (2011).

[38] Luño (1998).

[39] Pereira (2015), p. 144.

[40] Olvieira and Cipriano (2017).

suppliers or providers of services in the information society,[41] which truly unravel the life of the consumer without their authorization and even without their knowledge,[42] which is a serious violation of the right to privacy.

Privacy in Brazil is a constitutionally protected right, being part of the list of fundamental rights and guarantees of the citizen, inscribed in art. 5, item X, of the Federal Constitution,[43] expressly, making inviolable the private life of the people, as well as their intimacy, honor and image. In the same sense, item XI of the said article provides the secrecy of data, which is protected by the mantle of inviolability.

Because it is a fundamental right of constitutional origin, the preservation of the privacy in the personal, legal, commercial and relational scope it is a guiding principle to the other norms, explaining Doneda that "in the panorama of the Brazilian law, the recognition of data protection as an autonomous and fundamental right, does not derive from an explicit and literal diction, but from the consideration of the risks that the automated treatment brings to the protection of the personality in the light of the constitutional guarantees of substantial equality, freedom and dignity of the human person, together with the protection of intimacy and private life".[44]

In view of this, the Brazilian Civil Code,[45] Federal Law of general scope, brings privacy in the chapter of personality rights, allowing the prohibition of the disclosure of writings, transmission of the word or the manipulation of a person's image, should such acts reach him honor, good reputation or respectability. Specifically for the consumer market, the Consumer Protection Code addresses privacy by protecting and regulating databases and consumer registration.

In the digital field, the Brazilian Civil Rights Framework[46] of the Internet places the protection of privacy as a principle of Internet use in the country, being its guarantee the condition for the full exercise of the right of access to Internet, in the sense that any operation of collection, storage and treatment of records, personal data or communications shall respect the rights to privacy, protection of personal data and confidentiality in private communications. Furthermore, we have the recent General

[41] Pereira (2015), p. 190.

[42] Silva (2001), p. 212.

[43] Article 5. All persons are equal before the law, without any distinction whatsoever, Brazilians and foreigners residing in the country being ensured of inviolability of the right to life, to liberty, to equality, to security and to property, on the following terms: X—the privacy, private life, honour and image of persons are inviolable, and the right to compensation for property or moral damages resulting from their violation is ensured. (BRAZIL: Constitution of the Federative Republic of Brazil of 1988. Available at: http://www.planalto.gov.br/ccivil_03/Constituicao/Constituicao.htm. English version: <http://www.stf.jus.br/repositorio/cms/portalStfInternacional/portalStfSobreCorte_en_us/anexo/Constitution_2013.pdf>).

[44] Doneda (2011), p. 103.

[45] BRAZIL. Brazilian Civil Code. Federal Law 10,406 of January 10, 2002. Establishes the Civil Code. Available at: <http://www.planalto.gov.br/ccivIl_03/Leis/2002/L10406.htm>.

[46] BRASIL. Marco Civil da Internet. Lei Federal 12.965 de 23 de abril de 2014. Estabelece princípios, garantias, direitos e deveres para o uso da Internet no Brasil. Available at: <http://www.planalto.gov.br/ccivil_03/_ato2011-2014/2014/lei/l12965.htm>.

Law on Data Protection,[47] which establishes that the discipline of personal data protection is based on respect for privacy and informational self-determination, as well as stipulating innumerable rights and duties for all those involved in data processing, whatever it may be, with only few legal exceptions.

It is observed the plurality of laws that regulate the same case, with different intensities and foci. In this "legislative sea", which would best protect the rights and interests of consumers who use these new technologies? In Brazil, there is no choice of one or the other, since there is an orderly and coherent application of all the normative sources that regulate the concrete case. The essence of the theory is that legal norms do not exclude, supposedly because they belong to distinct legal branches, but instead complement each other. The theory was developed by Erik Jayme in Germany and by Claudia Lima Marques in Brazil.

Claudia Lima Marques affirms that the "recent doctrine is today seeking more harmony and coordination between the norms of the legal system (conceived as a system) than exclusion".[48] In this sense also "the theory of the dialogue of the sources arises to replace and to surpass the classic criteria of solution of the legal antinomies (hierarchical, specialty and chronological[49])". The groundwork of the theory lies in the fact that norms arise to be applied as a whole and not to be excluded or chosen, especially in the increasingly convergent field of all norms of law, with centralization in the human person.

It can be seen, therefore, that the theory of the dialogue of the sources allows the jurist not to be bound to one or another regulation, one or another microsystem, for which the application to certain cases was imagined, since the legal system is an unitary whole and thus must be applied, unlike the classical conception of legal antinomy solution. This conception has been written in the Brazilian Consumer Protection Code[50] and the national courts have embraced the dialogue of the sources for the protection not only of the consumer, but transforming it into a new general theory of Brazilian law and extending it to all legal matters.

It is also important to note that when it comes to technology, where the capacity for data agglutination by such instruments is invisible and abstract, there is no transparency about what is collected and how it is used, going beyond the purpose of use intended by the consumer, the consent to the use of such personal elements

---

[47] BRAZIL. Civil Landmark of the Internet. Federal Law 12,965 of April 23, 2014. Establishes principles, guarantees, rights and duties for the use of the Internet in Brazil. (Brazilian Civil Rights Framework for the Internet). Available at: <http://www.planalto.gov.br/ccivil_03/_Ato2015-2018/2018/Lei/L13709.htm>.

[48] Marques (2009), p. 90.

[49] Tartuce (2012), p. 66.

[50] Article 7 The rights provided for in this Code do not exclude others arising from international treaties or conventions to which Brazil is a signatory, of ordinary domestic legislation, of regulations issued by the competent administrative authorities, as well as those derived from the general principles of the law, analogy, customs and equity. (BRASIL, Code of Consumer Protection, Federal Law 8.078, dated September 11, 1990.) Provides for consumer protection and provides other measures Available at: <http://www.planalto.gov.br/ccivil_03/Leis/L8078compilado.htm>.

and as well their specific prohibition becomes obscure. That is, the consumer not knowing what information suppliers are using in exchange for the service provided (often free or at symbolic prices, referring to payment also with data), the subject becomes even more vulnerable.

Consumer vulnerability is a state of the person, inherent of risk or a sign of excessive confrontation of interests identified in the market, and can be permanent or provisional, individual or collective, unbalancing a certain legal relationship and weakening the subject of rights.[51] Note that although it is a presumption for the consumer natural person,[52] the concept of vulnerability fits perfectly into what happens in the Internet of Things dynamics, since the consumer is totally exposed to his supplier, who takes advantage of this to gain even more advantages in the consumer market.

It should be highlighted, among the species of vulnerability,[53,54] the informational vulnerability as characteristic of the consumer in IoT. This type of vulnerability consists in the lack of information on the part of the consumer (because of the previous "selection" of the information given) of the characteristics of a particular consumer relationship, or, on the contrary, but also detrimental, in the avalanche of information not always useful to the legitimate expectations of the consumer. In a world where information has more and more value and preciousness, paradoxically, the consumer has it less and less and the supply network has it more and more. In the virtual environment, Canto identifies that the inherent vulnerability to the consumer relationship changes with the situation promoted by the new technologies, which generates a worsening of the consumer fragility and arouses his distrust in the electronic field.[55] See a new way of vulnerability Brazilian law is developing: digital vulnerability. It concerns the technical,[56] informational,[57] legal,[58] factual[59] and

---

[51] Marques (2016), p. 323.

[52] Miragem (2016), p. 128.

[53] Claudia Lima Marques identifies technical, legal and factual vulnerabilities (Marques, 2016, p. 322).

[54] Paulo Valério dal Pai Moraes also categorizes other types of vulnerability, such as legislative or political, biological or psychic, and environmental (cf. Moraes, 1999).

[55] Canto (2015). *Passim*.

[56] Technical vulnerability: the consumer has no specific knowledge about the product or service, either in terms of its characteristics or in relation to its usefulness (cf. Marques, 2016).

[57] Information vulnerability: it stems from technical vulnerability, but must be treated autonomously, due to the dynamics that consumer relations have in the digital era, where access to information has been expanded in order to be decisive for the decision thus protecting the informational vulnerability of the consumer presupposes the control of the quality of the information transmitted by the suppliers (and not their quantity) (cf. Marques, 2016).

[58] Legal (or scientific) vulnerability: the legislator recognizes that the consumer has no legal, accounting, economics, financial mathematics and other knowledge, for example, about the interest charged (cf. Marques, 2016).

[59] Phatic (socioeconomic) vulnerability: the supplier is the holder of economic power, and is in a position of supremacy (cf. Marques, 2016).

situational[60] vulnerability of the consumer in a virtual environment, adding the vulnerabilities of the Internet of Things architecture, the lack of security in the processing of personal data, the possibility of fraud, the lack of security of the information and, also, the consumer's lack of knowledge about the destination given to his data.

The point to be reflected is that due the use of the tools which compose IoT, the possibility of consumer control of the data they want or not to be used or transmitted is made impossible, since the information gathering is constant and often unrestricted.

Finkelstein warns that there are entities called by the Brazilian Civil Rights Framework for the Internet—Federal Law n. 12.965/2014—Internet applications, that owns and control the technology needed to identify where, when, and how fast the individual accesses each site, documenting which stores they visited, which links they were interested in, in what order, and for how long.[61] This aspect of the concern for the protection of privacy stems from an active management of the consumer in the search for information on the Internet, when their data are collected and associated to create an advertising environment aimed at the supposed interests of the consumer.

On the other hand, the Internet of Things captures consumer information in a more subtle way, through the habits of location, food, leisure, health and consumption identified from the use of their devices. In this sense, Finkelstein also states that the commercialization of the data collected by the sites for other purposes, for commercial companies or services unrelated to the company that collected them deserves greater action of the Law in defense of the users and their privacy.[62]

Almeida says that even if the Internet is not a private place, it is necessary to maintain privacy, because any individual, when leaving the private sphere, does not leave their privacy available in the virtual environment.[63] Although the right to privacy is based on the notion of being left alone,[64] the use of IoT devices prevents the user from being completely *alone*, given the permanent connection with the servers that collect their data and information and constantly pass it on to the owners of the technology.

Facing these questions, Pentland, an MIT researcher, proposes a set of practices and principles that seek to define the ownership of data and control its flow in favor of individuals whose data is collected and called the "New Deal". The basis of the proposal is transparency, which is one of the basic principles of the protection conferred by consumer law, as set out in article 4, caput, of the Brazilian Consumer Protection Code. For Marques, "the central idea is to enable a closer and more

---

[60] Situational vulnerability: Consumer weaknesses are recognized in specific situations, such as e-commerce or the virtual environment (cf. Canto, 2015. *Passim*).

[61] Finkelstein (2011), p. 125.

[62] Finkelstein (2011), p. 125.

[63] Almeida and Almeida (2016), p. 32.

[64] Doneda (2006), pp. 7 ff.

sincere and less harmful contractual relationship between consumer and supplier",[65] meaning clear and correct information, also regarding the allocation of their own information and sensitive data.

The New Deal would give people the ability to see what is being collected about them, allowing them to choose whether or not to accept such capture, so that the "portrait profile" formed by their data is owned by the individual himself, and thus can dispose of it according to its autonomy, even if this transparency makes some business models unfeasible.[66] In other words, Pentland proposes that companies clearly inform their customers what data will be collected and how they will be used, not only in contractual conditions but in other more effective ways, and the final word of whether or not to accept this condition will fit exclusively the consumer. In the author's words, "I do not think companies realize that the costs of a 'get all the data' strategy are too high. They are taking on huge amounts of risk in the form of data breaches and damage to critical systems. In addition to being expensive to maintain security, violations will cost more and more."[67]

With transparency still at the heart of protection, Kranenburg presents another point of view to solve the privacy issues breached in the smart environment. An open approach is adopted here, i.e., "it would be good for society if the architecture of the Internet of Things were the same for everyone and the public had the same tools as those in power."[68] In other words, the author argues the best solution would be radical transparency, where all data (personal, business, government, etc.) would be public and everyone would have access to them. The author thus preaches an inverse logic based on equality: "If corporations and governments can make decisions based on the behavior, health and consumption patterns of their clients and citizens respectively, individuals would also have the same power of access to all corporate data, from the most obvious to the most secretive, and could interfere more effectively with business and government guidelines.[69]

Obviously, this model of total openness of all data is utopian, and even more, in almost in all countries of the civilized world privacy is a fundamental right of greater relevance. The contribution that Kranenburg's proposal brings is extremely important because it can serve as an alert to the government and to providers in general that the collection and use of data can not be unrestricted.

Although the radical transparency proposed by the author is impracticable, transparency as defined and defended by consumer law is adequate for the protection of data, but not sufficient in the face of the new technologies, deserving highlight, therefore, the Bill 3514/2015 in progress in the Brazilian Chamber of Deputies, which deals with the updating of the Consumer Protection Code with a special theme on electronic commerce, and specifically section XI, which should be added to

[65] Marques (2016), p. 745.

[66] Kadow and Camargo (2016), p. 158.

[67] Pentland (2014), p. 88.

[68] Kranenburg (2008), p. 9.

[69] Kadow and Camargo (2016), p. 159.

Article 6 of the Code, establishing as basic rights of the consumer "the privacy and security of the information and personal data provided or collected by any means, including the electronic, as well as free access of the consumer to these and their sources.".[70]

Another point that deserves to be underlined is that Article 45-A, which opens the new Section VII on e-commerce, places the protective spirit of consumers in relation to new technologies: "Art. 45-A. This section sets forth general rules for consumer protection in e-commerce and commerce by distance, in order to strengthen their trust and ensure their effective protection by reducing information asymmetry, preserving security in transactions and protecting self-determination and privacy of personal data."[71]

We realize that transparency and the manipulation of personal data are of enormous economic, social and, above all, legal relevance, because there must be a legal response to technological development regarding the preservation of all plural aspects of privacy in a postmodern life. With the Internet of Things, the challenge of protecting this privacy is renewed, given that we live in an age of surveillance, of control, where all are observed at all times through information data.

4 Final Considerations

It is true that the Internet of Things will bring countless benefits to consumers through the rising of the smart environment, that integrates almost every device and gadgets in everyday life. The burden of this technological evolution, however, is the data of the millions of consumers, who will feed the supply chain—which now also has the purpose of flowing and processing information.

In community life, citizens will gave up part of their privacy in favor of community living. It turns out that this part, in the structure of the IoT, is very significant, since through it the exact profile of the consumer, his preferences, his health, his banking data and an infinity of sensitive information is created and, in theory, these should be of only to the consumer or to whom he has gave them.

The value and principle that should govern these consumer relations (continued over time, not in the form of services per se, but in the form of uninterrupted data provision) must be that of transparency. This apparent duality, transparency versus

[70] BRAZIL. Chamber of Deputies. *Bill 3514/2015*. Available at: <http://www.camara.gov.br/proposicoesWeb/prop_mostrarintegra;jsessionid=A2DC470DFC9B40ABF4C59DED84632163.proposicoesWebExterno1?codteor=1408274&filename=PL+3514/2015>. Accessed 26 Aug 2021.

[71] BRAZIL. Chamber of Deputies. *Bill 3514/2015*. Available at: <http://www.camara.gov.br/proposicoesWeb/prop_mostrarintegra;jsessionid=A2DC470DFC9B40ABF4C59DED84632163.proposicoesWebExterno1?codteor=1408274&filename=PL+3514/2015>. Accessed 26 Aug. 2021.

privacy, will also ground and guide the consumer's right to new technologies that deal with personal data.

Therefore, it is of great importance to approve Bill 3514/2015, which is pending before the Chamber of Deputies, so the constitutional provisions of the Civil Code and of the Consumer Protection Code are strengthened even further, to promote a true dialogue of the sources and the consumer life integrity and privacy is preserved and extolled, not forgetting in this regard the necessary (and inevitable) balance that must exist between technological development and the protection of the human being in relation to new risks coming from, in particular, the Internet of Things in particular.

References

Almeida, J. E., & Almeida, D. E. (2016). A ditadura do algoritmo e a proteção da pessoa humana: uma análise do controle do si eletrônico. *Revista de Direito Privado, 17*(69), 29–43.

Atzori, L., Iera, A., & Morabito, G. (2010). The internet of things: A survey. *Computer Networks, 54*(15), 2787–2805.

Bauman, Z. (2003). *Amor Líquido*. Jorge Zahar Editor.

Bergstein, L. G. (2017). Conexidade Contratual, Redes de Contratos e Contratos Coligados. São Paulo. *Revista de Direito do Consumidor, 109*, 159–183.

Bolger, P. (2017). The Internet of Things: Cayla doll is banned in Germany over privacy and security concerns. *Technology and data privacy*. E-book. Retrieved August 26, 2021, from <https://www.lkshields.ie/>

Canto, R. E. (2015). A vulnerabilidade dos consumidores no comércio eletrônico: reconstrução da confiança na atualização do Código de Defesa do Consumidor. *Revista dos Tribunais*.

Cognizant. (2015). *The rise of the smart product economy*. E-book. Retrieved August 26, 2021, from <https://www.cognizant.com/InsightsWhitepapers/the-rise-of-the-smart-product-economy-codex1249.pdf>

Davies, R. (2015). *The Internet of Things: Opportunities and challenges*. European Parliamentary Research Service. Retrieved August 26, 2021, from <http://www.europarl.europa.eu/RegData/etudes/BRIE/2015/557012/EPRS_BRI(2015)557012_EN.pdf>

Doneda, D. (2006). *Da privacidade à proteção de dados pessoais*. Renovar.

Doneda, D. (2011). A proteção dos dados pessoais como um direito fundamental. *Espaço Jurídico Journal of Law, 12*(2), 91–108.

Dutton, W. H. (2013). *The Internet of Things*. Retrieved October 21, 2017, from <https://papers.ssrn.com/sol3/papers.cfm?abstract_id=2324902>

Evans, D. (2011). The internet of things: How the next evolution of the internet is changing everything. *CISCO White Paper, 1*(2011), 1–11.

Finkelstein, M. E. (2011). *Direito do Comércio Eletrônico*. Elsevier.

Freud, S. (2011). *O mal-estar na civilização*. Editora Companhia das Letras.

Galvão, A. P. (1999). A informação como commodity: mensurando o setor de informações em uma nova economia. *Ciência da Informação, 28*(1), 67–71.

Helberger, N. (2016). *Profiling and targeting consumers in the Internet of Things–A new challenge for consumer law*. E-book. Retrieved August 26, 2021, from <https://papers.ssrn.com/sol3/papers.cfm?abstract_id=2728717>

Jayme, E. (2003). Direito Internacional Privado e Cultura Pós-moderna. *Cadernos de Programa de Pós-Graduação em Direito da Universidade Federal do Rio Grande do Sul. – PPGD/UFRGS*. Edition with selection of texts of the work of Erik Jayme, in homage to the delivery of the title of Doctor Honoris Causa / UFRGS to the jurist, *1* (1). Porto Alegre: UFRGS.

Kadow, A., & Camargo, C. (2016). Internet das Coisas: vulnerabilidade, privacidade e pontos de segurança. *Revista Competência, 9*(1), 153–161.
Kranenburg, R. (2008). *The internet of things: A critique of ambiente technology and the all-seeing network of RFID*. Insitute of Network Cultures.
Kuneva, M. (2009). Roundtable on online data collection, targeting and profiling. *European EC Rapid Press Release*, SPEECH/09/156.
Luño, A. E. P. (1998). *Impactos Sociales y Jurídicos de Internet*. Retrieved August 26, 2021, from <http://www.argumentos.us.es/numero1/bluno.htm>
Marques, C. L. (2016). Contratos no Código de Defesa do Consumidor: o novo regime das relações contratuais. *Revista dos Tribunais* (8th ed.).
Marques, C. L. (2017). A nova noção de fornecedor no consumo compartilhado: um estudo sobre as correlações do pluralismo contratual e o acesso ao consumo. *Revista de Direito do Consumidor, 111*.
Marques, C. L., Bessa, L. R., & Benjamin, A. H. (2009). Manual de direito do consumidor. *Editora Revista dos Tribunais*.
Miragem, B. (2016). *Curso de Direito do Consumidor*. RT.
Miragem, B. (2017). *A internet das coisas e os riscos do admirável mundo novo*. Retrieved August 26, 2021, from <https://www.conjur.com.br/2017-mar-29/garantias-consumo-internet-coisas-riscos-admiravel-mundo>
Moraes, P. V. P. (1999). *Código de Defesa do Consumidor: o princípio da vulnerabilidade no contrato, na publicidade e nas demais práticas comerciais*. síntese.
O Globo. (2015). *Samsung adverte: Cuidado com o que você diz em frente a sua TV inteligente*. Retrieved August 26, 2021, from <https://oglobo.globo.com/sociedade/tecnologia/samsung-adverte-cuidado-com-que-voce-diz-em-frente-sua-tv-inteligente-15286181>
Olvieira, A. F., & Cipriano, A. C. M. (2017). *Consumo digital e hipervulnerabilidade do consumidor*. 2017. Retrieved August 26, 2021, from <https://www.conjur.com.br/2017-jul-19/garantias-consumo-consumo-digital-hipervulnerabilidade-consumidor>
Pentland, A. (2014). Com os grandes dados vêm grandes responsabilidades. *Harvard Business Review Brasil, 92*(11).
Pereira, M. C. (2015). *Direito à Intimidade*. Juruá.
Rifkin, J. (2016). *Sociedade com custo marginal zero: a Internet das Coisas, os bens comuns colaborativos e o eclipse do capitalismo*. M. Books do Brasil.
Rodotà, S. (2008). *A vida na sociedade da vigilância: a privacidade hoje. Organization, selection and presentation of Maria Celina Bodin de Moraes*. Translation: Danilo Doneda and Luciana Cabral Doneda (pp. 145–146). Rio de Janeiro: Renovar.
Silva, J. A. (2001). *Curso de Direito Constitucional Positivo*. Malheiros.
Stolpe, M. (2016). The internet of things: Opportunities and challenges for distributed data analysis. *ACM SIGKDD Explorations Newsletter, 18*(1), 15–34.
Tartuce, F. (2012). *Manual de direito civil*. Rio de Janeiro: Forense, São Paulo: Método.
Thierer, A. D. (2014). The internet of things and wearable technology: Addressing privacy and security concerns without derailing innovation. *Richmond Journal of Law & Technology, 21*(2).
Tidor, B. (2015). *The internet of things*. Oxford: The MIT Press. apud Kadow A., Camargo, C. (2016). Internet das coisas: vulnerabilidade, privacidade e pontos de segurança. *Revista Competência, 9*(1), 153–161.
United States Federal Trade Commission. (2015). *Internet of things: Privacy and security in a connected world*. Retrieved August 26, 2021, from <https://www.ftc.gov/system/files/documents/reports/federal-trade-commission-staff-report-november-2013-workshop-entitled-internet-things-privacy/150127iotrpt.pdf>
Xataka. (2016). *Roomba aspira tu casa pero tambien la espia*. E-book. Retrieved August 26, 2021, from <https://www.xataka.com/privacidad/roomba-aspira-tu-casa-pero-tambien-la-espia-y-crea-un-plano-de-ella-que-podria-vender-a-terceras-partes>

Youyou, W., Kosinski, M., & Stillwell, D. (2015). Computer-based personality judgments are more accurate than those made by humans. *Proceedings of the National Academy of Sciences, 112*(4), 1036–1040.

Guilherme Mucelin PhD Candidate in Law at the Federal University of Rio Grande do Sul. Master's degree in Law from the Federal University of Rio Grande do Sul. Specialist in French and European Contracts Law by the Université Savoie Mont Blanc. Specialist in Consumer Law and Fundamental Rights and in Public and Private International Law and Integration Law at the Federal University of Rio Grande do Sul. Member of the Research Groups "Mercosur, Consumer Law and Globalization" and "Private Law and Market Access", both from UFRGS. CAPES scholarship holder. E-mail: mucelin27@gmail.com.

Part IV
Legal Tech

Legal Tech and Computational Legal Theory

Burkhard Schafer

Contents

Abstract The paper looks at the past, present and future of AI in the legal sector through the lens of the public reaction to Alpha Go's victory over a human opponent. It discusses the merit and limits of the "game metaphor" for an evaluation of legal AI, and asks in particular if empathy and creativity in law will become pushed aside should more and more legal tasks get performed by machines. Or can we instead think of different futures, either because of our evolving understanding of what it means to be a "good lawyer", or our evolving understanding of the capabilities of AI?

B. Schafer (✉)
The University of Edinburgh, Old College, Edinburgh, UK
e-mail: B.Schafer@ed.ac.uk

G. Borges, C. Sorge (eds.), *Law and Technology in a Global Digital Society*,
https://doi.org/10.1007/978-3-030-90513-2_15

1 Legal Tech Is (Alpha) Go

1.1 Playing Games with AI and Law

When DeepMind's AlphaGo beat the South Korean Go master Lee Sedol before the eyes of a captivated worldwide audience numbering in the hundreds of millions, it catapulted artificial intelligence into the public limelight. Widely deemed a litmus test for AI research, Go had for a long time eluded all attempts by computer scientists to develop gaming software that was better than a mere beginner. The social and political impact of AlphaGo's victory is difficult to overestimate. To give but one example, the South Korean Prime Minister announced that Korea would invest $863 million in AI research over five years, identifying it as a key technology to maintain competitiveness. In her speech, she emphasized the benefits that AI will bring to society, the "fourth industrial revolution", which will as radically transform how we work and live as the invention of the steam engine did in the eighteenth century. The announcement went on to say: "Above all, Korean society is ironically lucky, that thanks to the 'AlphaGo shock', we have learned the importance of AI before it is too late."

Interest in legal applications of AI benefited from the same phenomenon, with several reports in the technical media evoking explicitly AlphaGo to bolster the case that Legal Tech is going to revolutionise legal service delivery.[1] The Law Society for England and Wales referenced AlphaGo in its Horizon Scanning report *Artificial Intelligence and the Legal Profession*,[2] as did Neota Logic's Michael Mill in his report, *Artificial Intelligence In Law: The State Of Play 2016* for their Legal Executives Institute.[3]

This paper will ask if these references to a computer game in the context of legal AI are merely an example of "story telling", a means to capture the public imagination, or if there are more substantive lessons that can be learned by approaching law and legal adjudication through the perspective of games and gameplaying. Is the connection between AlphaGo and the law ultimately just a metaphor, or worse, misleading hype, or can we think of it as a useful heuristic to study the way in which technology might transform the practice of law? I will argue that indeed, we can draw from it some inferences about the scope and limits of AI in the justice system, and with that also about the most promising strategy on how to change legal education and make law students "future proof".

The focus of this chapter is not so much what type of technology is best suited for legal knowledge representation, for this issue there exists already extensive literature from the AI and Law community.[4] Rather, by using the public reaction to AlphaGo and its interpretation by legal writers, it will question some often taken-for-granted

---

[1] See e.g. Holmes (2017) and Taylor and Osafo (2018).

[2] The Law Society (2018).

[3] Mills (2016).

[4] For an overview see Bench-Capon et al. (2012).

assumption about law, the legal profession, how lawyers think within and about the law, and what this can tell us both normatively and descriptively about the future of law in an age of AI.

Through such an analysis of the underlying assumptions and metaphors, and in particular our intuitions about "mechanical machines versus creative humans", we can develop a better understanding of what we can and should demand from technology in the justice system. As important as this is though, we can also begin to understand better what (if anything) is intrinsically human in the quest for justice.

Using mechanistic metaphors for law in this exploratory way is by no means a new idea. The opposing views of law as a system of rules, to be applied mechanically, unfeelingly, and in a machine-like manner, and the vision of law as a vehicle of substantive justice that also requires empathy and, where appropriate, mercy, has a long tradition. The "mechanical jurisprudence" of eighteenth and nineteenth century legal formalism deployed metaphors of "machine like" legal decision making long before computers or robots existed—so e.g. Julien de La Mettrie in his "L'homme machine" where he argued that distinguishing right from wrong is as much a mechanical process as distinguishing blue from yellow, and both within the capacity of deterministic machines.[5]

By contrast, his contemporary, Luc de Clapiers, wrote in his *Reflections and Maxims* of 1746 that "one can not be just if one is not humane."[6] His competing vision of justice sees it as irreducibly tied to the human condition and human experiences. Only through the exercise of empathetic judgement with our fellow creatures can we achieve a measure of justice in this world. These opposing views were to shape the development of modern legal systems ever since. The "computational legal theory" that this paper will outline will also interrogate and challenge this old dichotomy, which structured in the past much of jurisprudential debate.

1.2 From Games to Law

AI is increasingly seen as capable of transforming substantially the way legal services are delivered, both by academic commentators and by practitioners.[7] To realise this potential, it is often deemed necessary however that on the one hand, the justice system and its regulatory regime adapt appropriately, and on the other hand, legal education responds to the challenge and trains a new type of technologically competent and digitally literate professional of the future.[8] These questions are obviously connected, as regulatory bodies in many jurisdictions proscribe the

[5] De La Mettrie (1748), see also Campbell (1970) and Thomson (2016).

[6] Clapiers' optimism regarding human nature put him at odds with his fellow moralists including Mettrie. On the impact this has on his legal and political philosophy see also Meagher (1994).

[7] See e.g. Newman (2017) and Askin (2018).

[8] See e.g. Cornell Tech (2016).

minimum skills necessary to practice law competently, and often also influence the law curriculum through their power to accredit law degrees. In Scotland for instance, the Law Society has created a new recognised specialism of "legal technologist", supported by an appropriate training and evaluation regime. In England, the University of Swansea is offering an entire LLM programme on Legal AI since 2018, and "Coding for lawyers" courses are increasingly offered as an elective subject across jurisdictions.[9] In parallel, discussions by the regulatory bodies on how to respond to legal tech are starting to raise issues also about possible dangers and how to respond to them, with e.g. the Law Society of England and Wales setting up a *Technology and the Law Policy Commission* chaired by the Society's president. This body, which released its report[10] in June 2019, carried out a wide-ranging consultation on the issues of algorithmic decision-making in the justice system.

Many of the first-generation academic researchers in AI and law will look at the current interest with a mix of bemusement and concern. Some of the technologies now promoted as AI enabled game-changers are based on approaches and methods developed in the early days of legal AI, in the late 1980s and 90s. Neota Logic, one of the success stories of commercially deployed legal AI, describes some of its solutions as being based on "Rules: If/then mappings, decision trees, or decision tables". From a logic perspective this is a revival of the early rule-based expert systems such as the TAXMAN,[11] Divorce Advisor[12] or ADVOCATE.[13] ROSS, another headline grabbing legal AI, is essentially an information retrieval system similar to those that power all large legal databases. And while most of the current excitement about AI is driven by developments in machine learning, this too has been trialled in legal settings long before the current interest, e.g. in the hybrid SPLITUP system developed by Zeleznikov and Stranieri.[14] SplitUp combined a rule based element with machine learning that used a neural network to detect patterns in decision from 20,000 first instance family court decisions. This historical reminder is by no means meant to downplay the ingenuity of these modern developments, which succeed where earlier systems failed to reach the market and to have measurable impact. But if we want to anticipate the impact of AI on law, and also want to develop regulatory responses, it is important not just to remember the lessons from the past. We also have to understand what exactly has changed since the 1990s, and which aspects of this change, if any, pose genuinely new challenges.

[9] Fenwick et al. (2018).

[10] The Law Society (2019).

[11] McCarty (1976).

[12] Duguid et al. (2001).

[13] Schafer and Bromby (2005).

[14] Zeleznikow and Stranieri (1998).

1.3 The Law and Ethics of Legal AI

Not only are the technologies that drive the current interest in legal AI rooted in ideas developed during the 1980s and 90s, the discussion on the social, economic and ethical implications of the use of AI in the justice system also precedes current concerns about the "black box society" and algorithmic accountability. The UK introduced fully automated legal decision making as early as 1998, through the Social Security Act, whose section 2 on 'Use of computers' provided that:

> any decision falling to be made by the Secretary of State may be made or issued not only by an officer of his acting under his authority but also '(a) by a computer for whose operation such an officer is responsible.

According to the government introducing the legislation, this meant that "They [sic] will be able to calculate and award benefit in straightforward cases without the need for human intervention".[15]

The measure was controversial, and led to an in-depth discussion of the ethical and social implications of leaving legal decision-making to machines. Oliver Letwin, then newly elected to Parliament, spoke for many:[16]

> What is important here is . . . the fact that the complainant feels even more strongly that he is not dealing with a group of human beings who in any sense understand the human problems that he may be experiencing. He will feel that he is dealing with a monolith that will take on the appearance of a large network of computers producing pieces of paper–sometimes correctly, sometimes not–with never a human hand behind them.

Then and now, there is concern about the factual or legal correctness of the decision, how to identify and prove mistakes, and how to challenge them through a process of judicial review. However, this was only one part of the discussion, and not its end-all. Even perfect computational decision makers, so the opposition, would still pose significant ethical concerns, dehumanising citizens who were already marginalised. The concern was that it is inappropriate for the state to interact with its citizens in this way, even in cases where hypothetically, the quality of decision making is as good as that of human decision makers, or even surpasses them. Or again quoting Letwin, and noting in passing the chess metaphor that likens citizens to mere pawns:

> The long-term effects of this increasing depersonalisation of the Department of Social Security are, and will become, even greater than many imagine. Many individuals receiving or seeking benefit undoubtedly find that this impersonal approach, which contrasts strikingly with the way that they are dealt with by other Departments, including the national health service, or by commercial entities, makes them continue to feel that they are pawns. The ethic of all of us . . .should be to re-integrate as fully as possible into the wider community individuals who receive benefit, by making them feel that they are not pawns in the hands of a monolith. Clause 2 moves significantly in the opposite direction, and runs counter to our perception of the proper way to proceed, as well as to Government policy.

---

[15] Baroness Hollis of Heigham (1998), see also Le Sueur (2016).

[16] Letwin (1997).

By contrast, the contemporary debate surrounding algorithmic decision making in public administration and the justice system focusses almost exclusively on the reported mistakes that AIs have been found to make, in particular when these are perpetuating social biases and prejudices. The proposed solutions, such as the FAT principles for Fair, Accountable and Transparent computing, frame the societal and ethical question of the use of automation in the justice system as a technological problem that can be resolved through better algorithms and more data. As we will see, the public reaction to AlphaGo, within and outside the legal technology community, shows that such an approach might not go far enough. We begin therefore our substantive comparison of AlphaGo and the law with a quote that puts the game into a wider social context;

> In China, Go is not just a game. It is also a mirror on life. We say if you have a problem with your game, maybe you also have a problem in life[17]

These words by the European Go Champion Fan Hui explain some of the extraordinary reactions to AlphaGo's victory over a human opponent. They also help to understand better some of the perils of legal AI. First, we note that there are two different ways the above statement can be read. In the less problematic version, winning at Go is merely a *diagnostic tool* for certain skills that are necessary, but not sufficient to succeed also in other aspects of life. Both aspects are important: first, it is merely diagnostic of an underlying skill, not the skill itself, and second, even the underlying skills are only necessary, but not sufficient for success. If understood like this, there would be little or maybe even no point to train aspiring administrators, judges or kings in Go, even though there might be a case for using Go during their entrance examination.

There is a different way the quote can be interpreted though. In this reading, the skills one learns while acquiring mastery of Go transfer directly to other domains. As Cho Chikun, a Go 9-Dan, put it:

> studying Go is a wonderful way to develop both the creative as well as the logical abilities of children because to play it both sides of the brain are necessary.[18]

Understood like this, the game board can represent anything, from armies to rivers to legal documents. This idea found its artistic expression in *The Battles of Coxinga*, a Japanese Kabuki play by Chikamatsu, in which two Immortals are playing a game of Go. Sankei, the main protagonist, observes them, marvelling how they can be so engrossed in the game. They reply that while to him, the game is but a game, they could see it for what it truly is: the world itself, the opposition of yin and yang in the black and white, with the 361 sections corresponding to the 360 days of the lunar calendar, and the tactics and strategies of the game the same as for the war that is the background of the play: "The ordinary man, confused of mind, takes it for a mere contest between Go stones . . . the fish swimming in the water . . . mistakes

[17] Quoted in Gibney (2016).

[18] https://www.usgo.org/notable-quotes-about-go (accessed 26 Aug 2021).

it for a fish hook ... the bird soaring above the clouds ... is frightened, thinking it a bow ..."[19]

Why does this matter for the discussion of legal AI? First, it points at a crucial aspect of human intelligence. We are indeed able to transfer skills, including intelligence, across domains in often surprising ways. Whilst learning Go alone is not going on its own to turn a student into a good lawyer or administrator, the argument is made that it nonetheless hones certain abstract skills such as pattern recognition, problem solving under stress, planning and self-control.

Artificial Intelligence by contrast remains highly domain specific, "idiots savants" rather than "general" AI. Even the most accomplished Go computer program can't write poetry, drive a car or analyse a court decision. Even the most advanced automated car can't play chess or walk, and AIs for facial recognition would fail simple text comprehension tasks. The same holds true for legal AI. Some of the underlying techniques are to a degree generic—ROSS is an information retrieval system and in this sense agnostic towards the field of law that is researched. But information retrieval is all that it does. Other systems are even more domain specific. The DoNotPay app helps its user to challenge parking tickets, and nothing else, its knowledge of the (limited and highly constrained) relevant law may be exhaustive, but that is all that it knows That includes a lack of understanding that there are other fields of law, and with that comes also an inability to detect if a user is using it for the wrong purpose and needs a different type of support.

This ability to transfer skills across subject areas is also one enabler of human creativity—seeing on an abstract, symbolic level similarities between otherwise radically different domains underpins our ability to generate metaphors and similes, and helps us to find new ways of using old solutions. The greatest danger when using computer tools in legal practice might then be that we project our own capability to move between domains onto the AI—especially if, as is increasingly the case, they come with avatars that mimic human facial expressions and demeanour. The danger is not the "singularity", AI developing beyond the point where we cannot any longer control it, but the inappropriate use of ultimately very simple (but deceptively human-like in outward behaviour) tools for tasks they are not suited for. Because *for us* being good at Go is indicative of deeper skills that transfer to other domains, we assume, wrongly, that this must also hold true for computers.

2 A Case Study in Legal AI

One example where the above tendency to inappropriately project into AI human-level reasoning skills can be seen in the responses to a recent report of a successful experiment to develop an AI that predicts decisions by the European Court of

[19] Keene (1961), p. 130.

Human Rights.[20] This story was also widely circulated in popular science news, and generally generated considerable public interest.[21] Similar ideas have since been developed for French Supreme Court cases,[22] the US[23] and Thailand.[24] Inevitably, while some news outlets reported correctly that the authors did not see this as a first step to replace judges, many others framed theor story through the question if systems like this could one day replace human decision makers,[25] or by using the term "Robojudge" indicated that despite the disclaimer, this was an important step towards emulating judicial decision making. This however highlights a number of misconceptions, which invite us to probe more deeply into the study.

The first, and more minor misconception, is the problematic inference from "being able to predict X" to "being able to do X". I can predict the outcome of a tennis game between Andy Murray and a tennis novice. That does not mean I could replace Murray and achieve the same result. Nor is the fact that outcomes of sport games can to a degree be predicted a reason not to play the game anymore. The playing itself is of as much importance as reaching the decision. Similarly in law, the trial may or may not be an efficient way to establish the truth and come to a correct decision. But as a rule-governed procedure that is open and transparent, and that involves in various forms the citizenry as communicative interlopers (as witnesses, jurors, observer-journalists etc.) it fulfils a function that an Oracle that merely gives the right solution cannot replace. Giving the accused the ability to respond in a structured way to the accusations against them is more than a mere tool for fact finding, it is crucial for a public "holding to account" that at least as an ideal shapes and changes the perception of all those who participate in it.[26]

Next, by framing the technical task as a binary classification problem, Aletra's system predicts either a "complaint upheld" or "complaint not upheld" verdict, but it does not predict which line of reasoning the court is going to deploy. From a legal perspective however, this can of course be the most important aspect of the decision. Indeed, it is often easy to predict *if* the court will accept or dismiss the claim, the interest focusses solely on which grounds the judges are going to do so. For the future development of the law, this can be crucial—a dismissal due to lack of standing for instance will have a significantly different impact on future cases than an analysis that reaches the merits of the case. We can see this even more clearly in the jurisprudence of another court, the US Supreme Court, that also publishes opinions that concur in the verdict, but which cite very different reasons for doing so than the majority. These concurrences can be utilised in future litigation to distinguish a new case, or to entice the court to abandon its jurisprudence. By the

[20] Aletras et al. (2016).

[21] Sayer (2016), Vincent (2016) and Wakefield (2016).

[22] Sulea et al. (2017).

[23] Katz et al. (2017).

[24] Kowsrihawat et al. (2018).

[25] Crichigno (2018).

[26] For a discussion with application to international criminal law see e.g. Wringe (2010).

same token, the AI under consideration does no distinguish between a resounding success for one party and a mere "Pyrrhic Victory" that while technically ruling in favour of one party awards remedies that render the case a de facto loss.

Second, we learn from the paper that the accuracy of the prediction was "79%"—but without any context, this information is all but meaningless. Is this better or worse than an average lawyer? A good lawyer? A layperson? This will become an issue also for the regulatory bodies. At the moment, we are not only lacking standards that tell us how accurate is "accurate enough" for an AI to replace human decision makers, we are even lacking the methodological framework to determine how to formulate such standards. "Being on average as good as or better than a human decision maker" seems initially a plausible criterion, but on closer inspection immediately fails. It would mean for instance to use an AI that in most cases is slightly better than a human, but will make on rare occasions terrible mistakes that no human would even contemplate. Would a responsible law firm hire a job applicant who through their brilliance outperforms all competitors during four days of an assessment week, but on the fifth day, realising their lead is now unassailable, does not show up for work at all, and thus not only loses a case due to a missed deadline, but also exposes the partnership to litigation? This dilemma has also been recognised in the context of Go. Toby Manning Treasurer, British Go Association, commented on the game:

> The thing that struck me, playing through the games you couldn't tell who was the human and who was the computer. With a lot of software you find the computer makes a lot of sensible moves and suddenly loses the plot. But here, you couldn't tell which was which.[27]

Superficially, this reads like a version of the Turing test. But something different is at stake here. It is not so much important that the computer surpasses the human player over a set number of games. Rather, the absence of "catastrophic mistakes" is flagged up as the decisive factor. There are some types of mistakes that are unforgivable, even if the performance otherwise is stellar. For the regulatory bodies, and also for future courts dealing with malpractice cases involving lawyers (mis)-using AIs, this raises several, and so far under-researched, challenges. How proficient must a computer be to count as a permissible replacement (or maybe even an obligatory one) for a human decision maker, and over what period of time should we evaluate its performance? What counts as "good", what as "good enough", and how should we in law classify those cases where a computer that was judged to be proficient enough on average nonetheless gets the wrong result in a specific case?

A look at the academic literature on AI and law immediately points to a number of significant problems. The overwhelming majority of proposed software tools has not been empirically validated, and those few cases where there is at least an attempt being made, it is done in an ad hoc fashion, typically comparing the computer's decisions with those of friends and colleagues of the author, or of their students. There is hardly any discussion of the methodology of evaluation legal AI. Law poses

[27] Cited in Gibney (2016).

here unique challenges. In a medical diagnostic system nature provides "hard" feedback—the patient gets better or not. Law is more malleable than that. If a computer systematically disagrees with human judges, does this necessarily mean the computer is "wrong", and what does this, in law, mean? Or could it be that the computer found the "correct" solution, even though it fails to predict the answer given by the judiciary?

Here a significant role for jurisprudential reflection is needed, since the answer will depend at least in part on what we consider "right" legal advice to be. This in turn depends at least in parts also on the legal culture for which the answer is given. German legal culture e.g., with its formalist leaning and its systematic juxtaposition of "dominant opinion, judiciary" vs "dominant opinion, literature" in education and training might be amenable to the idea that the AI discovered the correct solution through its grasp of the heaven of legal concepts. In the common law tradition with its realist leanings and focus on pragmatically sound advice to paying clients by contrast, this could still be seen as "bad" advice. In some conceptions of the common law, judges create in an ontological sense the law, and therefore can't be "wrong". Rendering a decision is constitutive, not merely descriptive of the law. In this interpretation of the common law, the very concept of a correct legal answer that nonetheless diverges from the one given by the courts may not compute, leading to a very different evaluative methodology.

This is also a concern for Aletra's paper. In its evaluation of the system, it treats all mistakes as ultimately the same. The success rate of 79% does not tell us if any of the cases that were *not* classified correctly were of such a nature that any sufficiently competent lawyer would have predicted the outcome, or if they are all the type of borderline cases, maybe with significant disagreement between the judges, that would have taxed the human expert too. This matters obviously if it comes to issues of liability for incorrect predictions. To remedy this problem, additional analysis would have to be carried out to determine how contested the decision was. If the jurisdiction in question publishes dissenting opinions, or is open about whether the court was unanimous in its decision or split, this could be relevant data to decide just how contested a decision was. Failing that, an analysis of the academic legal literature before the decision was delivered could give indications of the degree to which the outcome was deemed obvious. AIs failing to predict obvious outcomes could then be distinguished from those that only called contested decision wrongly.

Even more problematic is the understanding of "prediction" that informs the paper, and also the subsequent reporting in the press. A layperson hearing that a computer predicts court decisions could be forgiven to think it means something like this: The clients comes to the software operator with a legal problem, express the problem in their own words, and then ask the AI if they will prevail. The AI then predicts an outcome, the case reaches the court, and the AI is then either vindicated or proven wrong.

But this is not how the program presented in the paper works. First, it is not trained on all possible outcomes that can arise if a submission is made to the ECHR. Only decisions that reached the final stage of adjudication, were decided on merits, and published, were considered.[28] This already excludes the majority of ECHR disputes from the training set, as most cases that reach the court are rejected on procedural grounds, without the reasons ever getting published. But this means the system is suitable, at best, only for a small subset of all possible cases—those where the fact constellation is such that they require and permit a discussion of the merits of the case. For all other cases, it should not be used at all. They are not the type of fact constellation the system has been trained for. But this means that a user has to predict (significant parts of) the outcome of the court's likely reasoning in order to decide if they can use the system to predict the outcome of the (rest of) the court's reasoning—not a particularly convincing value proposition for real world use.

Nor is it quite correct, or rather again overly anthropomorphising, to say that the AI learns from the statement of facts in past cases and from this calculates how likely a specific outcome of a future case is. Rather, the sections of text from the decision that are marked as "fact" are treated as "bags of words" (BOW), that is multisets that disregard grammar and word order but keep multiplicity, or in other words, a count of the number of times a word appears in a text. The "bag of words" of Art 3 of the ECHR, "No one shall be subjected to torture or to inhuman or degrading treatment or punishment", would look like this:

{no:1, shall:1, degrading:1, or:3, inhuman:1, punishment:1, to:2, treatment:1, torture:1, one:1, be:1, subjected:1}

The order of the entries does not matter, and the same BOW would be created had Art 3 said instead "...subjected to inhuman or degrading treatment or punishment or torture...". To avoid giving frequent but uninteresting terms such as "to" and "or" too much weight, these are normalised, that is they are assigned a weight that is inverse to its document frequency, i.e. its occurrence in the entire corpus. The paper uses a specific, somewhat more sophisticated form of bags of words, n-grams. N-grams preserve to a degree word order when n is larger than 1 ($n = 1$ yields the outcome above). For $n = 2$ for instance, the text of Art 3 would create a BOW: { "no one":1, "shall be":1, "subjected to":1, "torture or"....}where again order of the elements does not matter, if $n = 3$ the BOW becomes {"one one shall":1, "be subjected to":1, ...}. The paper uses groups that are up to four words long. To see the importance of this, simply consider that "John killed Peter" and "Peter killed John" produce the same BOW if $n = 1$, but not if also $n = 2$ is considered. This means that a lawyer, reading the facts of a case, will understand considerably more of its content than the analysis carried out by the AI. How much more the lawyer understands depends also on the decision which value is chosen for "n".

[28] Even this is not quite true. Only cases were selected where the decision is available in English. This excluded a significant number of cases where only French rulings are available. It is difficult to say if this affects the validity of the result—this would in particular be the case if the choice is not random, but driven by features of the case, eg. In which jurisdiction the matter first arose. It would equally be a possible source of distortion if the choice of language were dependant on the composition of the court, the nationality of the judges who hear a specific complain and who may bring the methodology of their home jurisdiction to the table.

The analysis then asks which, if any, of these n-grams correlate positively with either "violation occurred" or "violation did not occur". If we now look at the list of n-grams that the program correlates with judicial outcomes, some unexpected entries can be found. For Art 3 for instance, terms such as "overcrowding", "police officer" and "beating" unsurprisingly correlate with finding a violation. More unexpected is to find in this list also "June" and "Ukraine". For lawyers, it would seem immediately clear that it *should* not matter, as a normative or legal-doctrinal statement, whether the violation occurred in June or April, in the Ukraine or the UK. So what could we make of this finding?

The first, and probably most intuitive, approach would be to consider this simply a mistake, a statistical fluke and further proof of the old adage that correlation does not entail causation. Maybe of the relatively small number of cases on Art 3 that were decided on their merits, disproportionally many originated in the Ukraine. Or maybe, Ukrainian cases are more likely to be published in English. Of course, we would assume and hope that if a similar constellation of facts were to arise in the UK, the outcome of the trial would be just the same, and the correct AI-prediction too would be the same as if the case had come from the Ukraine. This indicates a significant difference between machine learning in tasks such as playing computer games. Positions and moves in chess or Go are reducible without loss of information, to mathematical structures. "Qa2 to a3" tells me all there is to know about the move of the Queen. This is very different for text, here significant information content is lost, and what would be obviously and immediately identifiable as an absurd outcome for a human is still a possible outcome for the AI.

On closer analysis, things get even more complicated. Interpreting the connection between "Ukraine" and "winning the case" as a statistical fluke is not the only possible interpretation of what happened here. This issue brings us back on one hand to the issue of "correctness" of legal AI applications, on the other it prepares the ground for the discussion below on computer creativity. Maybe the AI spotted a real, if from a doctrinal perspective extremely worrying, pattern. Maybe there is indeed hostility towards some countries by the judges. Maybe, as "digestive jurisprudence" suggested, even the month a case is heard can matter—maybe judges are more plaintiff friendly if the weather is warm outside and they want the deliberation to be over quickly for a nice round of golf. In this case the AI would have found a novel and unexpected pattern, it was creative rather than merely mistaken.

One difficult with legal AI is that it might be more difficult than one would think to distinguish these two possible interpretations. Should a "good" lawyer advice a client that even though abstractly, their case is sound, they should not make the significant investment in terms of money and time, because they are not Ukrainian and hence their chances limited? Or should the lawyer advice only on what the law, to the best of their knowledge, demands?

In practical terms, both interpretations are worrying were such a system to be used in practice. Even if we do not replace judges, but merely use AI to help lawyers in advising clients about their success chances, or as a triage for the court to allocate resources (as the authors of the paper suggested), this would mean that fewer cases from countries other than the Ukraine would be heard. Apart from the highly problematic violation of the principle of equality before the law, this would also mean that the training set gets progressively more skewed in favour of Ukrainian cases, enhancing the predictive value of "Ukrainian" in the statement of facts simply because the very same system is now used to further filter cases.

The final, and arguably biggest problem with the paper though is that the computer program did not really predict the outcome of cases, not even of those that had been admitted to a full hearing, though this was frequently the way it was reported in the popular press. What it did was to learn and analyse the fact statement at the beginning of a written judgement, and from that fact statements it "predicted" the outcome of that very same decisions. Now, this would look like a genuine "prediction" if and only if in a court deliberation, the statement of facts is written first, and independently of any legal analysis, and the judges then use all and only these facts and their legal knowledge to reach the conclusion. But this is of course not how court decisions are drafted. When the judgement was written, the decision had been already taken. This means in particular that mainly those facts that support the decision will be listed or at least emphasised, others will be dropped. Nor are the party submissions neutral statements of facts—by the time they reach the ECHR after a long procedural journey, they will have been recast and written specifically with an eye towards the applicable law, based on the legal theory of the lawyers for their respective parties.

Aletras et al. recognise this problem and acknowledge it as a potential shortcoming, but ultimately consider it unlikely to have a bearing on their analysis. In their view the statement of fact in the court's decision simply reports the pre-trial submissions by the parties. They argue that the duty of the court to be impartial is enough to ensure that the statement of facts is an objective rendition of the problem space of a legal dispute. They also argue that the limited ability of the ECHR to engage in its own fact finding will make the statement of fact a good proxy for the facts as submitted by the parties. But it remains unclear why either should prevent dropping or emphasising/selecting those facts that align eventually with the verdict. The problem is not that the judges are dishonest or cheating. Rather, the statement of fact is inevitably theory-laden, Aletras' analysis assumes an ontology whereby some "objective fact" out there is represented unambiguously in the language of the court. This ignores the work that goes into the "framing" of the question, how facts are constructed in the legal process, and the skills and knowledge required for what German lawyers call the "subsumption" of facts under rules.

Take the classical example of Hart's concept of "penumbra of doubt", the rule that prohibits vehicles in the park. Assume the legal issue is if an electric wheelchair qualifies as vehicle for the purpose of the law. The following are different statements of the same fact:

(1) A powered wheelchair was driven into the park
(2) The defendant drove his wheelchair into the park
(3) The defendant visited the park in his wheelchair.
(4) The disabled defendant visited the park with the aid of his wheelchair

There is no "fact of the matter" that makes any one of them "more true" than the other. Yet intuitively, we might expect that if we find 1 or 2 in a statement of fact, the verdict is likely to go against the defendant, whereas 3 (and even more so 4) looks like an opening of an argument that is likely to distinguish vehicles from wheelchairs.

Judges writing the decision will at the time they formulate the statement of facts already know how they decided the case, and select accordingly between the possible formulations. This threatens to make one of the main claims of the paper, that decisions are highly fact specific, trivial. The AI "predicted" something that was "jury rigged" from the word (alpha)Go. The authors of the study acknowledge most of this explicitly in the paper, however, the subsequent news reporting consistently did not pick up on this crucial point, including the reporting directed at legal practitioners.

From this we can take an important lesson. One key skill lawyers of the future will have to acquire is to critically interrogate systems that purport to assist them in doing their job. What does a success rate of 79% mean? What is it benchmarked against? What types of mistake does the algorithm make, and what are the implications for each type of mistake? What data was it trained on, and how was that data selected? In the Aletras paper, only three articles of the ECHR were analysed, Art 3, 6 and 8. Should the user feel entitled to then also apply the same approach to any of the other provisions? Does it matter that only decisions published in English were used, does it matter that cases that were rejected on procedural grounds were not used at all?

Most importantly, and most difficult to teach, this requires also a degree of self-reflexive detachment. The reason that many legal commentators were impressed by the capabilities of the ECHR algorithm is that writing and interpreting court decisions had become second nature to them. This means that all the separate cognitive steps that are necessary to resolve a legal dispute had become habituated and internalised. In the case at hand, these were the crucial decisions that went into "framing" the question and selecting the relevant facts from of the totality of all submitted information. The mistake then became to think that an AI that seemingly performed the same task as a human lawyer must have done so using the same cognitive steps. In reality though, the AI simply piggy-backed on the cognitive labour of its users, making the human input invisible.

With this we are back at the problem that started this section. Just as we tend to project domain independent intelligence onto an algorithm that is good at Go or Chess—because humans who are good at Go or Chess typically have generic cognitive skills—there is a danger that we also project generic legal skills onto an algorithm that seemingly replicates a specific legal reasoning task merely because solving that task, *if performed by humans*, would be an indicator of generic legal skills.

3 Symbolic Meaning, AI, HCI and the Law

We can now return briefly to the "founding myth" of Go, and the way the game is used in the story above to symbolize the art of war. Seeing the game as more as it is can also be understood as a testament to our ability to imbue everything with symbolic meaning, and ascribe in this way also intrinsic value to otherwise meaningless activities. We can project battles and landscapes onto a game board, and playing the game then acquires a different type of significance. In human cognition, seeing things through symbols seem to be intrinsically linked to advanced cognitive tasks—something that in our evolutionary past seems also to have played a key role in the emergence of normative systems such as the law.[29] The practice of law consequently is imbued with symbols and symbolic meaning. We cannot entirely separate these symbolic performances, the robed judges or the solemn oath, from the substantive content of the law.

Trials in particular, as a public holding to account, are about more than merely getting the right decisions, they are performative acts where the community reaffirms publicly its values, gives the accused a forum to explain themselves, and hold them if necessary, to account.[30] Replacing human decision makers by algorithms changes also the performative role of legal procedure. This I argue was also at the heart of the earlier concerns about algorithms in public administration that we saw in the UK. Whether the software took the same decision as a human adjudicator, or maybe even a more consistent one, is secondary to the change in symbolic meaning that comes with automation. When a petitioner is only interacting with a machine and not any longer with humans as representatives of the community, a statement is also made about the importance the government gives to that person.

We do not yet know how exactly, if at all, our symbolic understanding of legal decision making will change if performed by machines. We might end up with the world of La Mettrie that intentionally and visibly strips some of the "excess meaning" from legal transactions and processes. The machine-like nature would then not be a bug, but a feature, and possibly converge with other trends to replace traditional legal intervention with science-based ones. One could think in this context for instance of the medicalisation of crime and the mooted replacement of criminal sanctions with medical interventions, accelerated by the recent surge in "neurolaw" and advances in neuroscience.[31] AI in medical diagnosis has avoided many of the pitfalls that have made legal AI a challenge, with Mycin one of the early medical expert systems that was developed at the same time as early rule based legal expert systems.[32] The current machine learning based approach too counts medical diagnosis as a success story. La Mettrie's dictum that distinguishing right from

[29] See e.g. Alcorta and Sosis (2005); Holloway (1981); on the role in the evolution of normative order see Arnhart (1998).

[30] See e.g. Duff (2003).

[31] See e.g. Zeki et al. (2004).

[32] Buchanan and Shortliffe (1984).

wrong should be as easy as to diagnose something as blue and not yellow could then become true, but not because AI improves to the level that it can replace human judges *as we currently conceive their role*, but because we changed our expectation of what we want the legal system to achieve. Legal AI then becomes the stalking horse of a very specific conception of justice, turning what should be a contested public debate about the values of law into a technocratic decision of what is computationally possible.

In another possible future, we might learn to project the same symbolic meaning that we ascribe currently to human action also onto a machine. The avatar with wig and robes or the virtual replication of courtroom design could then create a continuity in symbolic perception between the old world and the new. This however would further enhance the danger of overestimating the capabilities of the AIs. It would also need a public debate to decide if this type of deception is legitimate.

Finally, we might develop an entire new "legal mythology" that creates new forms of symbolic meaning to help us integrate automated decision making more seamlessly into its societal context. Already, we see how some of these "new mythologies" borrow explicitly from old narratives—it is not by coincidence that we reuse words such as "avatars" or "oracles" in computer applications, rather, these are ways in which technological change can be mediated and imbued with shared symbolic meaning. Staying with the specific example, we could foresee a future where the robojudge does not mimic a present-day judge, but the Cassandra of old, delivering judgements from the (computational) cloud. The ubiquitous "Minds" in Ian Bank's Culture novels and similar fictions could provide other blueprints.

The field of study that concerns itself with the way in which humans make sense of technology with a view on designing tools better suited for the way we think is Human Computer Interaction (HCI). HCI began to emerge as a distinct field of computer science research in the 1970s, the same time when we also find first attempts to build legal AIs. It became a household name in 1983 when Card, Newell, and Moran published their book *The Psychology of Human–Computer Interaction*,[33] which remains the standard work in the field. HCI combines insights from computer science, design, media studies, anthropology and sociology among others to elucidate how individual users interact with digital computational artefacts. If we look however at the topics discussed at the two main conferences on AI and law, JURIX and ICAIL, over the past 25 years, HCI papers are noticeable only though their absence. Some of the more recent commercial applications will deploy some generic HCI techniques, simply because they are by now a standard component of most computer science degrees, but there is until now no research into HCI that is specific to legal applications. Such a field of study would draw on the rich jurisprudential and socio-legal literature that contextualises law in its social context, but with a view to determine how exactly human users give meaning to legal AI tools, and adjust their design accordingly. Is it for instance a good idea, or is it positively dangerous, to give legal AIs human looking avatars that openly display symbols we normally associate

[33] Card et al. (1983).

with the legal profession? Will users assume that the AI is subject to client confidentiality, for instance, and could that be in all cases guaranteed? This is the second lesson we can learn from the public reaction to AlphaGo: To educate a new generation of legal professionals that can work together with software developers, an understanding of the basics of HCI seems much more important than coding skills. Thinking about how people experience their interaction with law *now*, and how a new tool is likely to change this, brings a specific type of legal expertise to the table that lawyers are best placed to contribute.

4 Creative and Empathic Legal AI

Our discussion above already touched upon the different attitudes towards creativity and empathy as potentially "irreducibly human" features of legal decision making. A good lawyer, and most certainly a great one, must have more than competent knowledge of the rules, they must also have an ability to understand the needs of their clients (empathy) and the ability to find if necessary non-obvious or novel ways to achieve the client's objectives (creativity). One response to the perceived threat to jobs in the legal profession is therefore not just to introduce coding skills into the curriculum, but to focus on "soft skills" such as empathy and creativity as market differentiators for human legal advice. This assumes first as we noted above that these are indeed legal goods—and in the juxtaposition of Mettrie and de Clapiers, we have seen how for a long time, opinions on this have been divided. Second it assumes that AIs are indeed incapable of creativity and empathy, something we are now going to probe further.

4.1 The Robot as Legal Metaphor

As we discussed in the introduction, control over emotional responses has been historically identified as one of the skills that the teaching of Go can bring. Fan Hui, the first Go champion to be beaten by AlphaGo without a handicap,

> compared Alpha Go to Lee Changho, who is famous for not showing any emotion when he plays, just as AlphaGo does not attribute any emotional characteristics to the moves.[34]

This quote opens up the question of emotion and empathy in legal reasoning, and in cognition more generally. Proponents of the use of automated legal decision making have consistently extolled the virtue of neutral, unemotional and hence objective decision making by machines.[35] On the few occasions that AI and law

---

[34] Cited in Koshkina (2016).

[35] Notably and with further references Eggers (2007).

research has addressed the issue, it has echoed the sentiment expressed in the above quote. The SplitUp system mentioned above, and the online dispute resolution tools that it inspired, used the absence of emotional involvement by the AI as an important advantage. It is meant to facilitate non-adversarial conflict resolution by "emotional de-escalation", forcing the parties to become much more explicit about their desires, preferences and plans in their communication with the AI. This then generates the type of input with which a computer can work. Similarly, for automated sentencing support tools the often highly emotionally charged context of sentencing, can lead to inconsistent decision making that computer tools may be able to remedy.

Instructive in this context is a statement by Sonia Sotomayor, who in her testimony before the U.S. Senate Judiciary Committee stated that

> We're not robots [who] listen to evidence and don't have feelings. We have to recognize those feelings and put them aside. That's our job. I think the system is strengthened when judges don't assume they're impartial.

Here, the cold, robotic decision is an unachievable ideal, but as human judges, what we can do is to identify our emotions—and then transcend them and "put them aside".

Robots and AIs are indeed frequently depicted as emotion-free, cold and rational, a trope that also found its way into the law. In the US case of Keppel v. BaRoss Builders for instance, the decision states that:

> Above all, it showed that a judge is a human being, not the type of unfeeling robot some would expect the judge to be.[36]

In this quote, we find an echo of de Clapiers' dictum that one cannot be just without being humane. The evaluation now flips in the opposite direction, the emotion-free robot is a dystopian, not utopian decision maker. Justice requires the intuitive and empathetic recognition of oneself in another, a recognition of shared humanity. The golden rule is ultimately an expression of such a form of direct, personal relation with another human being, putting oneself into the position of another and in this way treating them as an equal moral agent.[37]

If empathetic judgement is a necessary, irreducible and desirable aspect of at least some legal decision making, several options for legal technology arise. One is again a separation of responsibilities. We should identify those fields of law where we value empathetic judgement, and reserve it for humans. For the law curriculum, this would mean a much stronger focus on "empathetic reasoning", emotional intelligence and similar soft skills that computers are less likely to replicate.[38] The importance to teach empathy in this way has been recognised for some time in medicine.[39] It remains however the exception in legal education. The need to compete with computers may be the reason needed to push law in the same direction.

---

[36] Keppel v. BaRoss Builders, Inc., 509 A.2d 51, 56.

[37] Veitch (1998) and Bańkowski and Schafer (2007).

[38] See for examples Douglas and Coburn (2009) and Rosenberg (2001).

[39] Batt-Rawden et al. (2013) and Stepien and Baernstein (2006).

This focus on soft skills will also entail an ability to listen patiently and attentively, to de-escalate conflicts and to create space for the parties to reflect about their objectives and desires, to "give them time". Again a quote, this time by Toby Manning, who refereed the AlphaGo v. Hui is instructive:

> The one thing that was not human was the way it managed its time. Fan Hui took longer over his moves than AlphaGo[40]

In legal AI too, the speed advantage of AI over human decision makers has been a key selling point. TrademarkNow for instance advertises its services by pointing out that a trademark search will now take seconds, where in the past it took human experts days if not weeks.[41] A recent competition between human lawyers and AIs on who is fastest in document review also had the AI as clear winner.[42] As justice delayed is justice denied, there is also a good ethical case to be made for using AIs to speed up the legal process.

However, reduced transaction costs and increased speed can often have unexpected consequences. The reduced costs and increased speed of email in comparison to letters increased considerably the use of written communication even for trivial issues, and eventually gave us spam.

We find first indicators of similar dynamics with technology-enhanced legal practice. "Speculative invoicing" is the practice to send threatening letters to a large number of merely potential violators of copyright law or other regulations, in the hope that as long as even a small number pays up, the law firm makes a profit. This means also that it does not matter any longer if the case against any individual in the targeted group is weak or the accusation even false. What makes this potentially profitable is the reduction in cost that automated document review and generation of legal letters brings.[43] while high cost and low speed of litigation can be an unjust access barrier to justice, especially for economically disadvantaged groups,[44] the experience with email shows that reduced costs can simply drive up demand, with even higher costs accumulating elsewhere in the system. Not making it too easy to sue one's neighbour could in that sense be a contribution to social peace.

Careful analysis of the regulatory environment, including the role of professional bodies and the enforcement of ethics standards, will be needed to address these side-effects of more efficient computational administration of justice, something that "training for patience" could assist with.

But if we want to teach empathy, patience and good listening practices, we need to rethink more than merely the content of the law curriculum. Shortened degrees, modularization and constant assessment which drive at least in the UK a desire for more "efficient" legal education are all inimical to such an understanding of the role

[40] Cited in Gibney (2016).

[41] https://www.trademarknow.com (accessed 26 Aug 2021).

[42] https://www.lawgeex.com/resources/whitepapers/aivslawyer/ (accessed 26 Aug 2021).

[43] Wakefield (2015).

[44] Rhode (2004).

of the legal profession. They ultimately force humans to compete with computers where the latter are at their strongest.

4.2 Legal AI with Soft Skills

4.2.1 When Time Is of the Essence

So far, we have taken it for granted that AI is intrinsically incapable to replicate soft human skills such as empathy, and hence a separation of responsibilities between AI and humans is possible. There is however another response possible. Staying for a moment with the question of speed, the potential problem that AIs are so fast that human decision makers get overwhelmed has been recognized in other context, and again gaming led the way. To give an example, the programmer's annotation to Apple's chess player revealed that the answer of the machine was artificially delayed:

```
// Paradoxically enough, moving as quickly as possible is
// not necessarily desirable. Users tend to get frustrated
// once they realize how little time their Mac really spends
// to crush them at low levels. In the interest of promoting
// harmonious Human - Machine relations, we enforce minimum
// response times.
```
[45]

This is indeed by now a common strategy, and artificial delays like this are ubiquitous in the design of chatbots, games and other interactive tools. Temporality and how it affects user engagement with technology has been studied extensively in HCI and give rise to a rich body of literature.[46]

For legal AI, Diver made the argument that an "affordance of delay" will be necessary to preserve the normative role that delay can play in the law—from "cooling off periods" in contract law to the requirement of trial separations before a divorce is granted.[47] This however goes well beyond small delays in response time to make interaction with a legal chatbot such as Josef[48] slightly more natural. Rather, it asks hard questions about the way in which legal AI design can support or undermine legal ideals in often less than obvious ways, through design features that are not directly related to legal knowledge representation, and through their cumulative effects on users rather than necessarily aspects of an individual application alone. How we as emotional and empathetic beings interact with legal

[45] https://opensource.apple.com/source/Chess/Chess-300.22/Sources/MBCEngine.mm.auto.html (accessed 26 Aug 2021).

[46] Hallnäs and Redström (2001), Marsh (2016), Odom et al. (2018) and Sebald (2020).

[47] Diver (2020).

[48] https://joseflegal.com/product/ (accessed 26 Aug 2021)—Josef too is marketed with the promise of "speeding up" legal services.

technology requires therefore a new type of interdisciplinary research that builds on HCI, but also draws on legal theory and socio-legal studies. Temporality and the "speed" of legal AI, taken as an undisputed benefit of legal technology, becomes in this vision an important aspect that will shape the way in which computers and human users will co-create justice in the future.

4.2.2 Empathetic Legal AI

Thinking about the temporal aspects of legal AI was a first step to recognize the often unexpected normative salience of design choices in Legal Tech. By suggesting to replace the logic of "efficiency through AI" with a paradigm of slow computing, it potentially creates space for human users to bring their soft skills and intuitive, empathetic and creative skills to the task, We can now ask if the alleged opposition between human and computational reasoning modes can't be diminished even further, discuss how an "empathetic" legal AI would look like, and if such a novel approach to legal tech design would be possible and desirable.

Research into computational empathy or "affective computing,[49] the correct reasoning about and response to emotions, has emerged as a key research theme especially in care robotics.[50] "Emotion mining" and sentiment analysis, identifying and classifying not just the content of a text, but also the quality and strength of the emotional attachment of the writer to the opinion expressed in it, has become an even more mainstream research field in information retrieval.[51] There has been so far little or no interaction between researchers in legal AI and affective computing research, explainable by the formalist paradigm of traditional AI and Law approaches that we encountered above. Some applications seem straightforward. We can e.g. analyze the correlation between emotional commitment in party briefs and success at court, the same way in which Lex Machina currently analyses the correlation between specific legal arguments and success.[52] A legal chatbot could analyze if the user is in so much distress that a referral to a human lawyer should be proposed, not because of the legal difficulty of the question, but because the client may need support other than a mere legal answer.

More adventurous would be the question if we should imbue legal advice tools with emotional skills. This could mean training them to recognize the emotional state of their user-clients and include this in their advice, warning e.g. against the overly aggressive wording of a complaint. Going even further, they could be trained to display an appropriate emotional response, showing compassion to the user-client, or express revulsion in a guilty verdict. Such strategies are explored in assistive

---

[49] Tao and Tan (2005).

[50] See e.g. Tapus and Mataric (2007).

[51] See e.g. Yadollahi et al. (2017).

[52] Moses and Chan (2014).

robotics research,[53] but have also raised considerable ethical concerns.[54] A similar discussion in legal AI is so far missing, but seems both necessary and desirable. It would force us to reflect what role and values we assign to emotions at various stages of the interaction between citizen and the justice system, and in that way also alert us to what we may lose though automation.

In section 1, we discussed the response of the UK parliament to the automation of social benefit decision making, and the danger that this could further alienate vulnerable people. If instead of a simple rule-based system with web interface, they had been given a human-looking avatar that in addition responds appropriately to their expression of anguish or concern, would this have been an appropriate technological solution to a social problem, or rather a form of deception that in reality enhances rather than reduces the message that (certain types of) claimants need not be taken serious as moral agents?

The question of empathetic judgement, creativity and analytical reasoning skills is complex and within the cognitive science community highly contested. Any discussion in detail would go beyond the scope of this paper.[55] It allows us however to recast traditional jurisprudential concerns about the role of compassion, mercy and forgiveness in legal reasoning. Certain modes of reasoning that are institutionally embedded in most legal systems, for instance the concept of amnesty, sometimes treat empathy as a "necessary deviation" from normal practice,[56] sometimes as an irreducible structural aspect of adjudication.[57] For our purposes, this means we can reconnect the discussion about empathetic legal AI with the difficulty to decide what it means for a decision or legal argument to be creative and novel, as opposed to being plain wrong. Similarly, once modelling of empathetic reasoning becomes a possibility, we need to ask whether empathetic legal judgement is a general good, a necessary but ultimately regrettable corrective, or an unacceptable deviation from the rule of law ideal. Even though the answer may influence the way in which we design legal AI, these are not decision we should leave to programmers and developers. They need a jurisprudentially informed discussion within the legal community and democratically validated decisions by legislators.

4.3 Creative Legal AI

Empathy and creativity are correlated skills in humans. Empathetic reasoners are also more likely to be creative.[58] As noted above, we can't assume the same for AIs

[53] See e.g. Pereira et al. (2010).

[54] See e.g. Bryson (2017).

[55] See e.g. Carlozzi et al. (1995); for legal reasoning see in particular Hu et al. (2015).

[56] Christodoulidis (1999).

[57] Nussbaum (2017) and Feigenson (1999).

[58] Carlozzi et al. (1995).

as highly domain specific tools, so techniques to imbue them with empathetic reasoning skills are not also going to lead to creativity. This means that to answer the question if teaching creativity in legal education will make lawyers of the future "automation proof, we also have to understand what it could mean for a computer to be creative.

On this topic too, the analysis of AlphaGo can prove helpful. What made many commentators argue that AlphaGo was indeed a watershed moment in AI was an apparent display of highly innovative and creative problem-solving skills.[59] In game 2, AlphaGo made its by now famous "move 37", a move that no human ever would have considered, wrongfooting its opponent and winning the game. Let us quote four immediate reactions, two from professional Go players who were commentators on the game as it unfolded, the other two by Fan Hui, the three-time European Go champion who lost five straight games to AlphaGo.[60]

> That's a very strange move,
>
> I thought it was a mistake,
>
> It's not a human move. I've never seen a human play this move,
>
> So beautiful. So beautiful.

The quotes, taken together, invite us to reflect on what it means for a computer program to make a mistake. AlphaGo had calculated a 1/10,000 chance that a human would make that move. From the observer perspective, this could mean two things: that the machine simply made a mistake, in fact acting "crazy". Or that it displayed the capacity for genuine creativity. In the narrow context of a rule-based game, what allows us to decide between these interpretations is ultimately a post-factum analysis. The move was successful. It did not break any rules. AlphaGo won, hence it cannot have been a mistake. What makes the move "right" is therefore not an intrinsic characteristic of the move, but of the move together with the response of the human player to it. A least to a certain degree, we see here again a problem of projection—we make sense of an unexpected move by thinking of it as creative rather than random or plain wrong, and the difference between randomness and creativity is very much in the eyes of the beholder.

This same issue can be found discussed in the literature on computational creativity. Some of the oldest methods to create new and unexpected works used a random element. Indeed, the mechanical or automated generation of works of art using a combination of random throwing of dice together with simple composition rules preceded computers by several centuries.[61] In his seminal paper of 1950 that gave us the idea of the Turing Test, Turing speculates briefly about the capacity of random processes to combine with rules in the production of true novelty, and also anticipated the objection that we just saw—that the perception of novelty under these

---

[59] See e.g. Metz (2016a).

[60] All quotes from Metz (2016b).

[61] Schafer et al. (2015).

conditions is an interpretative act of the human user rather than an intrinsic feature of the machine or its output. More recent computational creativity research has moved beyond identifying randomness, or merely unexpected outcomes, with proper creativity. The target is not to give up on the very idea of computational creativity, but rather a aspiration to come up with two things, a better understanding of what we mean with "creativity" and also better computational tools that model this new understanding of creative endeavours.[62]

For legal AI, this poses a number of challenges. One, as we saw above, is the issue of liability and fault. To the extent that legal AIs have been evaluated at all, this was typically by comparing the outcome to that predicted by human experts.[63] Systems that make a "Move 37" would inevitably be deemed faulty, not ingenious, in this approach.

A resulting lack of creativity as a design feature may or may not be a desired result. In law, concepts such as the idea of binding precedent express a sentiment similar to the underlying philosophy of machine learning, that the past is a reliable guide to the future. Lon Fuller, in his *Eight ways to fail to make law* makes a broad jurisprudential argument for the need of stability in law.[64] In this sense, there is convergence between legal values and properties of the technology.

However, the inherent conservativeness of law would be further amplified and entrenched if significant amounts of decision making were delegated to systems developed within this paradigm, consistently reinforcing the past rather than challenging it. We encountered this problem above in our discussion of "Ukraine" as possibly misleading predictor of success at the ECHR, and how the otherwise irrelevant fact that most of the past successful cases originated there could mean a reduced chance of future successful cases from elsewhere.

More realistically, we are likely to restrict legal AI to those applications and contexts were creative problem solving is unnecessary or unwanted. Traditional legal AI systems typically targeted "low cost, high volume, low difficulty" scenarios. Richard Susskind's influential study on the future of the legal profession follows the same logic: the future will see legal routine tasks commodified and carried out by AIs, there will also be room for "bespoke" services that require, and allow for, creative problem solving.[65]

There are substantial implications of this model for legal education and training that have not yet been fully appreciated. If "creativity" is going to be one of the unique selling points for bespoke, human-delivered legal services, then it should also play a much more dominant role in legal education.[66] In exchange, all memory-

[62] See e.g. Colton (2008).

[63] See e.g. Stranieri and Zeleznikow (1999).

[64] Fuller (1964).

[65] Susskind (2017).

[66] Which is not straightforward, see Menkel-Meadow (2001). There is a substantial body of knowledge on teaching creative problem solving in the natural sciences, and of course long experience with creative writing education, but comparatively little on legal education.

intensive tasks, should be drastically reduced. It is unlikely that the lawyers of the future will ever find themselves cut off from instantly searchable knowledge repositories, including databases of precedents, so wasting time memorising these does indeed seem an increasingly pointless exercise, and may have to go the way of other skills once crucial for a lawyer, such as preparing a swan feather for writing.

So far, we have discussed two reasons why legal AI may be "non-creative by design". Yet as we saw, AlphaGo was successful precisely because it (seemed to have) found a genuinely novel solution to a problem. Even more so than computer games, advances in computational creativity research call the assumption of the necessarily non-creative computer into question. There has been very little systematic exchange between computational creativity research and legal AI, arguably also for the jurisprudential reasons outlined above. But by putting computational legal creativity at the centre of a future AI and law research agenda, we may not only be able to develop new tools, it can also help us to rethink the relation between creativity, correctness and randomness in law.

Randomness as noted above has been used to mimic artistic creativity, and also enabled AlphaGo to learn more efficiently. Could we conceive of legal AIs that use randomness strategically? Neil Duxbury has documented the use of random processes in the law,[67] including jury selection in common law systems, so the idea might not be as far-fetched as it initially seems. But random jury selection only impacts on the way the decision maker is constituted, not the decision itself. Could we imagine however a sentencing hearing for example where the law prescribes merely an upper and lower limit of punishment for the offence in question, but the precise value is determined by a lottery? Would we accept if the validity of a contract were to be determined by the toss of a coin?

One possible justification for the use of randomly generated decisions would be based on an argument inspired by the theory of evolution. Humankind survived pandemics such as the Spanish flu because not everybody's immune system followed the same script, and random mutations, mostly harmful but sometimes beneficial, created eventually a resistant pool. To prevent machine learning systems to entrench the past and as a consequence deliver sub-optimal solutions once the environment changes, we may need the equivalent of random mutations that can temporarily unsettle the system, only for it to achieve through selection, a new, stable equilibrium. Introducing a random element in the AI decision making process could achieve such a more dynamic and adaptable system. Still, the very notion that the outcome of a legal case could depend on random elements raises issues of justice that would need further analysis. In particular, it would mean to sacrifice for the benefit of the *system* (i.e. its resilience and adaptability) justice in *individual* cases, where a random element would create both "bad" and "good" mutations".

[67] Duxbury (2002).

4.4 Embracing Bias?

A very different way to think about creativity distinguishes it at least partly from notions of novelty and unpredictability of outcomes. A great artist will succeed in externalising subjective, internal states. These in turn will have been shaped by personal experiences that will have to a certain degree been typical for their time, class, social environment, upbringing etc. They are not random the way a random number generator is, but "unique" in the sense that numerous unplanned individual events create life histories which in turn have shaped individual characteristics and traits. We are all different also because we each experienced different things. We may not be able to *predict* the painting an artist is going to create from the knowledge of their social environment and upbringing, but we can *explain*, to a degree, why and how specific choices they made were influenced by these experiences.

This idea echoes Sotomayor's notion that good judges are aware of their humanity and the way their personal experiences have shaped their world-view—but can self-reflectively cast them aside when this is appropriate. AIs do not have personal histories, no social context that formed their individuality and personality, so no need for this self-reflective correction—or so it is often portrayed. However, recent experiences with machine learning systems in the justice sector have called this assumption into question. The "racist sentencing algorithm"[68] whose bias against defendants from ethnic minorities made headlines in the US gave a major impetus to research into algorithmic bias more generally.[69] The reason that the Compass algorithm discriminated against black defendants was in part that the human-generated data had in turn been biased, reflecting the views of past human decision makers but stripped of their historical, social and political context.[70]

We face as a result a dual problem: AIs can acquire implicit biases from the human-generated data that they learn from, but they are lacking the self-reflexive ability to identify their biases and "cast them aside". At the same time, external observers still tend to imbue computers with super-human neutrality and detachment, making them less likely to challenge the outcome of machine decision making.[71]

Currently, "de-biasing" AI is an important and active field of research.[72] However, it is also an approach particularly prone to the danger of proposing simple technological solutions for complex social problems, and in the process potentially hiding rather than preventing bias.[73] But what would happen if rather than eliminating, we were to embrace bias, as a necessary condition for an adaptive legal system and legal creativity? By bringing diverse viewpoints to the table, based on different

[68] Angwin et al. (2016).

[69] Larsen et al. (2016); see also Huq (2018).

[70] Osoba and Welser IV (2017).

[71] Citron (2007).

[72] Bolukbasi et al. (2016) and Mehrabi et al. (2019).

[73] See e.g. Gonen and Goldberg (2019).

personal experience, new perspectives can be opened up. This positive role of "biases" or "subjectivity" mirrors how we characterised artistic creativity above. In an interesting way, we can see this also being played out in law, an in particular the way in which law has deployed the metaphor of the "robo-judge". In Pennsylvania v. Local Union, the issue of potential bias by the judge was raised who, like the plaintiff, was a person of colour. In dismissing the objection, judge Higginbotham ruled:

> Thus by the subtle tone of their objection, they demonstrate either that they want black judges to be robots who are totally isolated from their racial heritage and unconcerned about it, or, more probably, that the impartiality of a black judge can be assured only if he disavows, or does not discuss, the legitimacy of blacks' aspirations to full first class citizenship in their own native land. Thus, the critical issue is, what conduct by black judges will assure their impartiality? Should they be robots? Should they demean their heritage by asking for less than first class citizenship for other blacks? Should they not tell the truth about past injustices?[74]

In this quote too, the "robo-judge", the algorithmic lawyer, is seen as inherently decontextualized and lacking the personal life history that grounds its decision in personal experience. If as we argued, creativity is closely connected to the ability to transform personal experience into external representations, then one could argue that machines, as perceived by judge Higginbotham, will indeed be lacking a crucial aspect of human creativity.

However, computational creativity research has challenged this view. Simon Colton's "Painting Fool" is a particularly interesting example that aims to capture also this aspect of human creativity, by creating short "life contexts" for the creative endeavour that the AI develops through its exposure to news item: "I don't even give it the notion of a person or a topic. It will wake up in the morning and look at the newspaper headlines."[75] There is still a random element, when the software runs its own web searches and trawls through social media websites to identify which news items to read. This replicates the vagaries of personal life histories which nonetheless result from a situated social experience. On the basis of what the Painting Fool finds, it autonomously develops a "mood", which then finds its external expression in the art that it creates—too many bad news, and the style will be distinctly dark. Eventually, the hope is that by giving AIs "life experiences", they too will create with an "authentic voice".[76]

If we were to transfer this approach to legal AI, a radically different approach to the current attempt to debias legal machine learning could emerge. We could try to

[74] Commonwealth of Pennsylvania v. Local Union 542, International Union of Operating Engineers, 671 F.2d 495 (3d Cir. 1981).

[75] Colton cited in de Lange (2012).

[76] Colton et al. (2018).

replicate the reasoning style of individual judges, in the same way in which some computational art replicates the writing[77] or painting[78] style of individual artists. Or we could train the algorithm on selected sub-sets of data, to recreate the equivalent to generic group perspectives. Sotomayor's quote above came in reply to questions about an earlier comment of hers, from a speech in 2001 at the University of California, Berkeley, law school:[79]

> I would hope that a wise Latina woman with the richness of her experiences would more often than not reach a better conclusion than a white male who hasn't lived that life,

If we understand bias in this contextualised sense, we may be able to use computational legal thought in entirely new ways, possibly getting closer to imbue them with some of the "soft skills" that currently are still seen as the exclusive domain of human decision makers. At the very least it would give us a testbed to analyse further what exactly we mean with soft skills like empathy and creativity in law, what they are and under what conditions, if any, they are to be amplified.

5 Conclusion

The aim of this paper was to use the lens of technology, and in particular gaming technology, to reflect and revisit some long-standing jurisprudential disputes about the "mechanic" nature of law, and what makes a good lawyer. This in turn should inform where we go from here, and how we want AI to shape the future of law. There is a danger of an unholy alliance between technological determinism and political decisions to underfund the justice system. In this environment, "cheaper, faster, more accessible" justice through AI seems like a winning idea. But what this obscures are the political choices that are taken when we follow this route, the hidden assumptions about what makes for a "good" legal system, the systemic and long term consequences for the administration of justice, and also some ill-informed assumptions about the power and limits of computation. But the cold, objective precise, predictable and emotion free robot-adjudicator is not a preordained fate. If it comes about, then because of the decisions we make, and neither the nature of law nor the limits of technology necessarily lead us down that path.

If as Oliver Wendel Holmes famously said, "The life of the law has not been logic; it has been experience", and if creativity is closely linked to life experiences so that truly creative computers need to find a way of replicating them, as Colton argues, then we might countenance the idea to build AIs that intentionally approach law from individualised, personal perspectives. Rather than having one legal AI, we might end up, just as we do in most courts, with several, each representing a unique

[77] Deutsch (2016).
[78] Schwab (2018).
[79] Sotomayor (2009).

world view that they will have to deliberatively reflect upon, justify, and sometimes but not always discard. Building a wise LatinAI as robo-judge would be a far cry from the current paradigm of legal software engineering, but one that could help us explore further the role of creativity, empathy, personal experiences, and biases, the role of collective decision making versus single judges.

This paper used comments on the victory of AlphaGo as a lens for the analysis of the social understanding of legal AI, and as a springboard for some provocations. Legal AI, from its metaphorical origins in the eighteenth century to the expert systems of the 1990s and the more recent machine learning approaches, has been wedded to a specific conception of justice. In the binary logic of the law, it was associated with consistency, not creativity; rationality, not empathy; objectivity, not subjectivity. The paper tried to outline a research project that challenges these juxtapositions, first, by querying if existing AIs really embody these ideals, or if they are rather the result of misguided projection by their observers. Second, by looking at developments in other fields of AI research to ask how creative, empathetic and subjective legal AI could look like, and if it might meet an as yet unsatisfied need. Answering these questions requires both, a conceptual analysis that draws on the rich tradition of jurisprudential thought, and empirical testing with methods from the HCI community.

Acknowledgments Work on this paper was supported by the UK Research and Innovation "Creative Informatics" grant and the UKRI network grant on Trusted Autonomous Systems.

References

Alcorta, C. S., & Sosis, R. (2005). Ritual, emotion, and sacred symbols. *Human Nature, 16*(4), 323–359.

Aletras, N., Tsarapatsanis, D., Preoţiuc-Pietro, D., & Lampos, V. (2016). Predicting judicial decisions of the European Court of Human Rights: A Natural Language Processing perspective. *PeerJ Computer Science, 2*. Retrieved August 26, 2021, from https://peerj.com/articles/cs-93/

Angwin, J., Larson, J., Mattu, S., & Kirchner, L. (2016). Machine Bias: There's software used across the country to predict future criminals. And it's biased against blacks. Retrieved August 26, 2021, from https://www.propublica.org/article/machine-bias-risk-assessments-in-criminal-sentencing

Arnhart, L. (1998). *Darwinian natural right: The biological ethics of human nature*. Suny Press.

Askin, S. (2018). AI trends driving the legal industry. *Law Technology Today*. Retrieved August 26, 2021, from https://www.lawtechnologytoday.org/2018/03/ai-trends/

Bańkowski, Z., & Schafer, B. (2007). Double-click justice: Legalism in the computer age. *Legisprudence, 1*(1), 31–49.

Batt-Rawden, S. A., Chisolm, M. S., Anton, B., & Flickinger, T. E. (2013). Teaching empathy to medical students: An updated, systematic review. *Academic Medicine, 88*(8), 1171–1177.

Bench-Capon, T., Araszkiewicz, M., Ashley, K., Atkinson, K., Bex, F., Borges, F., et al. (2012). A history of AI and law in 50 papers: 25 years of the international conference on AI and law. *Artificial Intelligence and Law, 20*(3), 215–319.

Bolukbasi, T., Chang, K.-W., Zou, J. Y., Saligrama, V., & Kalai, A. T. (2016). Man is to computer programmer as woman is to homemaker? Debiasing word embeddings. *Advances in Neural Information Processing Systems, 29*, 4349–4357.
Bryson, J. J. (2017). The meaning of the EPSRC principles of robotics. *Connection Science, 29*(2), 130–136.
Buchanan, B. G., & Shortliffe, E. H. (1984). *Rule based expert systems: The MYCIN experiments of the Stanford Heuristic Programming Project*. Addison-Wesley.
Campbell, B. (1970). La Mettrie: The robot and the automaton. *Journal of the History of Ideas, 31*(4), 555–572.
Card, S. K., Moran, T. P., & Newell, A. (1983). *The psychology of human-computer interaction*. Lawrence Erlbaum Associates.
Carlozzi, A. F., Bull, K. S., Eells, G. T., & Hurlburt, J. D. (1995). Empathy as related to creativity, dogmatism, and expressiveness. *The Journal of Psychology, 129*(4), 365–373.
Christodoulidis, E. A. (1999). The irrationality of merciful legal judgement: Exclusionary reasoning and the question of the particular. *Law and Philosophy, 18*(3), 215–241.
Citron, D. K. (2007). Technological due process. *Washington University Law Review, 85*, 1249–1313.
Colton, S. (2008). Creativity versus the perception of creativity in computational systems. In D. Ventura, M. L. Maher, & S. Colton (Eds.), *AAAI spring symposium: Creative intelligent systems* (Vol. 8). Retrieved August 26, 2021, from https://www.aaai.org/Papers/Symposia/Spring/2008/SS-08-03/SS08-03-003.pdf
Colton, S., Pease, A., & Saunders, R. (2018). Issues of authenticity in autonomously creative systems. In F. Pachet, A. Jordanous, & C. León (Eds.), *Proceedings of the Ninth International Conference on Computational Creativity*. Association for Computational Creativity. Retrieved August 26, 2021, from http://computationalcreativity.net/iccc2018/sites/default/files/papers/ICCC_2018_paper_64.pdf
Cornell Tech. (17.2.2016). 3 reasons why tech companies need a new kind of lawyer. *Law Tech Blog*. Retrieved August 26, 2021, from https://tech.cornell.edu/news/3-reasons-why-tech-companies-need-a-new-kind-of-lawyer/
Crichigno, B. (24.9.2018). Are we ready to accept an AI judge?. Retrieved August 26, 2021, from https://mastersofmedia.hum.uva.nl/blog/2018/09/24/are-we-ready-to-accept-an-ai-judge/
De Clapiers, L. (1746). *Reflexions & Maximes. New edition 2013*. CUP.
De La Mettrie, J. O. (1748). *L'Homme machine*. Elie Luzac.
De Lange, C. (19.1.2012). Artificial: The virtual virtuosos redefining creativity. *New Scientists*. Retrieved August 26, 2021, from https://www.newscientist.com/article/mg21328471-900-artificial-the-virtual-virtuosos-redefining-creativity/
Deutsch, M. (12.7.2016). How to write with artificial intelligence. *Medium*. Retrieved August 26, 2021, from https://medium.com/deep-writing/how-to-write-with-artificial-intelligence-45747ed073c
Diver, L. (2020). Computational legalism and the affordance of delay in law. *Journal of Cross-disciplinary Research in Computational Law, 1*(1).
Douglas, K., & Coburn, C. (2009). Students designing role-plays: Building empathy in law students? *Legal Education Review, 61*, 55–63.
Duff, R. A. (2003). Penance, punishment and the limits of community. *Punishment & Society, 5*(3), 295–312.
Duguid, S., Edwards, L., & Kingston, J. (2001). A web-based decision support system for divorce lawyers. *International Review of Law, Computers & Technology, 15*(3), 265–279.
Duxbury, N. (2002). *Random justice: On lotteries and legal decision-making*. Oxford University Press.
Eggers, W. D. (2007). *Government 2.0: Using technology to improve education, cut red tape, reduce gridlock, and enhance democracy*. Rowman & Littlefield.
Feigenson, N. R. (1999). Merciful damages: Some remarks on forgiveness, mercy and tort law. *Fordham Urban Law Journal, 27*, 1633–1650.

Fenwick, M., Kaal, W. A., & Vermeulen, E. P. M. (7.8.2018). *Legal education in a digital age: Why 'coding for lawyers' matters*. Lex Research Topics in Corporate Law & Economics Working Paper No, 2018-4. Retrieved August 26, 2021, from https://doi.org/10.2139/ssrn.3227967

Fuller, L. L. (1964). Eight ways to fail to make law. *Philosophy of Law*, 13–17.

Gibney, E. (27.1.2016). Go players react to computer defeat. *Nature News*. Retrieved March 15, 2021, from https://www.nature.com/news/go-players-react-to-computer-defeat-1.19255

Gonen, H., & Goldberg, Y. (2019). Lipstick on a pig: Debiasing methods cover up systematic gender biases in word embeddings but do not remove them. In J. Burstein, C. Doran, & T. Solorio (Eds.), *Proceedings of the 2019 Conference of the North American Chapter of the Association for Computational Linguistics: Human Language Technologies* (Vol. 1, pp. 609–614). Association for Computational Linguistics.

Hallnäs, L., & Redström, J. (2001). Slow technology – Designing for reflection. *Personal Ubiquitous Computing, 5*(3), 201–212.

Hollis, P. L., Baroness Hollis of Heigham, House of Lords Debate 30 March 1998, col 54.

Holloway, R. L. (1981). Culture, symbols, and human brain evolution: A synthesis. *Dialectical Anthropology, 5*(4), 287–303. Retrieved August 26, 2021, from https://link.springer.com/content/pdf/10.1007/BF00246207.pdf

Holmes, M. (2017). *Are we underestimating the impact of AI?* Retrieved August 26, 2021, from https://www.lawtechnologytoday.org/2017/06/the-impact-of-artificial-intelligence/

Hu, Y., Strang, S., & Weber, B. (2015). Helping or punishing strangers: neural correlates of altruistic decisions as third-party and of its relation to empathic concern. *Frontiers in Behavioral Neuroscience, 9*. Retrieved August 26, 2021, from https://doi.org/10.3389/fnbeh.2015.00024

Huq, A. Z. (2018). Racial equity in algorithmic criminal justice. *Duke Law Journal, 68*, 2019. Retrieved August 26, 2021, from https://scholarship.law.duke.edu/cgi/viewcontent.cgi?article=3972&context=dlj

Katz, D. M., Bommarito, M. J., II, & Blackman, J. (2017). A general approach for predicting the behavior of the Supreme Court of the United States. *PLoS One, 12*(4), e017469.

Keene, D. (Ed.). (1961). *Four major plays of Chikamatsu*. Columbia UP.

Koshkina, D. (27.7.2016). Fan Hui 2P: "AlphaGo is a new Lee Changho". *Amercian Go e-journal*. Retrieved August 26, 2021, from http://www.usgo.org/news/2016/07/fan-hui-2p-alphago-is-a-new-lee-changho

Kowsrihawat, K., Vateekul, P., & Boonkwan, P. (2018). Predicting judicial decisions of criminal cases from Thai Supreme Court using bi-directional GRU with attention mechanism. In *2018 5th Asian Conference on Defense Technology (ACDT)* IEEE 50–55. Retrieved August 26, 2021, from https://ieeexplore.ieee.org/document/8592948

Larsen, J., Mattu, S., Kirchner, L., & Angwin, J. (23.5.2016). How we analyzed the COMPAS Recidivism Algorithm, *Propublica*. Retrieved August 26, 2021, from https://www.propublica.org/article/how-we-analyzed-the-compas-recidivism-algorithm

Le Sueur, A. (2016). Robot government: Automated decision-making and its implications for parliament. In A. Horne & A. Le Sueur (Eds.), *Parliament: Legislation and accountability* (pp. 3–21). Bloomsbury Publishing.

Letwin, O. (1997). Standing Committee B, 28 October 1997 (Morning session).

Marsh, T. (2016). Slow serious games, interactions and play: Designing for positive and serious experience and reflection. *Entertainment Computing, 14*, 45–53.

McCarty, L. T. (1976). Reflections on TAXMAN: An experiment in artificial intelligence and legal reasoning. *Harvard Law Review, 90*, 837–893.

Meagher, P. (1994). Vauvenargues' quest for a new moral philosophy. *Lumen: Selected Proceedings from the Canadian Society for Eighteenth-Century Studies/Lumen: Travaux choisis de la Société canadienne d'étude du dix-huitième siècle, 13*, 125–136.

Mehrabi, N., Morstatter, F., Saxena, N., Lerman, K., & Galstyan, A. (2019). A survey on bias and fairness in machine learning. *arXiv preprint arXiv:1908.09635*.

Menkel-Meadow, C. (2001). Aha--Is creativity possible in problem solving and teachable in legal education. *Harvard Negotiation Law Review, 6*, 97–144.

Metz, C. (16.3.2016a). In Two Moves, Alphago and Lee Sedol Redefined the Future. *Wired*. Retrieved August 26, 2021, from https://www.wired.com/2016/03/two-moves-alphago-lee-sedol-redefined-future/

Metz, C. (14.3.2016b). How Google's AI viewed the move no human could understand. *Wired*. Retrieved August 26, 2021, from https://www.wired.com/2016/03/googles-ai-viewed-move-no-human-understand/

Mills, M. (2016). Artificial intelligence in law: The state of play 2016. Retrieved August 26, 2021, from http://www.legalexecutiveinstitute.com/wp-content/uploads/2016/03/Artificial-Intelligence-in-Law-The-State-of-Play-2016.pdf

Moses, L. B., & Chan, J. (2014). Using big data for legal and law enforcement decisions: Testing the new tools. *UNSWLJ, 37*, 643–679.

Newman, D. (29.8.2017). Top 5 digital transformation trends in legal. *Forbes*. Retrieved August 26, 2021, from https://www.forbes.com/sites/danielnewman/2017/08/29/top-5-digital-transformation-trends-in-legal/#2610caac76f8

Nussbaum, M. (2017). Equity and mercy. In P. A. Tabensky (Ed.), *Judging and understanding* (pp. 21–60). Routledge.

Odom, W., Wakkary, R., Bertran, I., Harkness, M., et al. (2018). Attending to slowness and temporality with olly and slow game: A design inquiry into supporting longer-term relations with everyday computational objects. In R. Mandryk & M. Hancock (Eds.), *Proceedings of the 2018 CHI Conference on Human Factors in Computing Systems* (pp. 1–13). ACM.

Osoba, O. A., & Welser, W., IV. (2017). *An intelligence in our image: The risks of bias and errors in artificial intelligence*. Rand Corporation. Retrieved August 26, 2021, from https://www.rand.org/content/dam/rand/pubs/research_reports/RR1700/RR1744/RAND_RR1744.pdf

Pereira, A., Leite, I., Mascarenhas, S., Martinho, C., & Paiva, A. (2010). Using empathy to improve human-robot relationships. In H. Lamers & J. Verbeek (Eds.), *International Conference on Human-Robot Personal Relationship* (pp. 130–138). Springer.

Rhode, D. L. (2004). *Access to justice*. Oxford University Press.

Rosenberg, J. D. (2001). Teaching empathy in law school. *USFL Review, 36*, 621–658.

Sayer, P. (24.10.2016). Not robocop, but robojudge? AI learns to rule in human rights cases. *PCWorld*. Retrieved August 26, 2021, from https://www.pcworld.com/article/3134495/not-robocop-but-robojudge-ai-learns-to-rule-in-human-rights-cases.html

Schafer, B., & Bromby, M. (2005). Wie Tajomaru seine NemeSys fand: Expertensysteme zum Augenzeugenbeweis. In B. Schünemann, M.-T. Tinnefeld, & R. Wittman (Eds.), *Gerechtigkeitswissenschaft* (pp. 259–278). Berliner Wissenschaftsverlag.

Schafer, B., Komuves, D., Zatarain, J. M. N., & Diver, L. (2015). A fourth law of robotics? Copyright and the law and ethics of machine co-production. *Artificial Intelligence and Law, 23*(3), 217–240.

Schwab, K. (4.12.2018). This AI paints like the old masters. Can you tell the difference. *Fast Company*. Retrieved August 26, 2021, from https://www.fastcompany.com/90167584/this-ai-paints-like-the-old-masters-can-you-tell-the-difference

Sebald, G. (2020). "Loading, please wait" – Temporality and (bodily) presence in mobile digital communication. *Time & Society, 29*(4), 990–1008.

Sotomayor, S. (2009). A Latina Judge's Voice. *Berkeley La Raza Law Journal, 13*, 87–95.

Stepien, K. A., & Baernstein, A. (2006). Educating for empathy. *Journal of General Internal Medicine, 21*(5), 524–530.

Stranieri, A., & Zeleznikow, J. (1999). The evaluation of legal knowledge based systems. In J. Bing, A. Jones, & T. Gordon (Eds.), *Proceedings of the 7th international conference on Artificial intelligence and law* (pp. 18–24). ACM.

Sulea, O. M., Zampieri, M., Vela, M., & van Genabith, J. (2017). Predicting the law area and decisions of French supreme court cases. arXiv preprint arXiv:1708.01681.

Susskind, R. E. (2017). *Tomorrow's lawyers: An introduction to your future*. Oxford University Press.

Tao, J., & Tan, T. (2005). Affective computing: A review. In *ACII 2005: International conference on affective computing and intelligent interaction* (pp. 981–995). Springer.

Tapus, A., & Mataric, M. J. (2007, March). Emulating empathy in socially assistive robotics. In A. Tapus, M. Michalowski, & S. Sabanovic (Eds.), *AAAI Spring Symposium: Multidisciplinary Collaboration for Socially Assistive Robotics* (pp. 93–96). Retrieved August 26, 2021, from http://robotics.usc.edu/publications/media/uploads/pubs/533.pdf

Taylor, D., & Osafo, N. (2018) Artificial intelligence in the courtroom. *Lawgazette*. Retrieved August 26, 2021, from https://www.lawgazette.co.uk/practice-points/artificial-intelligence-in-the-courtroom-/5065545.article

The Law Society of England and Wales. (2018). AI: Artificial intelligence and the legal profession – Horizon scanning report. Retrieved August 26, 2021, from https://www.lawsociety.org.uk/topics/research/ai-artificial-intelligence-and-the-legal-profession

The Law Society of England and Wales. (27.6.2019). Technology and the Law Policy Commission. Algorithms in the Justice System. Retrieved August 26, 2021, from https://www.lawsociety.org.uk/policy-campaigns/articles/public-policy-technology-and-law-commission/

Thomson, A. (2016). French eighteenth-century materialists and natural law. *History of European Ideas, 42*(2), 243–255.

Veitch, S. (1998). Doing justice to particulars. In E. Christodoulidis (Ed.), *Communitarianism and citizenship* (pp. 220–234). Ashgate.

Vincent, J. (24.10.2016). AI program able to predict human rights trials with 79 percent accuracy. *The Verge*. Retrieved August 26, 2021, from https://www.theverge.com/2016/10/24/13379466/ai-judge-european-human-rights-court-prediction

Wakefield, J. (17.11.2015). Sky warning over 'cash for porn' letters. *BBC News*. Retrieved August 26, 2021, from https://www.bbc.co.uk/news/technology-34842863

Wakefield, J. (23.10. 2016). AI predicts outcome of human rights cases. *BBC News*. Retrieved August 26, 2021, from https://www.bbc.co.uk/news/technology-37727387

Wringe, B. (2010). War crimes and expressive theories of punishment: Communication or denunciation? *Res Publica, 16*(2), 119–133.

Yadollahi, A., Shahraki, A. G., & Zaiane, O. R. (2017). Current state of text sentiment analysis from opinion to emotion mining. *ACM Computing Surveys (CSUR), 50*(2)., Article 25. Retrieved August 26, 2021, from https://doi.org/10.1145/3057270

Zeki, S., Goodenough, O. R., Greene, J., & Cohen, J. (2004). For the law, neuroscience changes nothing and everything. *Philosophical Transactions of the Royal Society of London. Series B: Biological Sciences, 359*(1451), 1775–1785.

Zeleznikow, J., & Stranieri, A. (1998). Split up: An intelligent decision support system which provides advice upon property division following divorce. *International Journal of Law and Information Technology, 6*(2), 190–213.

Smart Contracts in the Civil Law Countries: The Legislative Analysis and Regulation Perspectives

Elizaveta Zainutdinova

Contents

Abstract Lawmakers and theorists around the world are debating the need for a new set of rules to support transactions in a distributed ledger. It is particularly important for civil law countries because they rely on specific legislation provisions in order to formalize what has already appeared in practice. The same situation applies to smart contracts. The article addresses the appropriate legal response to smart contracts as a type of new digital contractual relationship. The author analyses their characteristics in Belarus, in the EU countries (Portugal, Germany, and Italy), and Russia, where the smart contract term has not yet appeared in legislation. Based on a review of the current legal approach to smart contracts in these countries it is determined whether new e-commerce trends towards creating a fundamentally different environment require a new legal approach.

E. Zainutdinova (✉)
Novosibirsk National Research State University, Novosibirsk, Russia

G. Borges, C. Sorge (eds.), *Law and Technology in a Global Digital Society*,
https://doi.org/10.1007/978-3-030-90513-2_16

1 The Smart Contract Definition

A smart contract should be defined as a contract with an automated performance of contractual obligations. There is nothing new in concluding a contract because it has a written form, namely, it is an exchange of electronic messages on the Internet. However, the principal issue is a performance of obligations that, in a smart contract, is automated, independent from actions of an obliged party, simultaneous for both parties. Other authors also point out that a smart contract is concluded between the parties, and the performance of obligations of a smart contract is enforced by an unchangeable computer code.[1] Some mention that smart contracts are self-executable contracts[2] while others point out that a smart contract is a digital form contract (namely, in a code form) that is self-executable and self-enforceable.[3]

It seems that the self-executability and self-enforceability of a smart contract lie in the automated performance of obligations. Therefore, we may identify a smart contract as a contract, due to which obligations agreed by parties are performed automatically and expressed in the form of a software code in the blockchain information system. Smart contracts are to solve one of the main issues of business relations and civil law—due performance of obligations taken by parties and fixed in a contract.

Considering the emergence of this new type of performing obligations, it seems reasonable to reflect its peculiarities in civil legislation in order to avoid disputes in the future and foster the use of such a new instrument that is to boost the level of trust in parties' relationship and help to duly perform their contractual obligations. What legislation needs is a provision that the performance of obligations could be automated with the help of a smart contract if parties agree so.

2 The Smart Contract Regulation in Belarus

Smart contracts raise a lot of legal issues in civil law countries. Some of them have already enacted legal provisions, though others restrained from introducing legal provisions on smart contracts into law.

For example, in the Republic of Belarus Decree No. 8 of December 21, 2017 "On the development of digital economy" (hereafter—"Decree") was enacted. This act came into force on March 28, 2018 and legalized smart contracts in the Republic of Belarus. Beyond a smart contract's definition, the Decree also clarifies issues regarding the blockchain, namely the actors of the blockchain market, tokens, and cryptocurrency. Actually, this is a pioneering act because, for the first time, on the state level it regulates smart contracts and other new issues in the digital field.

[1] Szczerbowski (2017).

[2] Kerikmäe and Rull (2016).

[3] Cornell and Werbach (2017).

In this act, a smart contract is a software code that is to function in blockchain or other decentralized information systems for automated conclusion and (or) performance of contracts or performing other legal acts.[4] It should be noted that in the Decree contracts related to the digital economy are named but a smart contract is not among them. This could mean that a smart contract there is not seem a type of contract but is seemed probably as a type of conclusion or performance of a contract. However, a smart contract is defined in the Decree mostly using technical terms such as "software code" but which has the legal function, namely to conclude and (or) perform contracts or perform other legal acts automatically. Therefore, in this definition, we face both technical and legal terms and this seems to be incorrect. Such terms as "software code" and "blockchain" are technical ones and these should be defined by legal terms such as "electronic message", "in the form of software code", "blockchain information system". As for legal terms used in the definition, it is not clear what "other legal acts" are prescribed. Overall, these could be acts that have legal consequences like the initiation of legal action, receiving an inheritance. However, these cannot be performed yet with the help of a smart contract. At the present moment, smart contracts are used only in deals.

It could be noticed that the main shortfall of the analyzed definition of a smart contract is that the question whether a smart contract is a form, a type or a way of performing obligations was not answered. Therefore, the legal nature of a smart contract was not identified. Could we consider a smart contract as a notice, an electronic document, or an electronic message? In paragraph 5.3 of the Decree, there is a provision that "the one concluded or performed a deal with the help of a smart contract is deemed to be properly informed about its conditions".[5] Analyzing this definition, we conclude that, firstly, a smart contract is seen as a way to conclude or perform contracts. Using Russian terms of Information Technology Law, we may name a smart contract as an electronic document or an electronic message. According to the Russian Law "On Information, Information Technologies and Protection of Information", an electronic document is documented information in electronic form that is perceived by a person with the help of computers and transmitted over the Internet, and an electronic message is information of any kind transmitted and received by a user of the Internet.[6] Secondly, the Decree introduces the presumption that a person that concluded a contract with the help of a smart contract knew the conditions of the contract and cannot challenge it. This presumption is to facilitate stability of the economic turnover and limit abusive practices. The presumption is essential for parties that use smart contracts and for the advance of the use of smart contracts overall.

---

[4]Decree No. 8 of December 21, 2017 "On Development of Digital Economy".

[5]Decree No. 8 of December 21, 2017 "On Development of Digital Economy".

[6]Federal Law No. 149-FZ of July 27, 2006 "On Information, Information Technologies and Protection of Information", art. 2, (as amended up to Federal Law No. 327-FZ of November 25, 2017).

Moreover, the important point is also that, according to the Decree, only residents of the Belarusian Hi-Tech Park could conclude and (or) perform obligations with the help of a smart contract. So, the use of this contractual means is limited to particular subjects. In regards to this, legal regulation of a smart contract in the Belarus Republic may seem to be quite measured and detailed. The lawmakers in Belarus succeeded in shaping the definition of a smart contract and this is itself revolutionary because the smart contract term is still so controversial and not many doctrinal researches were made on this topic so it is quite complicated to define a smart contract. Therefore, this experience of legal regulation of a smart contract should be used by other countries of civil law system.

3 The Smart Contract Regulation in Portugal

We are to consider the following jurisdiction where the issue of legal regulation of smart contracts was on the agenda. Under Portuguese law, contracts are generally valid if legally competent parties reach an agreement whether they agree verbally, in a written form or electronically. This is regulated by the Portuguese Civil Code, Article 219 and the Electronic Signature Decree-Law, Article 3.[7] In this regard, a contract that is concluded in blockchain platform and that is a smart contract is admitted to exist, authenticate the parties that signed it and be validly acceptable in a court.

Additionally, from 2007 Portugal has had a specific provision dealing with contracts executed by means of computers without human intervention in its E-Commerce Law No. 7/2004. This provision applies contract law to these types of contracts and smart contracts are among them. This is crucial that provisions of contract law are applicable to programming errors, malfunctions, and distorted messages[8] as to a mistake in a contract. Therefore, smart contracts are not specifically regulated in Portugal, however, consequences of technical mistakes arising in smart contracts are regulated. In other jurisdictions, there are no such or equivalent provisions.

[7] eSignature Legality in Portugal, 2018, https://www.docusign.com/how-it-works/legality/global/portugal. Accessed 01 September 2021.

[8] The Financial Technology Law Review: Portugal, 2020, https://thelawreviews.co.uk/title/the-financial-technology-law-review/portugal. Accessed 01 September 2021.

4 The Smart Contract Regulation in Italy

In Italy, a smart contract was defined in the legislation. The Law No. 135/2018 enacted on February 07, 2019, has defined a smart contract as a computer program based on distributed ledger technologies (we could concern this as decentralized distributed information system) performance due to which is legally binding if parties agreed so. Despite this, it is also noted in the Law that smart contracts satisfy written form requirements and requirements for the identification of parties of a deal. And it is also noted that distributed ledger technologies allow verifying data that cannot be modified or compromised.[9]

Analysing this, we may resume that a technically-oriented legislative approach is used in Italy in regards to smart contracts. In other words, a smart contract is defined from the technical point of view (as a computer program). However, it is specified that smart contracts are in a written form and provide for identification of its parties.

5 The Smart Contract Regulation in the EU

EU Regulation (EU) No 910/2014 on electronic identification and trust services for electronic transactions in the internal market (the "eIDAS Regulation") that came into force on July 01, 2016, should be considered.[10] Article 25(1) of the eIDAS Regulation provides that an electronic signature shall not be denied legal effect and admissibility in courts solely on the grounds that it is in an electronic form. This provision, of course, is applicable to smart contracts that are used in contractual relationships because the identification of parties of a smart contract's transaction is fulfilled electronically.

Article 9 of the Electronic Commerce Directive requires EU member states to ensure that their legal systems allow contracts to be concluded by electronic means which is also the case of smart contracts because these are concluded by electronic means. the Member States shall in particular ensure that the legal requirements applicable to the contractual process neither create obstacles for the use of electronic contracts nor result in such contracts being deprived of legal effectiveness and validity on account of that they are made by electronic means.[11] Therefore, we can conclude that in the European Union there is a basis for the legal regulation of smart contracts.

Noticing this, it should be stated though that there is currently no legal framework in the European Union specifically tailored for smart contracts. Definition of a

---

[9]Distributed Ledger Technologies and Smart Contracts in Italy, 2019, https://blockchain.bakermckenzie.com/2019/02/28/distributed-ledger-technologies-and-smart-contracts-in-italy/. Accessed 01 September 2021.

[10]Regulation (EU) N°910/2014.

[11]Directive 2000/31/EC, Official Journal L 178, 17/07/2000 P. 0001 – 0016.

"smart contract" and whether or not such a contract is legally binding are to be decided at the member state level. However, the European Union is now increasing its efforts to foster use of blockchain and smart contracts. Technological companies in Europe also seek to raise the "regulatory and supervisory capacity and knowledge" around blockchain.[12] The Action Plan on how to harness the opportunities presented by technology-enabled innovation in financial services ("FinTech Action Plan") identified areas of possible development, including legal clarifications around smart contracts, initial coin offerings, and other issues raised by the tech. According to this document, the European Union has already revised its Payment Services Directive, it now requires banks in the region to open up their communication channels, such as API, to blockchain applications that require payment account integrations. So there are already some steps taken for the use of blockchain and smart contracts by businesses. The FinTech Action Plan shows that since 2018 the European Commission takes action for a more competitive and innovative financial market.[13] Importantly, the European Supervisory Authorities already publish reports on innovation facilitators (regulatory sandboxes and innovation hubs) with best practices for them, namely on distributed ledger technology, crowdfunding and smart contracts.[14]

6 The Attitude Towards Smart Contracts in Germany

Therefore, despite there is no specific legal framework on smart contracts in the European Union, smart contracts are valid in the European Union countries in general. The thesis that smart contracts are generally valid in Europe could be proved in the case of Germany.

However, it should be noted though that German law is quite traditionalistic and even a bit slow in this regard. In 2013 the Federal Court of Justice of Germany ruled that in case of uncertainty, expression of the will of parties arising out of contracts in electronic form should be interpreted in a way how a person would interpret contract's provisions.[15] Therefore, the situation is that software code is made to be replaced by people's interpretation. By the way, smart contracts are based on technical language, namely on programming language, and here is the issue—is the programming language to be analyzed and interpreted by a person? Actually, this could cause ambiguities and unclarity of language since people interpret the same provisions in different ways though machine interpretation is determined and

[12] Simont Braun (2018).

[13] https://ec.europa.eu/info/publications/180308-action-plan-fintech_en. Accessed 01 September 2021.

[14] https://www.eba.europa.eu/esas-publish-joint-report-on-regulatory-sandboxes-and-innovation-hubs. Accessed 01 September 2021.

[15] BGH, Urteil vom 16.10.2012 - X ZR 37/12 (= NJW 2013, 598).

unambiguous. From the logical point of view, smart contract's provisions should be interpreted only in one particular way as they are perceived by computers. However, even having in mind that there are some propositions to transform people's language both in legal terms and in executable software code, creating the Common Language for Augmented Contract Knowledge,[16] anyway, while interpreting smart contracts' provisions we would do it as people and not as machines.

Considering the validity of smart contracts, we imply their acceptability in a court. And in courts, as it seems from the German perspective, smart contracts will be interpreted based on a perception of terms in software code by humans. Consequentially, while smart contracts are in the form not directly perceived by persons, namely in the form perceived by machines, these would be interpreted by lawyers with programmers. However, it is believed still that use of software code even interpreted by people helps to avoid many ambiguities and doubts as there are a lot of them in legalese.

7 The Smart Contract Regulation in Russia

7.1 Bills on Smart Contracts

If we are looking through the current legislation of Russia, we conclude that a smart contract term is not used at all. There is a provision on the automated performance of obligations in the Russian Civil code though.

Initially in the Russian draft bill "On Digital Financial Assets" of March 20, 2018, a smart contract was considered as a new type of contract in an electronic form. Rights and duties resulting from such a contract are fulfilled due to automated execution of digital transactions in a particular order and in case of occurrence of certain circumstances.[17]

In bill No. 424632-7 "On Amendments to Parts One, Two and Four of the Civil Code of the Russian Federation" of March 26, 2018, which was subsequently enacted there is no definition of a smart contract but there is a provision on the automated performance of obligations. Such performance of obligations is defined there as performance under certain circumstances where parties of a contract do not express separately their consent on the performance of obligations that have been specified in a contract with the help of information technologies.[18]

This regulation seems revolutionary because it describes what is the automated performance of obligations (this is performance of obligations out of a smart contract). However, it is quite a conventional approach since the smart contract

[16] Siedler (2016).

[17] Draft Bill No. 419059-7 of March 20, 2018 "On Digital Financial Assets".

[18] Draft Bill No. 424632-7 of March 26, 2018 "On amendments to Parts First, Second and Fourth of the Civil Code of the Russian Federation (on Digital Rights)".

term is not used in the legislation. According to such an approach, contractual obligations of parties can be performed automatically and for this, no special expression of parties' will is needed other than entering into a contract with the automated performance of obligations which is a smart contract. Therefore, provisions of the Civil Code of the Russian Federation describe the nature of smart contracts—automated performance of obligations stating that parties could use contracts with such type of performance, e.g. automated performance.

It seems reasonable to add a new type of performance of obligations to existing ones since this has started to be used quite actively in practice in the digital field and is to ensure due performance of obligations and the appropriate level of trust in parties' contractual relationship. The term "smart contract" itself is not used but it is described with the help of contract law instruments.

A smart contract is to be concluded when parties agreed on terms inside the blockchain platform. A smart contract is to be performed irreversibly after parties concluded it and according to conditions that parties agreed upon. This performance of obligations out of smart contracts helps businesses to boost the effectiveness of their contractual relationship, reduce transaction costs, and avoid non-performance of obligations.

So, commenting Russian bills on smart contracts it should be noted that considering that a smart contract term is quite debatable it is not used there despite it is used, for example, in Republic Belarus' Decree. However, the Russian lawmaker tries to describe the legal nature of a smart contract which lies in the performance of obligations that seems to be rational.

What may be missed in Russian draft bills is a provision that a contract's validity cannot be undermined only on the ground that this contract includes a "smart contract" term. This way to regulate smart contracts was used in particular states of the US.[19] It recognizes smart contract's provisions as valid while Russian lawmakers prescribe incontestability of smart contract's provisions. Maybe the Russian approach is even more prudent since it eases off courts stating smart contracts' incontestability as a general rule and thus reducing the vulnerability of transactions.

Meanwhile, the pitfalls of the definition of a new type of performing obligations in the Russian Civil Code are:

- this performance is not named as the automated performance of obligations while its nature is that this performance is automated;
- instead of using previously expressed consent of parties the term separate expression of consent of parties is used—the term separate expression is not clear, usually in contracts any additional consent to perform obligations out of a contract is not required;

[19] Expert Report Prepared by Employees of the Alexeev Private Law Research Centre under the President of the Russian Federation for the Non-Commercial Organization Foundation for Development of the Center for Elaboration and Commercialization of New Technologies, 2017. https://sk.ru/foundation/legal/m/sklegal02/21551/download.aspx. Accessed 01 September 2021.

- the term blockchain information system or distributed ledger technologies or distributed and decentralized system are not used so any contract performed with the help of information technologies could be considered as a smart contract (not only a contract concluded and performed in blockchain information system).

7.2 Other Provisions Applicable to Smart Contracts

In Russian legislation, we could find some other acts that could be applied to smart contracts. These are on concluding smart contractual provisions. Russian legislation includes provisions on concluding contracts in an electronic form. Contract in an electronic form in Russian legislation is a contract that is concluded due to the exchange of electronic documents.[20] Russian Law on Information, Information Technologies, and Information Security regulates electronic documents. According to this Law, electronic documents are documented information in an electronic form that is perceived by a person with the help of computers and transmitted over the Internet.[21] This type of contract was made quite common in business practice and represents e-commerce. This Law also takes into consideration electronic messages that are information transmitted and received by a user of the Internet.[22] Smart contracts, it seems, could be described in terms of this Law as electronic messages because they are not actually documented but still they are information sent in the blockchain information system. However, legislation tailored to smart contracts will help to reach their potential and avoid unnecessary disputes between the parties. What is mostly needed is a reflection of the automated performance of obligations and issues related to smart contracts' modification and termination.

8 Specific Issues Arising in Connection with Smart Contracts

Smart contracts introduce major changes into contract law. Without decentralized consensus, one party often enjoys huge bargaining power and monopoly rent. By the use of smart contracts and blockchain technology, contractibility and enforceability

[20]Civil Code of the Russian Federation (Part One) No. 51-FZ of November 30, 1994, art. 434.

[21]Federal Law No. 149-FZ of July 27, 2006 "On Information, Information Technologies and Protection of Information", art. 2, (as amended up to Federal Law No. 327-FZ of November 25, 2017).

[22]Federal Law No. 149-FZ of July 27, 2006 "On Information, Information Technologies and Protection of Information", art. 2, (as amended up to Federal Law No. 327-FZ of November 25, 2017).

in contingent contracts are increased.[23] Blockchain platform creates an environment of trust for parties. Payment due to a smart contract could be executed automatically which helps to bypass traditional banks and other payment intermediaries. Smart contracts help to resolve the problem of undue performance of obligations that is a pitfall of the existing system of contract law. However, the implementation of smart contracts still raises a lot of issues. Among them are: how smart contracts are to be analyzed by enforcement bodies, what principles and provisions should govern them, how obligations out of smart contracts are performed, and how we can modify smart contracts and apply measures of defense and liability to obligations from them. These are the ideas that came up and are to be addressed with the use of smart contracts:

1. Since smart contracts are in the form of software code it is tricky for enforcement bodies and parties without a technical background to analyze them. However, conclusion and verification of smart contracts in blockchain information system make it possible for parties not to draft typical legal agreements in printed form and provides an opportunity to secure data.
2. Smart contracts concluded with the use of blockchain platforms may be seen as a type of electronic contracts or a type of written contract. The origin of smart contracts is interconnected with the appearance of a new way of performing obligations—automated performance of obligations. Therefore, we may define a smart contract as a contract with the automated performance of obligations. It seems that smart contracts should be governed by contract law with special provisions on smart contracts that enforce smart contracts' validity or provide for automated performance of obligations and consent of parties for such performance.
3. Smart contracts are performed automatically but with the parties' consent. With the help of civil law doctrine, we may introduce a category of parties' previously expressed consent that helps us to define who is to enter into a deal. The point is that a party that gave its previously expressed consent on the performance of obligations with the use of the blockchain information system cannot influence the performance of obligations. This is the main difference and advantage of smart contracts: performance of obligations from a smart contract makes undue performance impossible; performance of obligations from a smart contract is always due (except for cases such as a technical omission or glitch, hacking attack, or other interference to blockchain information system). Problems with smart contracts are inevitable because of the bounded rationality of coders and contracting parties, incomplete foresight, limited information, and opportunistic behavior.[24]
4. Smart contracts may be modified if parties agree so and how the parties agree. However, neither of the parties may unilaterally restrain from the performance of

[23] Cong and Zhiguo (2018).

[24] Calcaterra and Kaal (2018).

its obligations. The issue is what we could do if smart contract provisions violate the law. This is up to the operator of the blockchain information system that could be authorized to terminate such a deal. A court or an arbitrator also may be enabled to override smart contractual provisions in a case when they violate the law.

5. Cases when means of defense and means of liability shall be applied to smart contracts are not so common. However, if there is a technical mistake such a mean of defense as specific performance should be used. In cases when a party of a smart contract hinders its performance by, for example, making changes in the blockchain information system, it could be seen as a liability of a party. The other party of a smart contract may pursue damages for such actions of such party (it would be tort liability like interference into a contract). It seems also that a programmer who made a mistake intentionally or unintentionally shall be liable for this before the party (his or her client)—this is a liability out of a contract on providing programming services. If a third party interferes in blockchain and smart contract this is considered as tort liability of such a person before smart contract's parties.

9 Conclusion

Therefore, after making an analysis of smart contracts' regulation in civil law countries, namely Belarus, the EU, and Russia, we may conclude that these contracts are still very new and revolutionary to these systems of law. A lot of issues regarding smart contracts' interpretation, performance, modification, and termination are to be properly addressed in tailored legislation and judicial practice. However, some changes due to smart contracts' appearance and use are already undertaken in the legal field. Among them are provisions on contracts' execution without human interaction, opening API channels for blockchain applications and summaries of best practices in this field, and use of categories as a special type of performance of obligations and consent of parties for such performance.

In civil law countries, it seems that smart contracts are made similar to existing contracts since they have the same goals—to perform parties' obligations. In Russia, due to the appearance of special provisions on the performance of obligations smart contracts were introduced into the Civil Code. This is what distinguishes the Russian legislative approach towards smart contracts from the approaches of Portugal, Italy, and other countries of the European Union. In Russia, we see that special provisions are crafted to describe the legal nature of smart contracts, and how successfully these would be applied in the legal system is still a question of time.

References

Articles

Calcaterra, C., & Kaal, W. (2018). Crypto Transaction Dispute Resolution. *Business Lawyer*. Retrieved September 01, 2021, from https://papers.ssrn.com/sol3/papers.cfm?abstract_id=2992962

Cong, L. W., & Zhiguo, H. (2018). *Blockchain disruption and smart contracts*. NBER Working Paper No. w24399. Retrieved September 01, 2021, from https://papers.ssrn.com/sol3/papers.cfm?abstract_id=3138382

Cornell, N., & Werbach, K. (2017). Contracts ex Machina. *Duke Law Journal, 67*, 313–382.

Kerikmäe, T., & Rull, A. (Eds.). (2016). *The future of law and eTechnologies* (p. 233). Springer International Publishing.

Siedler, N. (2016). The legal magnitude of blockchain. *Bucerius Executive Education*. Retrieved September 01, 2021, from http://www.bucerius-education.de/home/news-termine/blog/artikel/the-legal-magnitude-of-blockchain/

Simont Braun. (2018). Our review of the EU Commission Draft Fintech Action Plan. Retrieved September 01, 2021, from https://simontbraun.eu/news/2018/02/21/our-review-of-the-eu-commission-draft-fintech-action-plan/

Szczerbowski, J. J. (2017). Place of smart contracts in civil law. A few comments on form and interpretation. In *Proceedings of the 12th Annual International Scientific Conference 'New Trends' 2017*. Private College of Economic Studies Znojmo (pp. 333–338).

Crowdsourcing as a Means for Participatory Legislation

Pascal Soepper

Contents

Abstract This contribution looks at different forms of internet-based crowdsourcing to engage the general public in the legislative process. While defining the requirements that make this particular application of crowdsourcing possible and evaluating three representative examples (the Finnish off-road traffic law, the

P. Soepper (✉)
Yokohama, Japan

G. Borges, C. Sorge (eds.), *Law and Technology in a Global Digital Society*,
https://doi.org/10.1007/978-3-030-90513-2_17

Icelandic constitution reform process, and the platform Madison in the United States), two main approaches are identified as the most promising forms of crowdsourcing in order to engage as many people as possible. These are Feedback/Commenting as well as Know-How Accumulation, both due to their ease of access. Furthermore, arguments for and against the use of crowdsourcing in the legislative process are presented. The overall conclusion, despite some hurdles, is positive in that citizen engagement via crowdsourcing, under the right circumstances, has the potential to lead to a more deliberative, representative, open and transparent process that also heightens governments accountability, increases the quality of newly drafted laws and improves the overall acceptance of democracy itself.

1 Introduction

The term 'crowdsourcing' originally was coined by Jeff Howe in 2006 in an article for Wired Magazine as a combination of 'crowd' and 'outsourcing'.[1] Howe again, on his website dedicated to the phenomenon, is the one giving us the definition of crowdsourcing as "...the act of taking a job traditionally performed by a designated agent (usually an employee) and outsourcing it to an undefined, generally large group of people in the form of an open call".[2] This definition still needs to be amended by the importance of the internet as a platform for crowdsourcing and we arrive at the dictionary definition of the term as it is used today:

> The practice of obtaining information or input into a task or project by enlisting the services of a large number of people, either paid or unpaid, typically via the Internet.[3]

In the beginning, crowdsourcing was a phenomenon mostly to be observed in the private sphere. On one hand, it created a new way in which people would engage online to mutually and freely create and share, thereby giving rise to the open source software movement as well as many examples of crowdsourced know-how platforms, such as the online encyclopedia Wikipedia. The benefits of the efforts of countless individuals may be used freely by everyone with access to the internet. On the other hand, corporations developed a whole range of new business models which rely heavily on input from the crowd. Platforms such as the video portal YouTube or social networks such as Facebook supply the platform and rely to a great extent on its users to fill it with content. Revenue is generated through advertisement or via marketing of other for pay content. Crowdsourcing has various other applications,

[1] Howe (2006).

[2] Howe (n.d.).

[3] Oxford Dictionary of English (n.d.-a).

such as in research and development, know-how gathering or problem solving, for example on platforms like InnoCentive.[4]

Recently, crowdsourcing is used as a tool to engage citizens in political matters as well. Various platforms as well as individual projects have been initiated, oftentimes by politicians or legislatures themselves. These range from simple problem mapping systems like SeeClickFix,[5] gathering input on the allocation of the budget for a local community,[6] the development of a region,[7] up to crowdfunding civic projects[8] and drafting legislation. All these new ways of participation can be categorized as measures in the sphere of direct democracy and a lot more can be said about using crowdsourcing as well as its monetary application crowdfunding in the democratic process, which would exceed the purpose of this paper.[9]

This paper will focus on the very specific application of crowdsourcing in the legislative process. It will first attempt to broadly categorize different forms of participation (Sect. 2) and then present a selection of examples in which crowdsourcing is used or has been used recently to engage citizens in the drafting process of legal documents (Sect. 3). The mechanisms will be described, and the results of such efforts discussed. The discussion will then turn to a demonstration of how the application of crowdsourcing can potentially harness what is called 'the wisdom of crowds', i.e. the notion that unorganized crowds under the right conditions can produce excellent results, which are as good or even better than what an expert or a group of experts may produce. The discussion will also look at the prerequisite to harness that same wisdom in the legislative process. Therefore, Sect. 4 will focus on the discussion of how this wisdom can be harnessed in general as well as to possible rules that may apply to the legislative process in particular. Finally, it follows a discussion of further drawbacks and benefits of crowdsourcing legal texts. For that purpose, Sect. 5. will present arguments for and against crowdsourcing the legislative process and will ultimately demonstrate how crowdsourcing has the potential to improve deliberation and participation in law making processes, educate the public on public matters and create a more open process in which stakeholders' share responsibility, and in which decision makers can be held more accountable.

[4]https://www.innocentive.com/, accessed 29 August 2021.

[5]Where citizens can use a mobile app or website to send requests to their local governments, for example when they see a broken streetlight that needs fixing; https://seeclickfix.com/, accessed 29 August 2021.

[6]As the city of Chicago did in 2011 (Aitamurto, 2012, p. 23).

[7]As the region of Charlotte located in North and South Carolina with the CONNECT Our Future initiative; http://www.connectourfuture.org/, accessed 29 August 2021; (Warner, 2014).

[8]As on the discontinued platform Citizinvestor; www.citizinvestor.com/, last accessed 22 October 2017 (now taken offline).

[9]For a broader discussion of the various applications of crowdsourcing and crowdfunding in the democratic process see Soepper (2018).

2 Different Forms of Participation

This section shall give an overview of the possibilities for engaging the public in the legislative process and describe those abstractly, before Sect. 3 will turn to specific examples and their analysis. The following description is not meant to be exhaustive but intended to give a general outlook. Various overlapping schemes or completely different ones can be imagined.

2.1 *Opinion Polls*

Opinion Polls are a method which has been used extensively in the past, not necessarily initiated by policy-makers, but oftentimes carried out by private organizations such as Gallup. Examples include polls regarding the agreement rate of any certain party or candidate in an upcoming election. Much more relevant for the discussion at hand, opinion polls may also be used to gather information on the approval rate of any given policy decision. With the rise of the internet, nowadays opinion polls on websites are used extensively. This method of input gathering has only a very limited relevance for the area of crowdsourcing, due to its limits and simplicity, which basically ends in mere yes/no results. Rightfully, the simple extension of opinion polls to websites has been criticized as not a new form of direct engagement of the public, but just a change of technology for something that has been around for a long time.[10] The same criticism holds true for other mere voting on proposal schemes. Polls in the sense of up- and down-votes for proposals may however be used in addition to other forms of crowdsourcing in order to measure the approval rate of any one proposal.

2.2 *Initiatives/Petitions*

Initiatives or petitions also have been used in the past, but the Internet gave rise to new ways of facilitation. New online platforms have been created in various countries around the world, on which anybody may post an initiative or petition. The specifics may vary from place to place, but generally, if a proposal gathers a certain amount of support, it has to be discussed in or even decided upon by the public assembly. In the UK for example, 10,000 supporters are necessary for a petition to receive a response from the government whereas 100,000 supporters are needed for a petition to be considered for debate in parliament.[11] A recent example for a successfully crowdsourced bill, which had been brought to the public assembly,

[10] Rushkoff (2003), p. 56.

[11] See https://petition.parliament.uk/, accessed 29 August 2021.

is the Finnish marriage equality law. After having been proposed on the Finnish Open Ministry platform and gathering more than double than the necessary approval of 50,000 supporters on the first day back in 2013, the law became effective as of March 2017.[12]

2.3 Feedback and Commenting

Platforms for petitions, feedback and commenting systems may be implemented on any given platform, oftentimes in combination with up/down voting mechanisms for either the original proposal as well as the comments. In the legislative sphere, it is for example possible to post an original draft of a new bill online and allow for commenting and voting by the crowd. The example of the Icelandic constitution which will be discussed under Sect. 3.1 below, is a good example of such a scheme, as is the platform Madison, discussed under Sect. 3.2.

2.4 Know-How Accumulation

Another option is to start out without a pre-drafted bill and engage the public in know-how gathering, as has been done in the example of the off-road traffic law in Finland (see Sect. 3.3). In contrast to posting a pre-drafted bill on the internet for the crowd to comment on, here the crowd is actively engaged in the pre-drafting stage, i.e. the initial phase of problem mapping and solution gathering for the problems identified. This method of crowdsourcing may of course be combined with the above described feedback and commenting procedure, here on the individual proposals by other users.

2.5 Commons-Based Peer Production

The most engaging form of crowdsourcing is the same model that is used on Wikipedia or GitHub and called ‘commons-based peer production’. Generally, the idea is that anybody may use the original proposal (e.g. the text of a Wikipedia article or the source code on GitHub) and alter it in any way he/she sees fit. On Wikipedia those changes are immediately visible to the observer, whereas on GitHub a new

[12] Dawson (2013), Dittrich (2017), Eördögh (2012) and Petitioners take gay marriage bill to Parliament (2013).

fork of the original code is created, which via a so-called push/pull request may be implemented back into the main fork.[13]

Applied to the legislative sphere, the text of a new law may be drafted jointly online by a large number of people, just like a Wikipedia entry. Just as well, the community working on a draft would be evaluating changes and either accept or reject them. One example for this approach is the attempt by Assemblyman Mike Gatto in 2014 to use the Wikipedia-like platform Wikispaces to crowdsource a change to the probate law bill of California.[14] Clay Shirky, in his 2012 TED talk "How the internet will (one day) transform government", gives a good outlook on the application of open-source software tools such as GitHub to the democratic process.[15]

Unfortunately, this most engaging form of participation is at the same time the one with the highest hurdles of participation. Depending on the method used, already a great portion of interested citizens may be left out due to technical challenges of engaging with a platform like GitHub. At the same time, writing a legal text requires a certain amount of background knowledge as well as time commitment. Writing down a proposal for a new law or commenting on other peoples' ideas is much simpler than writing the whole text or at least parts thereof oneself. Therefore, commons-based peer production faces the challenge that only few people participate in the process,[16] which in turn makes the whole crowdsourcing process less inclusive and likely also less 'wise'.[17]

2.6 *Different Methods—Different Goals*

Crowdsourcing may be used not only in various ways, but each method may be applied with a different outcome in mind. Commons-based peer production can be seen as the strongest form of participation, in which the crowd is involved in the actual drafting of a legal text, whereas simple opinion polls are the weakest form, only taking note of the support of any given proposal as a whole. The most common and most promising approaches of crowdsourcing for legislation currently seem to be feedback and commenting schemes on a pre-drafted proposal, as well as the knowledge gathering approach, or a combination of both. Both have been described

[13] A fork is a copy of the original code which anybody can create. Via a so-called pull request, changes to the new fork can be sent back to the original project. GitHub describes forking as being "...at the core of social coding on GitHub."; https://guides.github.com/activities/forking/, accessed 29 August 2021.

[14] Heaton (2014).

[15] Shirky (2012).

[16] Aitamurto (2016), p. 2787.

[17] For a discussion of the wisdom of crowds, see Sect. 4 below.

above under Sects. 2.3 and 2.4 and shall be further analyzed via the examples to be described in detail in the next section.

What all the following three examples have in common, is that the crowd is not directly writing or changing the text of any given proposal, but that this task is up to a council or moderator, who uses the crowd as a large knowledge base as well as opinion machinery to gather input from the crowd either beforehand (as in Finland's off-road traffic law) or during the drafting process (as in the other two examples, where an initial draft is presented to the crowd in order to comment on and debate). Both methods have a low threshold of participation due to their ease of access. More people are willing and able to participate, which leads to a more diversified and in turn wiser crowd, which may produce better results. From the examples described in the following section, it will become apparent that those methods, combined with some form of moderation, are currently the most promising form of engagement of the general public in the legislative process.

3 Specific Examples

This section will describe three representative examples of engaging the public in the legal drafting process. One of those deals with a dedicated platform available for elected officials to publish legal texts (the Madison Project), two more deal with singular examples where crowdsourcing has been used to draft new laws for a specific purpose (Iceland's constitution, which involved mainly feedback and commenting on drafts of the text, and Finland's off-road traffic law, which involved idea gathering in the pre-drafting phase). The selection is not supposed to be conclusive; there are many more examples which deserve attention,[18] though it would by far exceed the margins of this paper to describe them all in detail. The three examples have been chosen to highlight certain arguments, such as the quality of crowdsourced legislation (Sect. 3.1), the simplicity and interactivity of certain crowdsourcing models (Sect. 3.2), and the benefits of crowdsourcing in the pre-drafting stage as well as of moderating the process (Sect. 3.3).[19]

[18] For example the Open Ministry Platform in Finland (Dawson, 2013); or the constitution of Mexico City (Campoy, 2016).

[19] Also, all three examples will serve important functions in the final discussion at the end of this paper.

3.1 Iceland's Crowdsourced Constitution

The example of the Republic of Iceland trying to crowdsource a new constitution in 2011 has been discussed extensively.[20] The process was novel for a constitutional reform. First, a constitutional council consisting of 25 delegates was enacted. The delegates were not professional politicians but chosen from the general populace. Second, this council did not hold sessions behind closed doors, but tried to make use of the internet to engage the citizens of Iceland in the process. Besides a dedicated website,[21] the constitutional council made ample use of various online platforms and social media, such as Facebook,[22] Twitter,[23] YouTube[24] and Flickr.[25]

Input from the crowd was gathered by posting each new draft of the constitution online for people to comment on. Additionally, council gatherings were public and streamed live on the internet. Social media could be used for commenting and messages could directly be sent to the council. Messages in turn were published online for others to comment on. All documents produced by the council, such as minutes of their meetings, were published online. Finally, the whole campaign was advertised in the media in order to encourage participation by as many people as possible.[26]

The final draft was presented to parliament on July 29, 2011.[27] An English translation has been prepared and is freely available online.[28] This section shall attempt to look at the overall quality of the text as well as certain specific stipulations therein. The preamble reads as follows:

> We, the people who inhabit Iceland, wish to create a just society where every person has equal opportunity. Our diverse origin enriches our society and together we are responsible for the heritage of generations, our country and its history, nature, language and culture.
>
> Iceland is a free and sovereign state with freedom, equality, democracy and human rights as its cornerstones.
>
> The government shall endeavour to strengthen the welfare of the country's inhabitants, encourage their culture and respect the diversity of the life of the people, the country and its biosphere.
>
> We wish to promote harmony, security and happiness amongst us and coming generations. We are determined to work towards peace with other nations and respect for the earth and all mankind.

---

[20] See for example Gylfason (2013), Landemore (2014, 2015) and Siddique (2011).

[21] http://stjornlagarad.is/, accessed 29 August 2021.

[22] https://www.facebook.com/Stjornlagarad, accessed 29 August 2021.

[23] https://twitter.com/Stjornlagarad, accessed 29 August 2021.

[24] https://www.youtube.com/user/Stjornlagarad, accessed 29 August 2021.

[25] https://www.flickr.com/photos/stjornlagarad, accessed 29 August 2021.

[26] The Constitutional Council (n.d.).

[27] The Constitutional Council (n.d.).

[28] http://stjornlagarad.is/other_files/stjornlagarad/Frumvarp-enska.pdf, accessed 29 August 2021.

> In light thereof we set a new Constitution, the supreme law of the land that all must observe.

The preamble exemplifies the overall tone of the whole text, which focuses on diversity, freedom and individual rights, but also on the responsibility for each citizen of Iceland towards fellow citizens and the environment. Noteworthy is also the lack of any religious reference in the text.

The first chapter begins with basic elements, such as the stipulation that Iceland is a parliamentary democracy in Art. 1. Chapter two then focuses on human rights and nature from Art. 6 onwards. Art. 6 itself, which is the constitution's equality stipulation, is a noteworthy example for input from the crowd. It reads as follows:

> Equal rights
>
> We are all equal under the law and shall enjoy our human rights without discrimination, such as due to gender, age, genetic character, place of residence, economic status, disability, sexual orientation, race, colour, opinions, political affiliation, religion, language, origin, ancestry and position in other respects.
>
> Men and women shall enjoy equal rights in every respect.

Attention shall be drawn to the attribute of 'genetic character', which is one of the suggestions coming from comments during the crowdsourcing process, addressing a problem nobody else had identified in advance. The rationale is that due the possibility of analysis of the genetic fingerprint of any given person, conclusions may be drawn about a person's health or other characteristics, which may give rise to discrimination in the same way as the other characteristics.[29]

Further noteworthy stipulations are the responsibility of citizens for Iceland's nature and environment in Art. 33 as well as the preservation of natural resources, including those in the oceans, which may not be sold off or pledged, thereby basically preventing a change to private ownership of such publicly owned natural resources. The following chapters of the constitution deal with matters such as Iceland's national assembly, the president, ministers, the judiciary, local governments and foreign affairs. Overall, the analysis shows that the crowdsourced text holds up to juristic scrutiny and that the input of the people was a valuable source of ideas.

Unfortunately, the final draft of the constitution, which had been approved by popular vote in a public referendum in 2012, failed to find approval in the public assembly, which was a prerequisite for its enactment.[30] Nonetheless, the example clearly shows that even a grand scheme, such as the supreme law of the land, may benefit from the crowd's input.

---

[29] Aitamurto (2012), p. 19.

[30] Gylfason (2013) and Landemore (2014).

3.2 Madison

Madison is a platform developed by the Open Government Foundation, a U.S. based non-profit organization. The Open Government Foundation was co-founded by two U.S. congressmen, Seamus Kraft and Darrel Issa, with the aim of developing various tools for connecting politicians and the public in the policy making process.[31] Their website states as their mission:[32]

> Serving those who serve the people
>
> The OpenGov Foundation is a fiercely apolitical nonprofit dedicated to serving those who serve the people in America's legislatures. [. . .]
>
> We empower those who serve in our democracy with better technology, data, and human-centered design. Everyone wins. [. . .]

Madison was first used to fight two bills, the Protect IP Act (PIPA) and Stop Online Piracy Act (SOPA), by proposing another bill, the Online Protection and Enforcement of Digital Trade Act (the OPEN Act). Neither SOPA, PIPA, nor the OPEN Act were finally enacted.[33]

The software to run the Madison platform is freely available as open-source on GitHub under GPL-3.0 license.[34] The Open Government Foundation also created a dedicated website to run Madison, which could be used by legislatures to open up the drafting process of legal bills to comments and discussions by the crowd, but which has been discontinued as of February 2019.[35] The 'How it Works' description on that website describes with a few words its rather straightforward mechanism (original emphasis):[36]

1. Elected officials share policy documents on Madison for public collaboration.
2. The people can read the documents, vote on the documents, and leave feedback on the documents, and their comments are sent **directly to the officials**.
3. Officials respond to these comments and even make changes to the documents based on the feedback.

Looking at any bill published on Madison highlights how important ease of use and accessibility were to the developers. Registration on the platform can be done with only registering an e-mail address and one's name. After verifying the e-mail address via an automatically submitted link, one is free to engage with the

[31] Finley (2013).

[32] https://www.opengovfoundation.org/, accessed 29 August 2021.

[33] Finley (2013).

[34] https://github.com/opengovfoundation/madison, accessed 29 August 2021.

[35] https://mymadison.io/, accessed 29 August 2021. Though the website is still online, the Madison platform has been shut down as of 10 February 2019 (The Open Gov Foundation, 2019). However, since the source code is still freely available, the technology behind the platform could be used by anyone to set up a new platform.

[36] https://mymadison.io/#how-it-works, accessed 29 August 2021.

documents posted on the platform. Noteworthy is, that registration does not require one to be a U.S. citizen or to be located in the U.S.

Each document posted on Madison can be read, up- or down-voted via two buttons ('I support this document' or 'I oppose this document'). Highlighting any part of the text with the mouse cursor allows for commenting on that part. If a comment has been made, the text is highlighted and on the right side of the text one can find a small button with a numbering counter, indicating how many comments or responses to this comment have been made. A click on this button in turn opens the discussion. Here a reply can be made to the initial comment and any additional comments as well as the original one can be liked via a thumbs-up button or flagged. In those discussion sections one can witness how feedback by officials is handled as well. If the name of the commentator is highlighted, this is a response from a government official of the legislature that posted the bill. The affiliation will be shown at the end of the comment.

This simple and straightforward mechanism allows for a direct engagement of the officials with the public. Anybody may comment and engage in the discussions or just read the text and vote on any given proposal. Just reading the bill or the comments does not even require creation of an account.

3.3 Finland's Off-Road Traffic Law

The reform of the off-road traffic law in Finland, attempted in 2013, is an interesting case study as its process was implemented by two researchers, Tanja Aitamurto and Helene Landemore,[37] who already have a proven track record of research in the area of crowdsourcing in law-making, for example with regard to the above described constitution reform process in Iceland.[38]

The process was divided into four stages, the first two of which included crowdsourcing.[39] "...[U]sers could propose ideas on the platform, comment, and vote others' ideas up or down. All of the crowd-generated input was visible to the public online. In order to leave a comment, propose an idea, or vote on the platform, the user had to register on the site".[40] Furthermore, the process was moderated by the researchers. They "...analyzed the participants' input and, together with the civil servants in the Ministry, organized the ideas and comments into categories. Then the ideas were assessed by crowd evaluation and expert evaluation...".[41]

---

[37] Aitamurto and Landemore (2016), p. 178.

[38] Aitamurto (2012), p. 18; Landemore (2015).

[39] Aitamurto and Landemore (2015), p. 5.

[40] Aitamurto et al. (2014), p. 15.

[41] Aitamurto and Landemore (2016), p. 179.

What makes this case study further interesting, is the fact that crowdsourcing was used in the research and drafting stage of the law.[42] This means neither was there a bill posted online to begin with, as in the two examples above, nor did the crowd decide on the final implementation of the new law. Rather, crowdsourcing was used to gather input from the crowd on what kinds of issues there were regarding off-road traffic (1st phase: Problem mapping) and how to solve those issues (2nd phase: Problem solving: Ideation); the non-crowdsourced 3rd and 4th phases were evaluation and law-writing.[43] Originally, the first phase started out with ten main problem areas, which had been pre-defined by civil servants of the Ministry of Environment and the researchers. Due to demand from the crowd, a 'Propose your own topic' category was added later on. This exemplifies how the crowd was able from the beginning to give meaningful input beyond the expectations of the officials.[44] Overall, "[t]he two crowdsourcing phases generated about 500 ideas and 4,000 comments, and 24,000 up or down votes from about 700 users".[45]

Evaluating the proposals and the following arguments around individual proposals set forth by the crowd on the platform, the researchers conclude that "...the participants exchanged arguments with others in the dialogical and intersubjective manner typical of the ideal of democratic deliberation described by deliberative democrats".[46] One example put forward on the discussion of age limits for driving off-road vehicles is worth summarizing here again. It went from a first commentator, who put forth the idea that there should be no age limits at all, drawing from his childhood experiences, via another commentator putting forth the dangers that come with young adults riding fast vehicles and another pointing out possible disturbances, to arguments put forward regarding distinctions between recreational and professional driving or different kinds of off-road vehicles (tractor or snow-mobile). Different commentators put forth different rationales and gave diverse input, all of which made the discussion broader and took into account former commentators' points of view. Furthermore, the tone of the discussion usually was respectful.[47]

Overall, it can be said that the process was very successful and via the online platform it was possible to gather a large amount of input, both regarding problems as well as their solutions with regard to off-road traffic. Furthermore, the crowd was able to interact respectfully with each other and no extreme polarization was observed. "Based on the evaluation analysis, it seems that the crowd preferred commonsensical and nuanced ideas, while rejecting vague and extreme ones."[48]

---

[42] Aitamurto and Landemore (2016), p. 177.

[43] Aitamurto and Landemore (2016), p. 179.

[44] Aitamurto and Landemore (2015), p. 5.

[45] Aitamurto and Landemore (2016), p. 179.

[46] Aitamurto and Landemore (2016), p. 182.

[47] Regarding the tone of the discussion it may be further noted that from the 4000 comments made on the platform, only 10–20 were removed due to inappropriate content or tone (Aitamurto & Landemore, 2015, p. 6; Aitamurto & Landemore, 2016, p. 187).

[48] Aitamurto et al. (2013).

After presenting these examples, the discussion now will turn to the drawbacks and benefits of crowdsourcing legislation, starting with the question of how to produce good results via the wisdom of crowds.

4 The Wisdom of Crowds

This section shall present and discuss the concept of what in collective intelligence research is called 'the wisdom of crowds'. First the initial idea shall be presented before the application to the legislative process shall be discussed.

4.1 Original Idea

The research on group dynamics and the good results that stem from aggregating the average of large amounts of peoples' input has been summarized by James Surowiecki in his book 'The Wisdom of Crowds', first published in 2004.[49] Surowiecki took the title for his book from an earlier study on crowd psychology by Charles Mackay, published in several volumes from 1841 onwards and entitled 'Memoirs of Extraordinary Popular Delusions and the Madness of Crowds'. Mackay presents the reader with numerous historical examples of group behavior gone bad, such as the crusades and witch hunts.[50] In contrast with the notion of mad crowds, Surowiecki presents the reader with numerous examples in which unorganized crowds have produced better results than experts or in which crowds were able to solve problems that experts could not solve.

The notion of unorganized crowds is contrasted with that of the classic group. On one hand, a crowd is not just a random gathering of people with the same purpose, as for example the crowd at a concert or a soccer match. A crowd in the sense of crowdsourcing needs something more, i.e. a form of cooperation to achieve a common goal. However, this cooperation is not as stringent as in a group, in which members are bound together and organized, usually with a leader at the top. The catchphrase that illuminates the contrast between crowds and groups is 'cooperation without coordination'.[51] Shirky illustrates the contrast by the different approaches to the development of open source software and proprietary software. Classic software production is based on a centralized approach in which tasks run together at a central entity to be combined into the finished product. Open source

[49] Surowiecki (2004).

[50] Mackay's work may now be accessed freely via the Project Gutenberg archive: https://www.gutenberg.org/ebooks/24518, accessed 29 August 2021.

[51] Shirky (2012).

software development, in contrast, is decentralized and shows a much more complex structure of interdependencies between individual contributions.[52]

A crowd may be unorganized in the sense that members may independently choose the tasks to engage with without being in need of justifying their decision. This however, does not imply a chaotic structure in which nothing works, but rather a structure in which everybody has a common goal and may voluntarily decide how much and in what way to contribute to this goal. In crowds, as Surowiecki points out, also two of the usual perils of groups, those of 'confirmation bias' (i.e. the tendency "...which causes decision makers to unconsciously seek those bits of information that confirm their underlying intuitions")[53] as well as 'group polarization' (i.e. the phenomenon, also observable in juries, that deliberation causes individual members to adjust their own opinions to the group's opinion)[54] are alleviated.

Surowiecki does not assume that crowds are wise all the time and under any given circumstances but identifies four ground rules for unorganized crowds to be wise, i.e. to produce as good or better results than groups or individuals. Those are:[55]

1. ...diversity of opinion (each person should have some private information, even if it's just an eccentric interpretation of the known facts),
2. independence (people's opinions are not determined by the opinions of those around them),
3. decentralization (people are able to specialize and draw on local knowledge), and
4. aggregation (some mechanism exists for turning private judgements into a collective decision).

The internet is the ideal crowdsourcing platform for harnessing the wisdom of crowds, as it by design fits all those four characteristics:

1. Although not (yet) every single person on earth is connected to the world wide web, the spectrum of opinions is as large as in the population as a whole (maybe with the noteworthy exception of certain non-tech-savvy minority groups).[56]
2. People on the internet, at least in societies with free and unrestricted access to the web and a guarantee of freedom of speech, are much less prone to be influenced by others around them. Of course, even on an online forum people do interact with each other and may be influenced by other users' comments. However, especially a person who would not speak out and utter his/her opinion in the real

[52] Shirky (2012).

[53] Surowiecki (2004), p. 178.

[54] Surowiecki (2004), p. 184.

[55] Surowiecki (2004), p. 10.

[56] Also, especially older people on average are not represented as much as younger generations and a lot of wisdom may be left out of the process. This concern may possibly be addressed by a mixed approach of crowdsourcing via public assemblies or paper questionnaires alongside internet-based platforms. However, even without such a mechanism, there currently is no other platform next to the internet, which allows for such a broad spectrum of opinions to be represented and aggregated by technical means.

world, may indeed do so in the more indirect and at times anonymous cyberspace.[57]

3. The internet's architecture is decentralized by design. Anybody living in a place with unrestricted access may participate in any endeavor on the internet, even in another part of the world.[58]
4. Different websites, forums or platforms offer various forms of aggregation. In some cases, it may be a centralized entity deciding on the best submission by a user, in other cases it may be the community at large deciding which submission makes it into the main fork of a project. The possibilities are almost limitless and new forms of cooperation without coordination spring up over time.

One illustrative example for the wisdom of crowds is that of prediction markets, among those most notably the Iowa Electronic Markets (IEM) run by the University of Iowa. "The Iowa Electronic Markets is a futures market run for research and teaching purposes. Traders can buy and sell real-money contracts based on their belief about the outcome of an election or other event. Using this 'wisdom of crowds,' the price of a contract at any given time is a forecast of the outcome."[59] Studies have shown that for example in predicting the outcome of U.S. presidential elections between 1988 and 2004, the political market of the IEM has on average outperformed predictions by polls 74% of the time as well as in every election when forecasts of 100 days in advance were analyzed.[60] Moreover, the results of the IEM in the presidential elections between 1988 and 2000 were, on average, off by only 1.37%.[61]

These arguments for the wisdom of crowds may not exactly be scalable to the legislative process. However, the legislative, or more generally the political process, may be able to harness that same wisdom, if done the right way. So far, there have been various experiments, some of which described above, that took different approaches in engaging the public in the legislative process. The next section will attempt to address the question of which ground rules apply to the wisdom of crowds

[57] Anonymity is not a prerequisite for crowdsourcing though. If it is an advantage or a peril in crowdsourcing legal texts is yet to be determined. In the above presented example of the Finnish off-road traffic law, possible anonymity did not hinder the respectful exchange of arguments (Aitamurto & Landemore, 2016, p. 187). Furthermore, Aitamurto/Landemore note that "...partial depersonalization of the exchange arguably allows it to be more fluid and enduring than actual face-to-face exchanges involving physical persons." (Aitamurto & Landemore, 2016, p. 186).

[58] It may be noted that of course other barriers than technical ones exist. One of the greatest barriers, despite the rapidly growing quality of translation services, is still language. It is unlikely that many users from other countries would be able to give input to an Icelandic or Finnish project. Also, the designers of any given platform may choose to limit access to the platform to people from a certain location (indicated by their IP addresses). This however would mitigate one of the benefits of crowdsourcing, namely that somebody from a different background and with a different point of view may be able to give valuable input to a project in which he/she has no direct stake.

[59] https://tippie.biz.uiowa.edu/iem/, accessed 29 August 2021.

[60] Berg et al. (2008), p. 5.

[61] Surowiecki (2004), p. 18.

in the legislative process and elaborate some of those arguments by using the examples discussed above.

4.2 Application to the Legislative Process

Aitamurto/Landemore take the example of the crowdsourcing process for the Finnish off-road traffic law as described above to extract five design principles for crowdsourced policymaking, including crowdsourcing in the legislative process.[62] Those are[63]

1. Inclusiveness ("...including a large number of people from a diversity of backgrounds...")
2. Accountability ("...justification of authorities' decisions and actions to the public, and their answerability to the citizens.")
3. Transparency ("...enables the participant to follow the policymaking process and peer communication...")
4. Modularity ("...enables the comprehensibility of a long policymaking process, enables participation in several parts of the process, and enables ad-hoc and long-term participation. It also fosters a sense of accomplishment.")
5. Synthesis ("...provides summarized outcomes and makes it possible to follow the developments in the process.")

These principles are not far off from the principles laid out above regarding wise crowds in general, but go a little bit further, taking into account the specifics of crowdsourcing public policy. Number 1. is comparable to Surowiecki's first (and partly second and third) principle, whereas numbers 2–4 are specifics for the legislative process (numbers 2 and 3 addressing questions of legitimacy and number 4 one of the process design). Number 5—Synthesis is a form of aggregation, as in Surowiecki's fourth principle.

One reason for the larger scope of these principles than the ones presented above is the difference between deliberation and aggregation. As described above, Surowiecki's version of the wisdom of crowds is mainly about aggregating the average of individual contributions to any given project, whereas not much focus is put on interaction and deliberation between the participants.[64] The legislative process however, as one of the strongest forms of democratic participation, highly benefits from a more interactive and deliberative design of the crowdsourcing platform, as the example of the off-road traffic law vividly highlights.[65]

---

[62] Aitamurto and Landemore (2015).

[63] Quotes taken from the overview table in (Aitamurto & Landemore, 2015, p. 10).

[64] Landemore (2013), p. 7.

[65] And this despite the fact, that the Finnish example was not designed for deliberation, but knowledge and idea search; deliberation just happened (Aitamurto & Landemore, 2016, p. 191).

These five design principles may not be the final solution to the question of designing crowdsourcing procedures for law-making, but they are a very good starting point. Future research and experimentation will have to address the design question and will likely find different answers for different goals, e.g. different designs for know-how gathering projects as opposed to commenting on a pre-drafted bill. What is evident from the case studies, however, is that interaction and deliberation between the users is a highly valuable source of the wisdom of crowds, that should be taken into account in future design models.

After making the point that crowdsourcing under the right conditions may be able to harness the wisdom of crowds and produce excellent results also for the legislative process, the next section will turn to a brief attempt at addressing further arguments for and against its application beyond the question of quality of the outcome.

5 Discussion

This section shall attempt to put forward further arguments regarding the usage of crowdsourcing the legislative process. Both arguments for and against its application shall be presented and discussed. The overall outlook of the discussion, however, is positive towards the idea of crowdsourcing legislation.

5.1 Openness and Transparency

Crowdsourcing generally is defined as an open call for anyone to participate in a task on the internet.[66] All crowdsourcing processes described here are open and transparent processes in the sense that they are first, open for all to participate in a free exchange of ideas, and second, transparent in the sense that anybody can see all the exchanges already present and engage with those as well. This is what Aitamurto/Landemore call "'horizontal transparency' of the deliberation (everyone's comments were equally visible to all)".[67]

Furthermore, the processes are open for participation by anybody, no matter if he/she is a citizen of that particular jurisdiction, a minor, stakeholder, using his/her real name or a pseudonym or any other possible characteristic about the background of the participant. This openness sets a very low bar for participation, which especially in the knowledge gathering approach to crowdsourcing serves the purpose of engaging as many and as diverse a crowd as possible, while nobody must fear being rejected or judged based on other factors than the force of argument.

---

[66] See definition of crowdsourcing in the introduction as well as Aitamurto and Landemore (2016), p. 177.

[67] Aitamurto and Landemore (2016), p. 186.

5.2 *Participation, Inclusiveness and Accountability of the Public*

As seen above at the category of commons-based peer production, the more complicated a procedure is, the less likely are people to participate. In order to maximize participation, the threshold has to be as low as possible. Even if the threshold is comparatively low, as is the case for example on the Madison platform described above, not every subject matter will gather a similarly large crowd. Quite the opposite, people will engage with those topics they care about and ignore others. As in the case of the Finnish off-road traffic law, "[t]he common denominator for the participants was that they had a reason to care about off-road traffic and the law regulating it".[68] Therefore, one can conclude that the more specific the area of the law to be crowdsourced, the more specialized (and smaller) the crowd will be.

This raises the question of how inclusive a process really is and what happens to the voices of those who just do not make themselves heard, although a new law may impact them as well, however remotely. As Aitamurto et al. put it for the Finnish off-road traffic law: "If the whole population were asked (in fact, they were asked, because the process was open for anybody to participate in) to generate ideas or evaluate those, would they bother to participate? If they don't care, should those who care not be heard at all? This is one of the core questions in participatory policy-making."[69] This is a universal question, which applies to any crowdsourcing of public policy. However, the answer can only be that this is what democracy also implies, that one has the right not to engage in the process.

At first glance it may seem a contradiction to propose on one hand that the larger and more diverse the crowd is, the better the results, and on the other hand to accept that the more specialized a subject matter, the smaller the crowd that will engage with it. Of course, crowdsourcing cannot meet its objective if the crowd is too small. However, this does not imply that a small and more specialized crowd cannot produce good results. What is more important in crowdsourcing, is the possibility to engage with the process, not necessarily the actual participation. A few great ideas are better than a whole lot of average ideas and much better than none at all. This line of reasoning can also be observed within the crowd under what is called the '1-9-90 rule'.

"The rule states that typically 90 percent of the visitors only observe the online activity and 10 percent participate more actively. Out of the 10 percent, 1% produce most of the content."[70] In the Finnish off-road traffic law example, out of 7000 visitors to the site only 700 registered and out of those only 175 users were the 'idea producers', out of which 121 produced only one idea and 46.44% of all ideas were

[68] Aitamurto and Landemore (2016), p. 181.

[69] Aitamurto et al. (2013).

[70] Aitamurto and Landemore (2015), p. 6.

produced by the ten most active users.[71] This shows that from an initially large group of people being interested in the process and visiting the site, only a very small portion actively engaged with it. This does not necessarily imply that the others only took a brief look and did not care about the process. It can also be interpreted as support for the ongoing discussion and the feeling not to be able to contribute anything else meaningful to it. Again, inclusiveness does not imply everybody has to comment, it only means everybody should be able to do so and that if he/she does, then the opinion has to be heard.

5.3 Deliberation

Depending on its design, the crowdsourcing process can be more or less deliberative. Strikingly, even in the off-road traffic example, deliberation took place despite the process design, which was not originally intended for deliberation.[72] Deliberation is defined as a "[l]ong and careful consideration or discussion".[73] The best example for this in the legislative process is again the off-road traffic law in Finland, in which the participants engaged in a civil and meaningful discourse regarding the proposals put forward on the platform (see Sect. 3.3). The researchers noted that the discussion focused not around the participants, but around the proposals and their surrounding arguments. "Deliberation takes place between viewpoints rather than persons themselves. Anyone can show up on the thread and take up a view or attack it. [. . .] The conversation can continue over several days, weeks, or months, with comments arriving at various times."[74]

This feature of an online crowdsourcing process is very distinct from the political arena in which proposals are oftentimes evaluated depending on the official or party it originated from. In this sense, one may argue that a more anonymous online platform allows for deliberation with less bias than the usual political stage.

The Madison platform also highlights the deliberative aspect of drafting legislation. Comments made by one person may be up-/down voted as well as commented on again by others, so that a discourse can evolve around every single comment. Furthermore, Madison allows for direct deliberation not only between users of the platform who are members of the general public, but also relies on direct interaction with public officials who are in charge of that particular bill. In that respect, Madison is a great tool for politicians to mitigate misunderstandings of any proposed bill and it challenges them to directly engage with individual comments and discussion threads, not only with the final aggregated result. The officials become part of the crowdsourcing process.

---

[71] Aitamurto and Landemore (2015), p. 6.

[72] Aitamurto and Landemore (2016), p. 191.

[73] Oxford Dictionary of English (n.d.-b).

[74] Aitamurto and Landemore (2016), p. 186.

Of course, any such discourse may not be able to resolve different points of view totally and there always will be some viewpoint that in the end is totally marginalized. It is likely impossible to create a process of perfect deliberation, at which end stays a result that everybody may support. Nonetheless, this is exactly what democratic deliberation is all about, that at least everybody has the chance to offer his point of view and make a strong case for it in order to convince others. The process does not have to chase some ideal form of perfect deliberation, it only has to be good enough to be accepted by the users (including public officials) and produce good results.

5.4 Representativeness

Connected to the questions of inclusiveness and deliberation is that of Representativeness. Will crowdsourcing create results, which are representative of the general public's point of view on the matter, or will it be only the opinion of a minority that engages in the crowdsourcing process? Or will the end-result even be a complete misrepresentation of public opinion?[75] One may further ask if this problem is heightened by the fact that people who are not affected by the law (e.g. foreigners from a different jurisdiction) may participate in the process.

This difficulty for once is alleviated increasingly, the bigger and the more diverse the crowd is. Apart from that, and although this may be hard to evaluate, lacking representativeness is not necessarily an argument against crowdsourcing. No matter how big the crowd may be to begin with, input even from a small crowd is more public input than without the use of crowdsourcing. To further heighten representativeness, it may become important to raise public awareness of the possibilities for crowdsourcing. Furthermore, input from a foreign crowd, where possible due the lack on language restrictions, may bring in a different perspective that nobody from the affected area has thought about before, therefore, as Sen rightfully points out, help to prevent 'local parochialism'.[76]

Another question in that respect is that of the voices of minorities. To deal with this issue, in the Finnish off-road traffic law example, the moderators used a clustering algorithm which was able to identify a minority cluster, which could then be taken into account more thoroughly. This kind of technology may help to make minority opinions heard and send the signal to minority groups that their efforts are not in vain.[77]

[75] Aitamurto (2016), p. 2788; Aitamurto et al. (2013).

[76] Sen (2010), p. 128.

[77] Aitamurto et al. (2013).

5.5 *Educational Approach*

Using crowdsourcing in the legislative process creates not only good results on the receiving end of the process, namely for the politicians, council or experts that employ crowdsourcing to gather input. It also is of benefit to the people participating and giving the input, as they learn from the process in various ways.

First, they learn from each other in a deliberative forum, by listening or interacting with one another and taking opposite points of view into account when engaging with the discussion. Thereby participants will learn how deliberation in the political sphere works, as well as enhance their knowledge of the subject matter.[78]

Second, they learn about the process of creating a new law itself, its challenges and hurdles and how to overcome those. For example, in the Finnish off-road traffic law, "[a]s stated by the Minister of Environment, the goal was to search for knowledge and ideas from the crowd, enhance people's understanding of the law, and attempt to increase the perception of the policy's legitimacy".[79] Crowdsourcing for legislation will help the public understand the law at stake as well as the general legislative process.[80]

Lastly, depending on the level of interactivity with public officials and the information provided by them, the crowd may even learn something about the political process as a whole and their role in it. The crowd in the Finnish example was "...hopeful, but realistic..." regarding their possible impact on the actual law, i.e. they were aware that one particular opinion would not end the debate and that they were not the ones directly creating the law, but only one form of knowledge accumulation among others, in order for politicians to make the final call in the end.[81] This is an important point which is related to the method used. In a crowdsourced bill employing commons-based peer production, the expectations of the crowd are very different, as they rightfully would expect that parliament will decide on that particular text without further changes. Any changes made need a special justification.

5.6 *Accountability of Lawmakers*

Legislators employing crowdsourcing in the legislative process raise their own accountability towards the public. If a proposal from the crowd gains enough traction, it will be much more difficult for the officials in charge to explain why such proposal has been rejected. Of course, the methods currently employed do not

---

[78] Aitamurto et al. (2013).

[79] Aitamurto and Landemore (2016), p. 178.

[80] Bautista (2015), p. 2.

[81] Aitamurto et al. (2013).

come with any measure that makes officials directly accountable to the crowd.[82] Still, democracy requires that officials be accountable to the public, which in turn requires proper justification for their decisions.[83] A large enough crowd may just be able to create enough pressure on legislators to hold them more accountable.

A problem in the Finnish off-road traffic law (and most likely also in the Icelandic constitution after the final draft was submitted to parliament) was, that law makers at large did not necessarily know how to handle the input from the crowd.[84] For the future of crowdsourcing in the public sphere, mechanisms have to found which alleviate such difficulty. The Madison platform again is a good example in that it provides a direct link between the officials making use of the platform to open it up to the crowd and the crowd itself. On Madison there is a direct dialogue between the users and the officials via the commenting possibility. Comments can be debated with the ones responsible for drafting the bill and officials can either accept new recommendations from the crowd or try to persuade the commentator of their position. If public officials are engaged in the process from the beginning, this may help to alleviate the problem derived from handing over a long report on the aggregated results of a months-long crowdsourcing process.

5.7 *Quality of Crowd-Input*

As described above, the quality of crowd-input can overall be categorized as good and valuable. The example of the Icelandic constitution shows that some aspects, such as the problem of discrimination based on genomics, had not been thought about before and only been put forward by the crowd. The discussion around the Finnish off-road traffic law exemplifies the deliberation that went into the problem-mapping and problem-solving phases and the vast amount of valuable input that could be gathered via crowdsourcing.

The wisdom of crowds suggests that the larger and the more diverse a group is, the better will the results be.[85] A great effort therefore has to be made to engage a crowd as large and as diverse as possible, which is one of the design principles for crowdsourced policy making (Inclusiveness).[86] The more inclusive and the better the original input, the better the actual outcome of the crowdsourcing process will be, assuming that in very large projects a form of aggregation can be implemented that allows for all voices to be taken into account.

[82] Unless one counts the support thresholds for discussion in the public assembly for initiatives and petitions as described under Sect. 2.2 above.

[83] Aitamurto et al. (2013).

[84] Aitamurto (2016), p. 2787.

[85] Aitamurto (2016), p. 2780.

[86] Aitamurto and Landemore (2015), p. 10.

5.8 *Sabotage Attempts and Interest Group Influence*

Wikipedia in the past saw several attempts by interest groups or stakeholders to edit the respective entry concerning themselves, in order to let them appear in a more positive light. Usually these attempts are recognized by the crowd as what they really are: sabotage or whitewashing.[87] The respective Wikipedia article can then be changed back to a more objective version. In cases where such whitewashing appears frequently, it is possible to limit the editing policies of any given article, so that special permissions are needed to change it.[88] Similarly, regarding the California probate law briefly mentioned above, the responsible assemblyman "Gatto explained that if a special interest group proposes a law or changes legislation in a blatantly biased way, people would see right through it".[89] Though both examples apply to commons-based peer production, the rationale holds true for other varieties of crowdsourcing as well.

In the off-road traffic law example, some participants openly participated as representatives of their stake-holding organizations.[90] This is not an argument against the process, but rather helpful, since of course also the stakeholders and special interest groups are members in the democratic process and should be able to have a voice during the crowdsourcing process. Considering that laws in the traditional legislative process are often highly influenced by special interest group lobbying, crowdsourcing may actually help to "...level the playing field..."[91] between the general public and such lobbyists.[92]

6 Conclusion

We have seen how with the rise of the internet and the rise of crowdsourcing in the private sphere, nowadays, those new technological means of engaging a large crowd of people for any given task can be used in the legislative process. Going way beyond what would be called a simple opinion poll, crowdsourcing is about engaging the public in the actual process of drafting a new law. There are various models, which have been presented above, the most promising of which are knowledge gathering in the pre-drafting stage and feedback and commenting on the actual draft of a new bill.

The knowledge gathering approach for the off-road traffic law in Finland has shown that especially people who are affected by the new law were the ones

[87] Singel (2008).

[88] https://en.wikipedia.org/wiki/Wikipedia:Protection_policy, accessed 29 August 2021.

[89] Heaton (2014).

[90] Aitamurto (2016), p. 2784.

[91] Bautista (2015), p. 2.

[92] Bautista (2015), p. 5.

participating. The tone of the discussion among the participants was respectful and people engaged with each other in a deliberative way, taking into account other users' points of view. The discussion evolved naturally and was neither one sided nor led to extreme positions. The process was divided into different phases and the moderators employed various techniques to identify and categorize the issues as well as to highlight minority groups' input. This was not done in a top-down approach, but the crowd was for example allowed to raise additional concerns that the moderators had not thought about, via the 'propose your own topic' section of the platform.

Feedback and commenting, exemplified by the Madison platform and Icelandic constitution, can create new points of view on a new law and help to correct errors or to give additional input, that no one in the drafting process had considered before. In the Icelandic example, this feedback loop evolved around the council posting several drafts of the constitution, each time trying to implement the input gathered beforehand.

Both approaches are most promising due to their low threshold of participation. The technical hurdles on Madison and on social media platforms employed for commenting on the Icelandic constitution are very low, so that even non-tech-savvy users may participate. Furthermore, giving input in the form of a proposal or a comment does not require the user to be either a specialist in the field discussed nor a legal professional who has to consider how to actually draft a legal text or which other existing stipulations have to be taken into consideration.

Additionally, both approaches may be combined in the future. Unlike in the Finnish example, in which phase 3 and 4, i.e. evaluation and drafting, were intended to be done by professionals without engaging the crowd, it is imaginable to design a crowdsourcing process in which first, the crowd will give input on the issues at hand and possible solutions, then a bill is drafted taking into account the crowd input and lastly, the crowd is asked for help again, by presenting the initial draft of the bill to the crowd for feedback and commenting. That way, the public has the opportunity to get involved in the whole process from the initial brainstorming up to the final text to be decided upon by the public assembly. One may even go one step further and implement a procedure by which a crowdsourced bill finally must be discussed in the public assembly, if it gathers enough support (like an initiative/petition). Furthermore, it is imaginable to organize a public vote on the final draft, as has been done with the Icelandic constitution. This heightened accountability would alleviate the problem that sometimes the crowdsourcing process was in vain, because no new law actually gets to the public assembly.

We saw that one of the greatest problems in that respect is the acceptance of crowdsourced laws by public officials, mostly based on the difficulty that those officials do not know how to handle the crowd input. Designing a process in which not only the crowd is engaged from the beginning, but also the politicians that are responsible for the final implementation, is a way of dealing with this difficulty. On Madison this is done by direct discussions between the users and the officials who put the draft online. In a combined process, public officials should be engaged from the very outset and take part in the problem mapping and solving phases as well.

This engagement may also help to let the participants know that their efforts will not be in vain, because their voices are heard by those finally deciding on a new law. The educational side of crowdsourcing would be heightened, since the crowd can get direct feedback on how realistic a proposal is to be implemented by the decision-making authority.

Aitamurto et al. propose that crowd input should be treated like any other form of input, e.g. input from interest groups or experts.[93] The proposal presented here goes beyond this approach. Crowd input cannot exactly be compared with input by interest groups or experts since those same interest groups or experts may be part of the crowd and may have engaged with the crowdsourcing process, thereby shaping the outcome. Furthermore, the democratic legitimacy that comes from support of a very large crowd must be taken into account. Therefore, in the most engaging projects, the crowd's voice should be given a stronger position than in others.[94]

We further saw that under the right circumstances, crowds are able to produce excellent results. This is not only true for the original research on aggregated averages, but still holds true in the deliberative forums of the legislative process. The practical advantages of employing crowdsourcing in the legislative sphere described above can help to produce better results than without crowd input. At the same time, giving people more power in the process heightens its legitimacy and holds officials more accountable. Certain dangers were pointed out, which can be mitigated either by the right process design or by the crowd itself. The here proposed crowdsourcing of the legislative process via know-how gathering and/or feedback and commenting is still supplemental to the classic process of legal drafting by public officials. This has two advantages. First, one may still experiment with the process and second, the dangers of extreme or populist positions is alleviated since the crowd does not finally decide on the measures to be implemented into a new law. If such danger really exists still has to be evaluated. So far, the examples clearly show that the opposite is true and that the crowd is able to regulate itself.

Overall, employing crowdsourcing in the legislative process bears the promise of creating an overall better process with better results. At the same time, as a supplement to the current democratic process, it poses very little risk. Legislators in the democratic world should consider using crowdsourcing more and reaping the benefits it may bring.

[93] Aitamurto et al. (2013).

[94] The Finish marriage equality bill, which gathered more than 100,000 supporters in just one day is a good example for this.

References

Aitamurto, T. (2012). *Crowdsourcing for democracy: A new era in policy-making.* Publications of the Committee for the Future, Parliament of Finland 1/2012. Helsinki, Finland, Available at SSRN. Retrieved August 29, 2021, from https://ssrn.com/abstract=2716771

Aitamurto, T. (2016, January). *Collective intelligence in law reforms: When the logic of the crowds and the logic of policymaking collide.* 2780–2789. https://doi.org/10.1109/HICSS.2016.349

Aitamurto, T., & Landemore, H. (2015). Five design principles for crowdsourced policymaking: Assessing the case of crowdsourced off-road traffic law in Finland. *Journal of Social Media for Organizations, 2*(1), 1–20.

Aitamurto, T., & Landemore, H. (2016). Crowdsourced deliberation: The case of the law on off-road traffic in Finland. *Policy & Internet, 8*(2), 174–196. https://doi.org/10.1002/poi3.115

Aitamurto, T., Landemore, H., Lee, D., & Goel, A. (2013, October 30). *Seven lessons from the crowdsourced law reform in Finland.* The Governance Lab @ NYU. Retrieved August 29, 2021, from http://thegovlab.org/seven-lessons-from-the-crowdsourced-law-reform-in-finland/

Aitamurto, T., Landemore, H., Lee, D., & Goel, A. (2014). *Crowdsourced off-road traffic law experiment in Finland: Report about idea crowdsourcing and evaluation.* Retrieved August 29, 2021, from https://www.researchgate.net/publication/291074594_Crowdsourced_Off-Road_Traffic_Law_Experiment_In_Finland_Report_about_idea_crowdsourcing_and_evaluation

Bautista, M. M. (2015). *Crowdsourcing and local law making: Closer to the people?* Retrieved April 21, 2016, from http://sffwlaw.com/wp-content/uploads/2015/04/Crowdsourcing-and-Local-Law-Making_Marilyn-Marchello-Bautista.pdf (not available as of 29 August 2021).

Berg, J. E., Nelson, F. D., & Rietz, T. A. (2008). Prediction market accuracy in the long run (Working Draft January 2008, https://pdfs.semanticscholar.org/0c15/429ff05a1d5376cf262cccdfd1575dba756f.pdf, accessed 29 August 2021). *(Final Version Published in:) International Journal of Forecasting, 24*(2), 285–300. https://doi.org/10.1016/j.ijforecast.2008.03.007

Campoy, A. (2016, April 21). Mexico City is crowdsourcing its new constitution using Change.org in a democracy experiment. *Quartz.* Retrieved August 29, 2021, from https://qz.com/662159/mexico-city-is-crowdsourcing-its-new-constitution-using-change-org-in-a-democracy-experiment/

Dawson, R. (2013, June 5). *How Finland's open ministry is crowdsourcing legislation.* Retrieved April 21, 2016, from http://www.resultsfromcrowds.com/insights/how-finlands-open-ministry-is-crowdsourcing-legislation/ (not available as of 29 August 2021, accessible via the Internet Archive: https://web.archive.org/web/20170927142956/http://www.resultsfromcrowds.com/insights/how-finlands-open-ministry-is-crowdsourcing-legislation/, accessed 29 August 2021).

Dittrich, B. (2017, February 21). Finland to allow same-sex marriage. *Human Rights Watch.* Retrieved August 29, 2021, from https://www.hrw.org/news/2017/02/21/finland-allow-same-sex-marriage

Eördögh, F. (2012, November 5). In Crazy Open-Source Project, Finnish citizens propose laws for parliament to consider. *Slate.* Retrieved August 29, 2021, from http://www.slate.com/blogs/future_tense/2012/11/05/open_ministry_finland_s_open_source_project_to_let_citizens_propose_laws.html

Finley, K. (2013, October 21). Out in the Open: Hackers bring lawmaking into the 21st century. *WIRED.* Retrieved August 29, 2021, from https://www.wired.com/2013/10/madison-project/

Gylfason, T. (2013, June 19). Democracy on ice: A post-mortem of the Icelandic constitution. *openDemocracy.* Retrieved August 29, 2021, from http://www.opendemocracy.net/can-europe-make-it/thorvaldur-gylfason/democracy-on-ice-post-mortem-of-icelandic-constitution

Heaton, B. (2014, July 2). *Is crowdsourcing the future for legislation?* Retrieved August 29, 2021, from http://www.govtech.com/internet/Experts-Predict-More-Legislation-Will-Be-Crowdsourced.html

Howe, J. (2006, June 1). The rise of crowdsourcing. *WIRED*. Retrieved August 29, 2021, from http://www.wired.com/2006/06/crowds/

Howe, J. (n.d.). *Crowdsourcing*. Crowdsourcing website: http://www.crowdsourcing.com/cs/. Accessed 19 April 2016 (not available as of 29 August 2021, accessible via the Internet Archive: https://web.archive.org/web/20180918122138/http://www.crowdsourcing.com/cs/, accessed 29 August 2021).

Landemore, H. (2013). *Democratic reason: Politics, collective intelligence, and the rule of the many*. Princeton University Press.

Landemore, H. (2014, July 23). The Icelandic experience challenges the view that constitutional process must be exclusionary and secretive. *Democratic Audit UK*. Retrieved August 29, 2021, from http://www.democraticaudit.com/?p=6747

Landemore, H. (2015). Inclusive constitution-making: The Icelandic experiment. *Journal of Political Philosophy, 23*(2), 166–191. https://doi.org/10.1111/jopp.12032

Oxford Dictionary of English. (n.d.-a). crowdsourcing | Definition of crowdsourcing in English by Oxford Dictionaries. Oxford Dictionaries | English. Retrieved August 29, 2021, from https://en.oxforddictionaries.com/definition/crowdsourcing

Oxford Dictionary of English. (n.d.-b). deliberation | Definition of deliberation in English by Oxford Dictionaries. Oxford Dictionaries | English. Retrieved August 29, 2021, from https://en.oxforddictionaries.com/definition/deliberation

Petitioners take gay marriage bill to Parliament. (2013, March 20). Yle Uutiset. Retrieved August 29, 2021, from http://yle.fi/uutiset/petitioners_take_gay_marriage_bill_to_parliament/6544701

Rushkoff, D. (2003). *Open source democracy: How online communication is changing offline politics*. Retrieved August 29, 2021, from http://www.demos.co.uk/files/OpenSourceDemocracy.pdf

Sen, A. (2010). *The idea of justice*. Penguin.

Shirky, C. (2012). *How the Internet will (one day) transform government*. Retrieved August 29, 2021, from https://www.ted.com/talks/clay_shirky_how_the_internet_will_one_day_transform_government

Siddique, H. (2011, June 9). Mob rule: Iceland crowdsources its next constitution. *The Guardian*. Retrieved August 29, 2021, from http://www.theguardian.com/world/2011/jun/09/iceland-crowdsourcing-constitution-facebook

Singel, R. (2008, September 10). Wikipedia Sleuths Win Journalism Award for Wired.com. Retrieved WIRED. Retrieved August 29, 2021, from https://www.wired.com/2008/09/wikipedia-sleut/

Soepper, P. (2018). Crowdsourcing and crowdfunding: Democratizing societies in the 21st century. 法制理論 = *Hosei Riron; The Journal of Law and Politics, 50*(2), 154–186.

Surowiecki, J. (2004). *The wisdom of crowds* (First Anchor Books Edition). Anchor Books.

The Constitutional Council – General Information. (n.d.). Retrieved August 29, 2021, from http://stjornlagarad.is/english/

The Open Gov Foundation. (2019, February 4). *Say Goodbye to Madison*. The OpenGov Foundation. Retrieved August 29, 2021, from https://medium.com/@OpenGovFdn/say-goodbye-to-madison-84629ce013a0

Warner, C. (2014, April 9). North and South Carolina counties use crowdsourcing to plan region. Retrieved August 29, 2021, from http://www.govtech.com/internet/North-and-South-Carolina-Counties-Use-Crowdsourcing-to-Plan-Regions-Future.html?utm_source=related&utm_medium=direct&utm_campaign=North-and-South-Carolina-Counties-Use-Crowdsourcing-to-Plan-Regions-Future

www.ingramcontent.com/pod-product-compliance
Ingram Content Group UK Ltd.
Pitfield, Milton Keynes, MK11 3LW, UK
UKHW020232290926
15470UKWH00002B/501
* 9 7 8 3 0 3 0 9 0 5 1 2 5 *